Visual Representations in Science

W0234820

Visual representations (photographs, diagrams, etc.) play crucial roles in scientific processes. They help, for example, to communicate research results and hypotheses to scientific peers as well as to the lay audience. In genuine research activities they are used as evidence or as surrogates for research objects which are otherwise cognitively inaccessible. Despite their important functional roles in scientific practices, philosophers of science have more or less neglected visual representations in their analyses of epistemic methods and tools of reasoning in science. This book is meant to fill this gap. It presents a detailed investigation into central conceptual issues and into the epistemology of visual representations in science.

Nicola Mößner currently holds a position as a lecturer at the Department of Philosophy at the RWTH Aachen University, Germany. Between 2015 and 2016, she was a Junior Fellow at the Alfried Krupp Wissenschaftskolleg Greifswald, Germany. She received her MA in German Literature and Linguistics from the University of Hamburg and her PhD in Philosophy from the University of Münster, Germany.

In the philosophy of science her main interests of research comprise, on the one hand, Ludwik Fleck's theory of social dynamics and influences on epistemic processes in science and, on the other, the epistemic status of visual representations in processes of scientific reasoning and communication. She edited (together with Alfred Nordmann) *Reasoning in Measurement* (2017) and (together with Dimitri Liebsch) *Visualisierung und Erkenntnis – Bildverstehen und Bildverwenden in Natur- und Geisteswissenschaften* (2012). Another area of her specialisation is social epistemology. In this context she worked on the epistemology of testimony and published *Wissen aus dem Zeugnis anderer – der Sonderfall medialer Berichterstattung* (2010).

History and Philosophy of Technoscience

www.routledge.com/History-and-Philosophy-of-Technoscience/book-series/
TECHNO

Series Editor: Alfred Nordmann

Visual Representations in Science

Concept and Epistemology

Nicola Mößner

Routledge
Taylor & Francis Group

LONDON AND NEW YORK

First published 2018 by Routledge

2 Park Square, Milton Park, Abingdon, Oxfordshire OX14 4RN

52 Vanderbilt Avenue, New York, NY 10017

Routledge is an imprint of the Taylor & Francis Group, an informa business

First issued in paperback 2020

British Library Cataloguing in Publication Data
A catalogue record for this book is available from the British Library

Library of Congress Cataloging in Publication Data
A catalog record for this book has been requested

ISBN: 978-1-138-08993-8 (hbk)
ISBN: 978-0-367-48705-8 (pbk)

Typeset in Times New Roman
by Out of House Publishing

Contents

List of illustrations

Figures

(Due to the standards of the production process, images are reproduced in greyscale only. Colour versions are provided on websites mentioned by links in the captions of the respective images.)

Tables

Preface

This book presents my ideas concerning an epistemology of scientific images. My considerations on this topic were motivated by the pioneering work of Laura Perini. Discussing the topic with her at several conferences helped to shape my ideas and to find a reasonable way to analyse a topic that crosses disciplinary as well as philosophical boundaries.

Within the last few years, I have discussed ideas with many people. They have helped me greatly to disentangle the complexities of this topic and my thoughts on it. For their helpful comments, I particularly want to thank (in alphabetical order) Joachim Bromand, Michael Gordin, Philip Kitcher, Erik Koch, Michael Krämer, Dimitri Liebsch, Patrick Maynard, Maria E. Reicher and Alexander Wagner. Moreover, between 2011 and 2015, I presented some of my first ideas at the research colloquium of the Department of Philosophy at RWTH Aachen University. I want to thank the participants of this colloquium for discussions and their valuable suggestions.

Some of the following ideas have been published in journal articles. Endnotes in the relevant sections of this book will inform the reader about the passages in question. I would like to thank reviewers of these journals for their helpful comments on my papers. Of course, any errors remaining are mine.

I thank the German Research Foundation (DFG) for its financial support. The DFG provided funding for my project *Visualisierungen in den Wissenschaften – eine wissenschaftstheoretische Analyse* (MO 2343/1-1), pursued at the Department of Philosophy at RWTH Aachen University between April 2012 and June 2015. This financial support enabled me to do the relevant research on which this book is based.

Furthermore, I want to thank the *Stiftung Alfried Krupp Kolleg Greifswald*. Their awarding me a junior fellowship made it possible to finish this book in the inspiring working atmosphere of the *Alfried Krupp Wissenschaftskolleg* in Greifswald.

I dedicate this book to my father, in loving memory.

Taylor & Francis
Taylor & Francis Group
http://taylorandfrancis.com

1 Introduction

Although mankind has always shown a huge interest in pictures as is documented in prehistoric cave painting (see e.g. Figure 1.1) and has always used images for different purposes (see e.g. Assmann 2009; Bosinski 2009; Lewis-Williams 2004; Robin 1992), there has been a significant increase in the production, distribution and usage of visual representations in the last few decades.[1] The reason for this can easily be detected in the realm of media technology, which has undergone a no less astonishing great leap forward. Most media rely heavily on visual representations.

In particular, the developments in the domain of information technology (IT) have drastically simplified the task of producing and distributing images. An illuminating example is the shift from the verbal to the visual in manuals. Take for instance software; just a few years ago complex programs were delivered with lengthy books explaining their features. Today more and more of these products are accompanied by visual explanations, i.e. by video clips (made accessible on the internet or attached as CD-ROM) instead of a written manual. Thus, *moving pictures* have taken over the task which the written word formerly fulfilled. This is also true in the *scientific domain* which is the focus of the following analysis. Here, too, a similar ubiquity of visual representations can be recognised (see e.g. Downes 2012).

This book is a study in the philosophy of science. From this perspective I will tackle the following questions: *what scientific images are, which epistemic functions they fulfil in scientific processes (focusing on the empirical sciences), and what their epistemic status is in comparison to other representational means.* My aim is twofold, namely to present an investigation into both the *concept* and the *epistemology* of visual representations in science. To approach these topics, it might be easiest to start with a brief example.

In 2016, physicists involved in the LIGO[2] project announced the detection of gravitational waves, i.e. roughly speaking, vibrations of space-time due to the merging of two black holes (see Abbott et al. 2016). The existence of these phenomena was predicted by Albert Einstein's theory of general relativity. Until the LIGO report, however, no empirical validation of the phenomenon had been achieved. Consequently, their detection meant another significant corroboration of Einstein's theory and, thus, a major success for the scientists

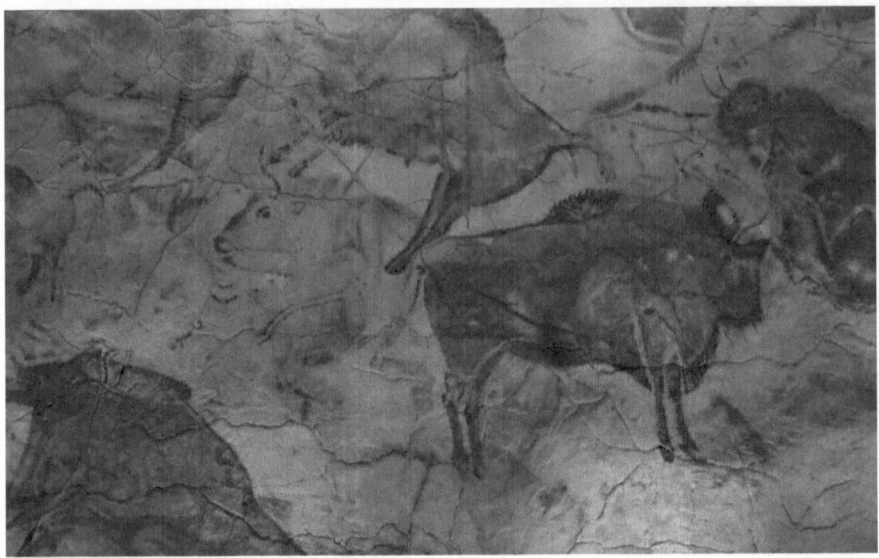

Figure 1.1 Picture of the reconstructed Altamira cave at the Deutsches Museum, Munich.

Source: photograph by Alexander Wagner.

involved in the LIGO project. Part of their reports about this important event consisted of many visual representations, in particular diagrams, one of which is examined as follows.

Figure 1.2 shows the measurement results of two detectors employed in the LIGO collaboration that registered the signal simultaneously on September 14, 2015. The first diagram shows the signal detected at Hanford, Washington, the second the detection at Livingston, Louisiana. Each of them is compared to the curve predicted by Einstein's theory. The third diagram presents a juxtaposition of the data registered at both Hanford and Livingston; that is, it makes visible the similarity of the data recorded.

Apparently, these images are of special importance to the scientists who have to report their detections. This becomes clear when taking a closer look at B. P. Abbott et al.'s paper (see Abbott et al. 2016), the first official announcement to the scientific community. This paper is 16 pages long; the main part of the announcement comprises roughly eight and a half pages, around two pages are footnotes, and five and a half are used to mention the authors of this article (133 people) and their affiliations. Furthermore, the article contains four complex figures, composed of different diagrams. So approximately half of the pages of the main part are filled by images.

Why are images that important to communicate the event? Is it because what is announced is an optical phenomenon so that it makes sense to show

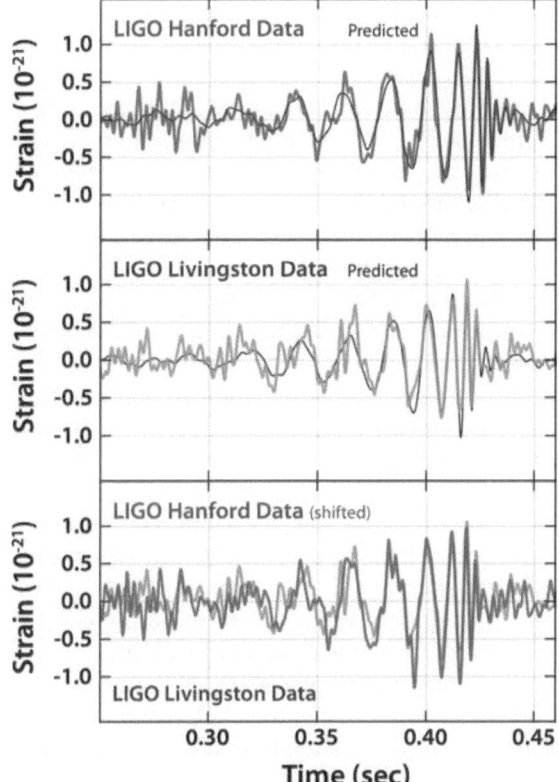

Figure 1.2 Signals of gravitational waves detected by the twin LIGO observatories at Livingston, Louisiana and Hanford, Washington.

Source: Caltech/MIT/LIGO Laboratory. For a colour version of the image see www.ligo.caltech. edu/image/ligo20160211a

the recipients what it looks like? Of course, this last question is only meant rhetorically. Gravitational waves in space-time, as predicted by Einstein's theory, are not in any way visible to the human eye. Visual events are only indirectly connected to those events. Moreover, as most of them (fortunately) occur in areas of the universe remote from us, their effects on our planet are very minor in scale, so minute that they are only detectable by instrumental means. Consequently, their detection requires a complex experimental set-up such as that realised by the LIGO collaboration.[3] Although the phenomenon they are investigating is not visual in any sense, the authors of this paper thought it relevant to distribute those images attached to their report. What roles do they play? What roles *can* they play in this context of information distribution? And what functional roles can visual representations gener- ally serve in epistemic processes in science? Should they be replaced by other representational means, such as linguistic expressions? Or can they play a role in such contexts that cannot be transferred to other vehicles of information transmission? These, then, are the central questions that we will investigate in the following analysis.

One major motivation to write this book consists in the astonishing mismatch between the obvious significance of visual representations in scientific practices and their long-lasting neglect in the analyses conducted by philosophers of science. Admittedly, there are many interesting case studies and investigations of visual representations in science,[4] but next to none of these studies has ever been compiled by philosophers of science, although theirs would be the traditional place to look for such an analysis when investigating the epistemic status of scientific images. As the example of the detection of gravitational waves shows, the contexts in which those images appear are often related to attempts at information acquisition and distribution, that is, they are *epistemic* in kind. The *philosophy of science* explicitly deals with questions of scientific knowledge, its justification and achievements (see e.g. Chalmers 1999; Ladyman 2003).[5]

The puzzlement about this mismatch between the dominance of the visual in scientific epistemic practices and the apparent lack of interest that philosophers of science have hitherto shown with regard to this topic I share with Laura Perini. Her pioneering analytical work on scientific images has been a major source of inspiration for this book. In her 2005 papers (see Perini 2005a; 2005b; 2005c), she not only makes us aware of the neglect of philosophers to deal with the topic, but also looks for explanatory reasons and an alternative handling of images by seriously taking them into account in the epistemic context.

To put it briefly, her diagnosis of why philosophers of science have not paid proper attention to visual representations rests on the following consideration: research results are presented as arguments in journal articles and the like. Here, scientists try to give reasons for why they think that their results are correct and, thereby, try to convince the members of their community of their findings. But, as Perini points out, it is still unclear what exactly is the kind of contribution that visual representations can make to scientific *arguments*.

The crux of the matter is that most philosophers think of argumentation as a merely verbal issue. This stance explains why visual representations are excluded as proper components of scientific arguments by these scholars (see Perini 2005b, 262): they simply belong to the wrong kind of representation. Following this line of reasoning, visual representations could only contribute to scientific arguments if they are transferred to the linguistic domain, i.e. by translating their content.[6] Scientists, however, do not seem to share this narrow perspective on the assumed cognitive capacities of visual representations. On the contrary, they use images in a variety of ways in their scientific work.

Combining these two observations allows Perini to extract the following characterisation of scientific communication that also highlights the essence of the dilemma:

> [w]hy do scientists include visual representations – why not defend ideas just with linguistic or numerical representations? Scientists do not act as if figures are 'mere illustrations' that are redundant expressions of information presented in the text, or convey information inessential to the argument. On the contrary, scientists treat figures as if they play integral roles in the arguments in which they appear.
>
> (Perini 2005c, 913)

It is not only this clash of opinions between scientists and philosophers that Perini highlights in her analyses. She also supports the scientists' assessment of the cognitive role of visualisations. From her point of view, it is essential to take visual representations seriously in epistemic contexts. It is only in this way that philosophers of science can do what they aim to do, namely to understand scientific reasoning.

> We cannot achieve a philosophical understanding of scientific reasoning until we understand the tools by which scientists present, defend and evaluate ideas; given the prevalence of visual representations in scientific communication, this means that we cannot understand scientific reasoning until we have the capacity to explicate reasoning involving scientific visual representations, in all their diversity.
>
> (Perini 2010, 132)

Although I do not agree with Perini's solution to the problem (see section 4.1.2 of this book), I am nonetheless sympathetic towards her diagnosis and I fully agree with her demand for a development of a serious epistemological analysis of visual representations in science. Thus, in this book I try to fill the gap that has grown out of the philosophical neglect of this topic.

1.1 Topics and methodology

This study aims at an investigation of the epistemology of visual representations in science. Since the role and status of images in scientific reasoning and cognition will be investigated, the focus of this book is on the *philosophy of science*. Nevertheless, to find answers to some of the above questions, I will make use of insights from other philosophical branches as well, most of all from *picture theory* and *social epistemology*.

Contrary to preceding attempts to analyse scientific images presented by scholars from Science and Technology Studies (STS) or the history of science (or the history of art), the focus of my investigation is not on case studies of

particular visual techniques, instruments or media. In contrast to such theoretical approaches, I will scrutinise the epistemic particularities of images
used in the empirical sciences on a more general level.[7] My approach is twofold in this respect. On the one hand, I will analyse the functional roles of
visual representations in the scientific processes of acquiring and distributing
information. On the other, I will investigate the epistemic status that can be
attributed to these images in comparison with other representational means
such as linguistic expressions.[8]

By taking a more detailed look at visual representations in different scientific contexts, I will try to solve the problem pointed out by Perini above;
namely why scientists apparently assess the epistemic capacities of images
differently in comparison with philosophers of science. Thus, the guiding
question will be whether there are epistemic reasons that support and justify
the formers' assessment. The following short guided tour through this book
is meant to highlight the line of reasoning by which I will proceed to meet
these aims.

It has already been noted that there is an astonishing diversity of visual
representations in science. To deal with this finding, we have to start with
the question of what the main characteristics of scientific images are (see
Chapter 2). Without doubt, all of these visual representations have their particularities, their special advantages and disadvantages as tools in scientific
practices. A first question, then, concerns the extension of the term, that is,
the scope of the concept of visual representations in science. Here, it will be
supposed that scientific images can best be understood as *artefacts*, that is, as
entities intentionally produced by using imaging technologies, instruments,
etc. to play a role in scientific cognitive processes. To find out more about
their particularities, I will begin my considerations by examining selected
paradigmatic instances. Approaching the topic this way will not only allow
us to find out how they are produced and, thus, how information is encoded
in the visual format, but also what functional roles they are supposed to play
in scientific processes and what their advantages and disadvantages as visual
tools are when serving these purposes.

A second question connected with a clarification of the concept of scientific images concerns its intension. Despite their diversity, can some common
characteristics nonetheless be extracted that also account for their epistemic
virtues and vices in scientific practices? To answer this question we will take
a closer look at insights put forward in *analytic picture theory*[9] and contemplate the possibility of gaining clues that might help us better to understand
the concept of scientific images. Proponents of picture theory deal with
questions about visual representations more generally (see Steinbrenner
2009, 284): *what are pictures? What is depiction? What does it mean to understand a picture? What is so special about the perception of pictures?* A variety of accounts has already been put forward to explain these phenomena
(see e.g. Gombrich 2004; Goodman 1976; Hopkins 2009; Hyman 2006;
Kulvicki 2006; Lopes 2006; Mitchell 2008; Newall 2011; Scholz 2009b),

yet is there also an approach that can tell us more about the concept of visual representations in science?

In Chapter 3, I will examine the fact that visualisations are not only diverse with regard to their appearances, but also with respect to their functional roles. The first aspect to be more closely examined concerns their main contexts of use in science. It will turn out that such a context-orientated approach will allow us to best explain both the common aspects and the differences in their use and design. In accordance with Felice C. Frankel and Angela H. DePace's suggestion (see Frankel and DePace 2012), I will propose a distinction between the *exploratory* and the *explanatory context*. Furthermore, we will compare this approach with another suggestion concerning the contextualisation of scientific images. The author of this second proposal adheres to Hans Reichenbach's famous distinction between the context of discovery and the context of justification.

The contextual analysis will not only permit a clearer grasp of some of the difficulties connected with scientific images, but will also suggest some solutions. For example, I will contemplate the possibility of drawing ontological inferences based on images with regard to the depicted objects of research, such as the announcement of the detection of gravitational waves being based on the diagrams discussed in the example above (see Figure 1.2). Here, it will be taken for granted that scientists broadly use visual representations in processes of information acquisition and distribution. One aim of the analysis is to systematise the functional roles played by those visual means in the two contexts indicated. Furthermore, we will tackle the question of whether scientists are justified in their reliance on those images in their epistemic endeavours. In what sense, for instance, can visual representations serve as evidence to support scientific hypotheses? Visualisations often make entities visible which are otherwise unobservable to the unaided eye. Here, scientific images touch on the core topics of the debate between scientific realists and anti-realists, such as whether we can have knowledge about the unobservable part of the world and also whether unobservable entities exist independently of the mind.

In recent years, it seems to have become somewhat fashionable to regard visual representations as supporting theses about the constructivist character of scientific achievements. Such approaches side with the scientific anti-realists' assumption that either the two questions above have to be denied or that both have to be discarded as nonsensical. Contrary to this, I will argue for a more moderate stance. Without denying that scientific images are artificially brought about and that instruments and other kinds of technologies play a crucial part – also entailing that theories are relevant to their handling – my initial investigations into the different contexts of usage will nonetheless pave the way to argue for a scientific realist approach to the epistemic capacities of scientific images.

Finally, focusing on scientific practices also highlights the social setting of science in which visual representations play a role. Teamwork in the laboratory and in communication processes to disseminate results in nationwide and international communities are paradigmatic examples in this respect. Taking this social component seriously will permit an explanation of the diversity of scientific visualisations much better than attempts relying on representational aspects alone. In developing this explanation, I will make use of Ludwik Fleck's theory of the relevance of social influences on scientific reasoning and cognition (see Fleck 1979; 1986a).

Chapter 4 finally constitutes the main part of my analysis. Here I will consider the question of the *epistemic status* of scientific images. In this chapter, Perini's initial suggestion to take visual representations seriously in the context of scientific arguments will be re-examined. The questions whether visual arguments are possible and how best to conceive of the role of scientific images in communicative contexts such as these will be analysed.

This topic is bound up with questions about the presumed cognitive content of visual representations and the possibility of transmitting information by its means to the observer. Do visualisations fulfil only supplementary tasks (as heuristic aids or illustrations, etc.) in cognitive processes or do they also make genuine and sometimes even indispensable contributions to these epistemic practices? If they can transmit information, what kinds of constraints (if any) have here to be taken into account? Are there, on the other hand, particular epistemic advantages to visual communication in science? An important stepping-stone to clarify these issues relates to the topic of propositional and non-propositional contents of perception. After all, it is by our visual senses that we can decode information presented in scientific images. Thus, if perception works (at least partly) non-propositionally, it can be assumed that visual representations can achieve the same epistemic ends. Yet, if it is admitted that scientific images can transmit non-propositional content, what are the effects on their epistemic capacities? Do visualisations, for example, enable or support the transmission of knowledge-how? Do they facilitate scientific understanding?

Notes

1 In the following text, I will use the terms 'visual representation', 'visualisation' and 'image' interchangeably.
2 This is the abbreviation of Laser Interferometer Gravitational-Wave Observatory, see www.ligo.caltech.edu/, accessed May 4, 2016.
3 Information about the precise set-up of the experiment is offered in Abbott et al. (2016, 061102–3f.), in Reichert (2016) and on the website of the LIGO collaboration: www.ligo.caltech.edu/page/ligo-gw-interferometer, accessed May 5, 2016.
4 In particular, art historians, historians of science and proponents of Science and Technology Studies (STS) contributed much to this topic (see e.g., in alphabetical order, Adelmann et al. 2009; Bigg 2009; Bredekamp, Schneider and Dünkel 2008; Bredekamp 2009; Burri 2008; Cartwright 1997; Carusi et al. 2014; Daston and Galison 1992, 2007; Daston and Lunbeck 2011; Hennig 2011; Hentschel and Wittmann 2000; Hentschel 2014; Kemp 2003; Vertesi 2015).

5 On the distinction between the philosophy of science, epistemology and metaphysics see Scholz (2013).

6 We will examine this problem in more detail in section 4.1 of this book.

7 Images are, of course, also used in the formal sciences such as mathematics. Readers interested in this topic might find a good starting point for their investigations in Brown (1997) and in Mancosu, Jørgensen and Pedersen (2005). Discussing visual representations used in those disciplines would, however, go well beyond the scope of this book. An important particularity of those images concerns their functional role to serve as mathematical proofs. The latter, however, are of a completely different epistemic quality than empirical data in science. Therefore, tackling both kinds under the same heading seems to be inappropriate.

8 However, I will not investigate the much broader topic of scientific representations in general. Readers interested in this and, in particular, in the topic of scientific models might find a starting point in Frigg and Nguyen (2016), in van Fraassen (2010) and, of course, in the works of Mary S. Morgan and Margaret Morrison (see e.g. Morgan and Morrison 1999).

9 As this book is written in the analytic tradition of philosophy, it seems natural also to adhere to similar approaches in picture theory. On the concept of *analytic picture theory* see Steinbrenner (2009).

2 What are scientific visualisations?

Taking a look at scientific practices, we encounter a great variety of different kinds of visual representations that are used by scientists for diverse purposes. For example, we find *diagrams* included in scientific publications and presentations showing measurement results and their relations. *Drawings* and *photographs* are used to illustrate books. Furthermore, photographs are regarded as evidence for certain phenomena, e.g. that there is a fifth satellite orbiting Pluto (see Figure 2.1). Climatologists use *images produced by computer simulations* to investigate the interaction of natural and anthropogenic factors with regard to their effects on global warming. Physicists take *graphs* as means for their error calculations. Physicians take *X-ray* and *PET-images* (positron emission tomography) as devices for diagnostic purposes. In the same way, the outputs of *magnetic resonance imaging* (MRI) have become indispensable evidence in medical reasoning.[1] This is just to give the reader some idea of the great variety of the appearances, distributions and functions of images in contemporary sciences.

Despite this diversity, we tend to think that the different phenomena have something in common. This assumption explains why we usually subsume them under the same label, namely 'scientific images'. This chapter is meant to find out whether such an assumed common core can indeed be identified.[2] How then can the term *scientific images* be best understood? To answer this question, the nature of scientific images will be approached in three steps.

Firstly, the *extension* of the term 'visual representation' in science will be considered. Here, four paradigmatic categories of scientific images will be discussed, namely photographs, imaging techniques, data visualisations and diagrams. This investigation into these paradigmatic instances of scientific images will provide the reader with some initial ideas about what kinds of visual representations are used by scientists and in which contexts. In particular, the epistemic capacities and limitations of these paradigmatic examples of scientific images will be analysed in more detail within their respective contexts. Does the depictive style chosen influence the usefulness of visual representations in scientific epistemic contexts?

Secondly, the visual element involved will be examined more closely and thus the *intension* of the term be addressed. As scientific images belong to the realm

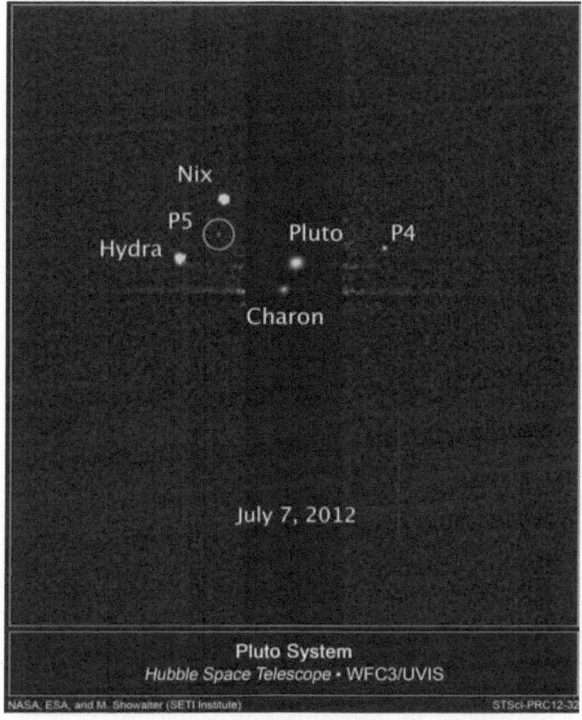

Figure 2.1 Pluto system: photograph taken by NASA's Hubble Space Telescope that shows Pluto and its five satellites. The circle marks the newly discovered moon, designated P5, as photographed by Hubble's Wide Field Camera 3 on July 7, 2012. (Credit: NASA; ESA; M. Showalter, SETI Institute).

Source: www.nasa.gov/images/content/666710main_p1232ay.jpg

of visual representations in general, it seems advisable to take into account what has been said about the latter in *picture theory*. Different proposals have been put forward in this context to explain (1) what *pictures* are, and (2) how *depiction* works. As, meanwhile, there has arisen a broad variety of theoretical approaches to these topics, it would go well beyond the scope of the current analysis to discuss all of them in detail. Yet what can be considered are *types* of theories, as Michael Newall (2011) has suggested. Approaching the topic this way, the following questions will be investigated: whether any of these explanatory attempts can help to reveal more about the nature of scientific images, whether one of those approaches in picture theory can be followed to achieve a better understanding of visual representations in science, and whether a common core of conditions will become evident when such definitional suggestions of the term are considered.

The last aspect to be taken into account in this chapter will be the question of whether there are any helpful correlations between insights gained by this analysis of paradigmatic instances of scientific images and any put forward in theoretical approaches in picture theory. Perhaps more can be discovered about the object under investigation by combining the aforementioned ways of approaching the topic?

2.1 Characteristics of visual representations in science

The list of examples of scientific images mentioned at the beginning of this chapter suggests that they are, broadly conceived, purposefully brought about to store and often also to transmit information. This characterisation motivates two initial theses of mine: firstly, visual representations can be regarded as artefacts.[3] As such they are not only produced, but also used for certain purposes. Moreover, these purposes can, but do not have to, coincide as the people (and their intentions) making use of one and the same image can change. It seems reasonable, furthermore, to regard scientific images as *signs*, if they are supposed to fulfil the tasks of information storage and transmission. However, it has to be added here that visual representations are not always used as signs in a meaningful way, that is, as representational means. Sometimes they are added for merely decorative purposes.

How a meaningful representational relation is achieved in the context of sign usage is most famously discussed by Charles Sanders Peirce in semiotics. Here, Peirce makes a distinction between signs as *icons*, *indices* and *symbols* (see Peirce 1983, 64ff.). This differentiation is based on assumptions about how the respective visual phenomena are related to their object of depiction: icons are characterised by resemblance relations to their object of representation. Indices are causally brought about. Symbols acquire their meaning via conventions. Now, it would be interesting to know whether visual representations in science can be subsumed under one of these three categories. This question will therefore be kept in mind in the subsequent discussion.

In the following, this investigation will commence with some short case studies on four selected paradigmatic instances of visual representations in science. These case studies are meant to further elucidate the scientific practices of utilising images of different kinds. The guiding questions will be about how they are produced, what they are meant for in scientific practices, and, from an epistemological point of view, what kind of information they can provide, what their epistemic limits are, and whether there are connections between these limitations and the way that information is visually presented. The discussion will involve descriptions of the respective imaging procedures and analyses of exemplary applications in science. It is not intended to offer a complete overview of image usage in scientific practices, the focus being rather on questions concerning the aptness of images for certain information processes in science and epistemological difficulties that might show up in those contexts.

A good starting point for this investigation is photography, which can be regarded as a prototype of both a measuring or detection device and a means of depiction in science.

2.1.1 Photographs

Photography utilised in the scientific context can best be understood along the lines of Patrick Maynard's theoretical account on this topic. He suggests regarding photography as "a branching family of technologies, with different uses, whose common stem is simply the physical marking of surfaces through the agency of light and similar radiations" (Maynard 2000, 3). What makes this approach particularly useful when analysing the role of photography in the epistemic processes of science is its *focus on the different functions* that this "family of technologies" can fulfil. Photography is, for example, used as a means of detection as well as of recording certain data. An example of the detective function of photography is supplied by Figure 2.1 above. Investigating the Plutonian system by means of a camera was part of the activities preparing the New Horizons space mission, aiming at the fly-by manoeuvre of a spacecraft in 2015. This mission presupposed background knowledge of potential hazardous asteroids and other objects in the vicinity of Pluto that might damage the spacecraft. As the position of such small celestial bodies could be revealed by recording their movements in comparison to the starry background and in relation to Pluto and its other satellites, taking a series of photographs of the system turned out to be a suitable way to discover the relevant facts. The astrophysicist Thorsten Ratzka points out that photography is a quite common tool in astronomy as, due to long-term exposures, cameras allow the collection of much more light than the human eye is able to detect and thus render even minute, remote objects visible (see Ratzka 2012, 246). In this case, using the photographic method[4] of investigation even allowed for the detection of a formerly unknown fifth satellite, now labelled Styx.

Even though scientists did not intend to take a photographic picture of Styx when using their camera to examine the system of celestial objects orbiting Pluto more closely, Maynard's account allows us to explain the functional role played by photography in this context quite easily by making use of his theses about its detective function. Admittedly, and due to this divergent starting point emphasising a multiplicity of functional roles, Maynard's theory is somehow at odds with many more common philosophical approaches to photography.[5] All of them commence theorising with the *photographic picture*, which narrows their focus to only one function that this technology is usually thought to fulfil, namely depiction (see Maynard 1989, 263). Contrary to this traditional point of view, Maynard's approach to photography makes clear that there are several functional purposes that this technology can serve. Scientists make use of this functional diversity of photography, as will be seen in due course.

From Maynard's point of view, which will be followed in the subsequent section, photography is predominantly a technological procedure, a "productive process" (Maynard 2000, 9). Three components of this process are then relevant to consider, namely *the input, the output and the nature of their relation.* Both input and output involve a diversity of phenomena, a fact which is also due to technological developments since the initial invention of photography in the nineteenth century.[6] To put it briefly, the input of this process is constituted by different kinds of radiation, whereas its output is created by varying kinds of media of inscription, so to speak.

In contrast to the diversity of input and output, the photographic process itself works in the same way all the time. It is a causal relation. How this causal process operates and how the notion of causality invoked here can best be understood is explained as follows.

Setting aside digital photography for the moment, the most traditional and simplest explanation is that light rays, reflected by a particular object in front of the camera, *cause* a photochemical reaction on a film or photographic plate and thus *cause* a recording of those emissions. In this ideal conception of photography,[7] any aspect beyond this triple of object, relational process and resulting image is withheld from investigation. In particular, all kinds of intentional decisions on both sides of the causal process are ignored. Attention is neither paid to intentional acts of how to select, arrange or otherwise influence the object posed in front of the camera, nor is it considered how the photographer might focus on this object, be it by means of a zoom, a flashlight or filters.[8]

Despite this ignorance in the ideal case of a photographic picture free from any intentional interventions, there are actually wide-ranging possibilities to *manipulate* or, expressing it more neutrally, to influence the photographic process. That is, the causal process can be intentionally interfered and, because of the relational connection between object and picture, a change of the latter can be effected by manipulating the former. Counterfactually speaking, this amounts to the following conditional statement: if the object (x) had been different in front of the camera, its photograph (y), as the result of the photographic process, would also have been different. It has to be added that, of course, not all changes affecting x will influence y. The sensitivity of the camera is crucial in this respect. If, for example, the camera can only produce black and white images, painting the object in front of it in another, but bright colour similar to the genuine one, will not affect the resulting image. Replacing a model made of wood by a visibly indiscernible one made of plastic will not make a difference to the resulting photograph either, and so on. The sensitivity of the camera restricts what kinds of interference are possible in the sense of making a difference to the resulting picture. The camera's sensitivity, which might change due to the different kinds of radiation that can be inscribed, as pointed out above, will therefore also disclose what kind of cause and effect are related by the very process.

A general theory of causality that can accommodate these particularities of the photographic process is James Woodward's suggestion of an *interventionist approach to causality* (see Woodward 2005). Although Woodward is not concerned about the causal process of photography in particular, his reflections on the concept of causality in general permit immediate recognition that the photographic process belongs to the set of causal phenomena. Especially, his definition of a "direct cause" fits neatly into our considerations above:

> [a] necessary and sufficient condition for X to be a direct cause of Y with respect to some variable set V is that there be a possible intervention on X that will change Y [...] when all other variables in V besides X and Y are held fixed at some value by intervention.
>
> (Woodward 2005, 55)

Changing the object in front of the camera will alter the respective image. Henceforth, I will take it that the photographic process can best be understood in Woodward's interventionist sense of causality. Remaining with causality for a moment, it can also be noted that it is the causal process on which photography as a technology relies that is also mentioned as a rationale to explain its particular *evidential status*. Lorraine Daston and Peter Galison's famous discussion of the *mechanical ideal* established by this technology serves as an example in this regard (see Daston and Galison 1992). In their analysis of the term 'objectivity' in the history of science, they highlight the fact that photography was particularly valued because of the assumption, arising in the late-nineteenth century, that it can be regarded as the paradigmatic instance of a "non-interventionist" ideal of scientific observations (see ibid., 82, 120). Paying close attention to the scientists' attitude towards photography in those early days of this technology, Daston and Galison introduce their concept of "mechanical objectivity". Describing what the characteristics of this conception of objectivity are, they state that "[n]onintervention, not verisimilitude, lay at the heart of mechanical objectivity, and this is why mechanically produced images captured its message best" (ibid.). Photography is thought of as being able to detect and record radiation without any human intervention.

Without doubt, Daston and Galison's concept of mechanical objectivity has informed common ideas about the *particular evidential status* of photographs.[9] Aaron Meskin and Jonathan Cohen, for instance, claim that "[...] photographs seem to have a distinctive epistemic status compared to other sorts of pictures. [...] we are inclined to trust them in a way that we are not inclined to trust even the most accurate of drawings and paintings" (Meskin and Cohen 2008, 70). The kind of trust that Meskin and Cohen point to is precisely based on the assumption of the apparent absence of human intervention that Daston and Galison highlight in their discussion of photography. Of course, it is objective evidence that is sought in science. If scientists analyse the photographs taken by the camera of the Hubble Space Telescope to find out about celestial objects which could be potentially hazardous to a future space probe fly-by manoeuvre in the Pluto system, they want to be sure that the objects depicted on those images are really out there. They want to be sure that those objects do have the orbits shown in those pictures so that reliable inferences about their potential interferences with the space probe can be made. If photography allows the reduction of human bias and other human errors in the course of observation, it would indeed be a very reliable kind of evidence in scientific investigations.

Roger Scruton, for instance, thinks that the ideal process of photography works exactly this way; that is, the camera can be regarded as enabling *automatic detection*. He points out that if we focus on the causal process of photography alone, what he terms the ideal case of photography, we can use the resulting images to infer

> [...] first, that the subject of the ideal photograph must exist; secondly, that it must appear roughly as it appears in the photograph; and thirdly, that its appearance in the photograph is its appearance at a particular moment of its existence.
>
> (Scruton 2008, 150)

Again, this would be exactly the kind of data that scientists at NASA are in need of when preparing the trajectory of their space probe. The question, then, is whether photography can indeed yield such a reliable kind of evidential data in science.

It might be objected that neither cameras nor other kinds of scientific instruments work automatically. No matter whether cameras, Geiger counters, ammeters or thermometers are analysed – all of these devices can be characterised as being sensitive to certain measurable parameters, as working via a causal connection. All of them show a certain result that is caused by the phenomenon to which the instrument is sensitive, be it visible light rays, radioactive radiation, electric current or temperature. However, none of these measurement processes works without human intervention, at least in the sense of setting up the experiment or working out the parameters to which the instrument has to be sensitive. Be this as it may, none of these measuring practices is regarded as particularly questionable concerning the objectivity of the results obtained. At least, in scientific practices we have become used to the interventionist part played by human observers,[10] but perhaps photography is wrongly put on a par with those other scientific instruments, since it is the photographic picture which is at issue here and not a detective function in general. I do not agree with this demurral. Responding to this issue, however, requires a more detailed look at the initial thesis that photography is a family of technologies. Without doubt, pictures are important in the context of photography, but decidedly no decisive feature. To explain this, let us come back to Maynard's theory of photography.

He makes us aware that the technologies belonging to this family can serve different purposes and fulfil varying functions. This is then the point where the diversity of the photographic output, indicated above, comes into play. Maynard explains that the same kind of technology which allows us to take family snapshots is also an essential part of the processes of photolithography, i.e. using light to pattern surfaces sensitive to this radiation, to produce, for example, solar cells or microchips (see Maynard 2000, 58, 118). Furthermore, the very same technology is utilised in reproductive practices to produce photocopies (see ibid., 18). In all of these instances, light rays are used to mark a surface, but rarely can a product of those processes be called a picture.

But even if the focus is narrowed to photographic processes in the vicinity of more traditional conceptions, that is, where photographic technologies are not used in the context of the production and reproduction of different goods, Maynard makes clear that photography does fulfil different functions. He discusses two such functions in detail, when drawing a distinction between photography as an *imagining technology* (see ibid., ch. 4)[11] and as a *detecting technology* (see ibid., ch. 5). Whereas the former function is essentially related to the capacity to produce images, the latter is not. How exactly do they differ?

About the detective function of photography Maynard states that:

> [i]ntended photochemical effects on specially prepared photosensitive surfaces enable reliable detection of the effects' causal situations, and have great importance for forensic, industrial, military, scientific, and many other forms of evidence. Though they no more depict nor constitute photographs of anything [...], plates and films recording the photochemical effects of electromagnetic radiation upon silver halides continue to play a very important role in modern scientific experimentation, in detection and in trace recording, via radiations.
>
> (Maynard 1989, 264)

As was pointed out above, cameras can record certain kinds of radiation that causally affect them.

As the example of the discovery of Styx already suggests, the detective function of photography can imply taking pictures in the traditional sense. Accordingly, Maynard states that the purposes of detection and image production can be intertwining processes (see ibid., 268ff.). He elucidates this functional interplay by explaining that:

> [...] whatever image (or part) constituted of photochemical effects is to function as depicting something, must also be a photochemical effect *of* that very thing. Thereby, the subject photographed will be, in principle, always detectable from the photograph. Detection then becomes a special *condition* on this kind of depiction.
>
> (Maynard 1989, 268, his italics)

Pluto's satellite Styx left its photochemical traces, so to speak, which allow us to recognise the photograph as a photograph of Styx, as its depiction. The particular advantage of this interplay between detection and depiction, Maynard describes by using the term "user-friendly interface" (see Maynard 2000, 131). That is, they meet the human condition of vision as our predominant sense to cognitively access our world: "[s]ince humans are visual it is little wonder that, where possible, access is through (enhanced) visual recognition, technically specialized: in other words, combinations in which depiction serves detection" (Maynard 2010, 30). We can simply evaluate the Hubble photographs of the Pluto system by using our eyes and take this as the reason to conclude that Styx does not belong to the background of fixed stars.

But even though the interplay between detection and depiction has undoubtedly many advantages for human observers, it is by no means a prerequisite of the detective function to result in images. Furthermore, as Maynard points out, "[e]ven where the detection/recording output is a figured image on a surface, that image need not be accessed as a *picture*. Our automatic devices for image scanning, fortunately, do not entail 'machine imagining'" (Maynard 2000, 123, his italics). Thus, the technical processing of information detected and recorded via cameras does not draw on its pictorial format. Yet it is not only in this mechanistic context that images are irrelevant to the detective function of photography. This point becomes most salient with regard to discoveries of new types of radiation such as X-rays (see Hentschel 2014, ch. 8.3) or other kinds of radioactivity via photography.

Kelley E. Wilder presents a fascinating case study on Henri Becquerel's discovery of radioactive radiation (see Wilder 2011). She explains how this discovery was "a matter of both accident and design" (ibid., 354). As a scientist, and due to the by then rather recent discovery of X-rays, Becquerel was aware of the fact that photography might reveal kinds of radiation not visible to the human eye. Actually, it was in the course of experiments to find out more about the connection between X-rays and luminescence, i.e. phosphorescent substances, that Becquerel discovered radioactive radiation. Wilder describes Becquerel's motivation to investigate this relation in the following way: "[g]iven Röntgen's discovery of a new invisible and penetrating ray, any number of other bodies might have been emitting similar and as yet undetected penetrating rays" (ibid., 355). In this sense, Becquerel's scientific inquiry was informed by Röntgen's former discovery. However, the results that he obtained differed in crucial respects from those of his colleague. Although in both instances it was the darkening of photographic plates that constituted the evidence of their respective discovery, in Becquerel's case this process took place in the absence of any light rays or electric current at all. "His [Becquerel's, N.M.] images were not made by 'exciting' any response by either light or the cathode ray tube. They appeared in the dark, merely in the presence of a substance [namely uranium salt, N.M.] considered inert, and thus 'spontaneous'" (ibid., 356). Moreover, contrary to X-ray images that were quickly brought to use as depictive devices, Becquerel's detections could not be called pictures in any sense of the word. They were evidence of the presence of a particular kind of radiation, but the result was not a photographic picture of a specimen emitting this radiation, let alone illuminated by it. "One of the differences [between X-rays and radioactivity] is the subject matter. Becquerel photographed the rays themselves, whereas X-ray photographers photographed objects with the rays. There was no real-world equivalent to Becquerel's photographs, and nothing like them had existed before" (ibid., 361f.), as Wilder points out. It has to be added that, of course, there was "a real-world equivalent", namely radiation itself. The modern dosimeter still works in accordance with the principle revealed in Becquerel's initial experiment, namely by darkening a specially prepared surface in the presence of and relative to the intensity of radioactive radiation. Wilder is, however, perfectly right in stressing the difference to picture production. Regarding radiation as the object of the photograph obtained is a rather uncommon way to think of the photographic process. Thus, Becquerel's photographs were not photographs *of* objects. Wilder's example makes perfectly clear how to conceive of the photographic process without the production of images.[12]

As a concluding remark on the different functions of photography, its depictive function as Maynard thinks of it should be considered. Of course, depiction is the functional role that is commonly associated with photography. Family snapshots, portraits, passport photographs, but also photographs in field guides for biologists or geologists are paradigmatic examples. They show the appearance of their depicted object and, thus, permit later recognition. Daston and Galison's discussion of the scientists' changing attitudes towards the essential characteristics of *objective depictions* (see Daston and Galison 1992) can illuminate the shifting emphasis on what photographs are meant to depict in the examples just mentioned. Whereas, in the former three instances above, the photographic pictures show particular people and their characteristics – a fact especially relevant to the evidential role played by photographs in passports – images of the latter kind are meant to depict typical features to be found amongst the members of a breed or kind. Thus, photographs in field guides, albeit of particular animals, are not regarded as images of individual beings, but as representations of typical characteristics.

Maynard uses the term 'imagining technology' to label photography that serves the functional role of depiction. Why he makes use of the word "imagination" in this context becomes clear when considering his theory of depiction. As I will discuss different approaches to the topic of depiction in section 2.2 in more detail, I will only summarise his main ideas about this topic here and postpone a critical analysis until later discussions.

First of all, Maynard makes us aware that photographic depiction apparently calls for an explanation for why it attracted and still attracts people in a way that handmade pictures do not. He states that:

> [...] it would surely not be enough to say that such surfaces [photographic pictures, N.M.] resembled, referred to, or carried information about their subjects. If they had only done those things [...], there might have been many important uses but hardly a popular market for photography, then or now.

(Maynard 2000, 114)

It seems to me that the question indeed arises why photography so attracted the public's attention and thus why it has been so successful in replacing other kinds of images in so many contexts. Obviously, there is something special to photography in this respect. Now, Maynard thinks that what fills this gap – and what is actually left out in the above quotation – is "inspiring vivid imagining of direct seeing" (ibid.) that photographic depiction enables its recipients to experience.

Thus, he takes it that the latter function of photography serves the enhancement of "our imagining or visualizing activities" (Maynard 2000, 83) and thus brings in Kendall Walton's theory of depiction.[13] Briefly expressed, Walton states that depiction belongs to the *games of make-believe*. He claims that "to be a 'depiction' is to have the function of serving as a prop in visual games of this sort" (Walton 1990, 293). In accordance with Walton's account, Maynard takes 'depiction' as being a "generic imagining enhancer" which he explains in the following way:

> [i]f its [the depiction's, N.M.] function is prescription of imagining, and its mode of access is visual inspection of surfaces, it may be thought of as converting the latter activity into part of the former. We look at the surface in order to imagine, and immediately imagine about that very looking as well.
>
> (Maynard 2000, 106)

Thus, depictions stimulate our imaginative capacities and allow us to imagine the represented situation or entity (see ibid., 107). Additionally, depictions trigger our imagination of looking at the latter; that is, we do not only image the entity depicted, but the direct seeing of this very entity. As he puts it:

> [w]hat we actually do of course is to look at a picture. What we imagine doing is seeing something else: whatever it depicts. But more: we imagine that our looking at the picture is the act of seeing what is depicted.
>
> (Maynard 1989, 266)

Maynard argues that it is precisely this second imaginative act that matters most to people when looking at pictures. It explains why, particularly in the popular domain, depiction implies more than merely transmitting information about the object of depiction. As an imagined observer, the viewer becomes mentally a (true) part of the represented situation, that is, she has the possibility of getting (mentally) involved, and it is also this particularity of depiction that feeds Maynard's criticism of mere resemblance-based accounts. "People look at pictures and go to the movies in order to imagine situations vividly, hardly to see what visual resemblances they can observe" (Maynard 2000, 103). Although this depictive function can also be realised with other visual representations as well, photography has made picture production more efficient by allowing more people to produce more images in less time, and it enables viewers to imagine more detailed things at a variety of places and times. It is in this sense that Maynard speaks of photography as "an engine of visualization" (Maynard 1989, 266).

Regardless of the fact that I am sceptical about this approach to depiction, Maynard definitely has a point in making us aware of the necessity to draw a distinction between the different functions that photography can serve. Without keeping them apart, epistemological concerns relevant in regard to depiction are transferred to the domain of detection and vice versa, although they genuinely do not pose a problem under those other circumstances. If, for instance, I am solely interested in detecting and recording a particular radiation, it does not matter whether its depiction appears naturalistic or not. Moreover, this differentiation also helps us to understand why photography is an important technological device in science. Its detective function can be regarded as the paradigmatic instance of a causally functioning measurement process. Combining this with its depictive function explains why it is so useful to human observers: we can simply assess the measurement result with our own eyes. As a technology, photography is also probably subject to developmental processes that have continuously broadened its scope of scientific applicability. To return at this point to the diversity of input and output of the photographic process, mentioned above: the basic principle of the recording mechanism, described by Maynard, allows not only visible light, but also for a variety of radiations as input. Depending on the sensitivity of the camera – the *detecting device* – ultraviolet (UV) and infrared (IR) radiation can be recorded as well as X-rays (which is also the basis of Computer Tomography, CT)[14] or radioactive radiation (often used as contrast media in medical imaging). Of course, contrary to visible light rays, those latter kinds of radiation are not reflected by objects in front of the camera, but emitted either by those sources themselves or, in the case of X-ray photography, by the diagnostic apparatus.

Similarly, the output of the photographic process can be of different kinds, too. Discussing this issue at the end of this section about photography will also allow us to take a look at some more cases of its application in science. As explained above, only some results can and should be called pictures. But even within this class, there are varying types of output. X-ray images, for example, can be regarded as resembling their objects of depiction, though they show completely different aspects to photographs produced by using visible light rays. X-rays are capable of passing through soft materials such as human skin and flesh, but are blocked by harder ones such as bones. Consequently, photographic film will record radiations that pass through the human body so that the inner parts will be visible in the respective image. Because of this divergent pictorial content, whose interpretation was nonetheless informed by the familiar practice of photography to produce images of the depictive object's visible appearance, the invention of X-ray images was accompanied by a number of questions concerning their interpretation (see e.g. Golan 2002;

Dünkel 2008), namely who could read those images correctly, whose opinion could be regarded as an authoritative expertise in this respect, could they be regarded as reliable evidence in the medical sciences and elsewhere?

Another variety of the photographic image in science is *composite photographs* – a method that combines detections of different radiations in the resulting image. An example can be seen in Figure 2.2 showing the different kinds of radiation emitted by the galaxy M51. Obviously, such composites are not produced automatically, but presuppose careful reasoning about what information is supposed to be combined in a picture and why. Combined images of different radiations, such as shown in the pictorial example of the Whirlpool Galaxy (see Figure 2.2), allow, among other things, inferences about hidden sources of radiation. If, for instance, cosmic dust covers a cluster of stars so that its emission of visible light is not directly detectable, infrared (IR)

Figure 2.2 M51 in false colours. Copyright: X-ray: NASA/CXC/Wesleyan Univ./ R.Kilgard et al; UV: NASA/JPL-Caltech; Optical: NASA/ESA/S. Beckwith & Hubble Heritage Team (STScI/AURA); IR: NASA/JPL-Caltech/ Univ. of AZ/R. Kennicutt. For a colour version of this image see www.jpl.nasa. gov/spaceimages/details.php?id=PIA10200

images can still show its presence. Combining the IR photograph with one taken by using the visible-light spectrum will show both the cluster of stars in the background and the cloud of cosmic dust in the foreground. Hence, composite images will often combine elements visible and invisible to the unaided human eye. But, of course, composite images can also be used to blend other aspects such as photographs of the same rural area taken at different time slots to detect changes in vegetation or an increase in population in cities, etc. Here, *false colours* are often added to indicate the different layers of information merged into one image.[15]

These last remarks about the different applications of the method of compositing make clear that the advantages of photography in science are not only due to its ability to detect and record a *variety of different radiations*, but that it can help to overcome the deficiencies of the human eye also in other respects. I will briefly discuss two more of such aspects in the remaining part of this section. These aspects are related to the dimensions of (1) *time* and (2) *space*.

Let us start with (1), the dimension of time. One apparent advantage of photography is that its ability to detect radiation is also combined with its capability to record it, thus making it available for later inspection. Moreover, this recording can take place at a speed much faster than discernible by the human eye. Consequently, one of its first applications in science was *chronophotography*, that is, the depiction of moving bodies.

Scientists regarded the possibility of slowing down or even stopping motion as an important achievement which enabled them to investigate certain processes in more detail. A famous example is offered by the photographs taken by Eadweard Muybridge. Figure 2.3 shows one of his picture series. Muybridge was interested in finding out the correct positions of the hooves of a galloping horse – a phenomenon not observable by the naked eye.

As the human eye is too inert to clearly discern particular sequences of motion at high speed, the possibility of chronophotography, of which Muybridge's studies furnish an example and which is commonly known from photo finishes in sports today (see Maynard 2000, 135ff.; 2017, 50ff.), is obviously of particular usefulness in scientific investigations. Yet it is not only photography's capacity to freeze time, so to speak, that makes it a useful device in scientific investigations. The photographic image is also a basic component of films, as moving images exploit a certain inertia of the human eye. Regarding the projection of 24 frames per second evokes the impression of movement. Thus, films are simply a fast projection of still images.

Combining this with (2), the aspect of space, guides us to another important domain of application in science. Often scientists make use of photography's capacity to collect radiation emitted or reflected by even faint entities that are either remote or minute by combining cameras with other scientific instruments, in particular telescopes or microscopes. Auto-tracking allows the recording of radiation during several hours (days, or even weeks) that enabled, for instance, Hubble's discovery of Styx (see Figure 2.1).

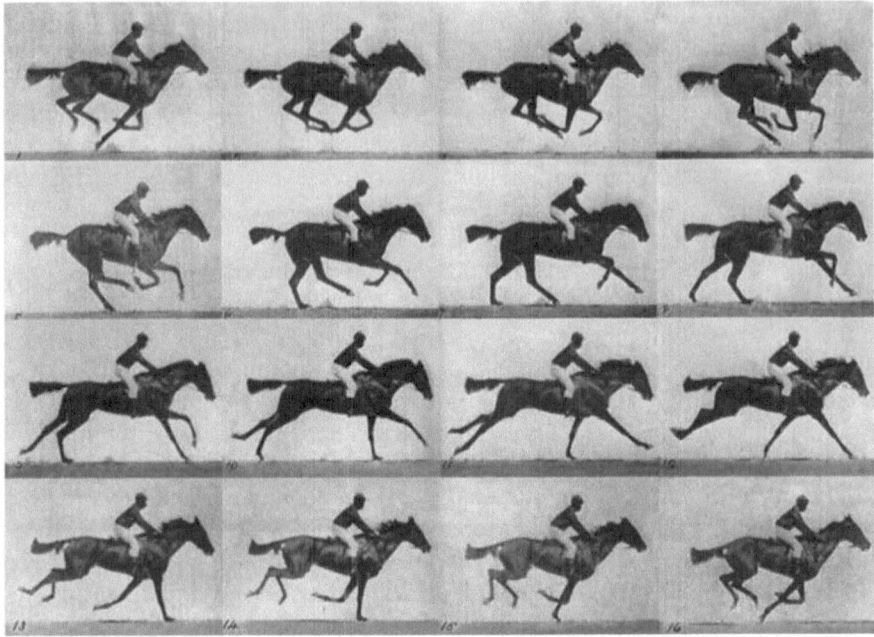

Figure 2.3 Photograph by Eadweard Muybridge, Human and Animal Locomotion, plate 626, thoroughbred bay mare "Annie G." galloping.

Source: http://commons.wikimedia.org/wiki/File:Muybridge_race_horse_gallop.jpg

Moving from the remote to the minute, another illustrative example of the extension of the scope of scientific investigation by photography is video microscopy. Here, Dieter G. Weiss shows that sometimes we need motion to discern particular properties of a research object (see Weiss 2012). Within the context of his discussion of the history of microscopy, he not only points out an interesting alteration in this observation technology, namely the shift from electron microscopy to video microscopy in the 1980s, but also makes clear that this shift came along with a significant conceptual change concerning the human cell in microbiology.

The investigation of the human cell with the aid of electron microscopy guided scientists in their ontological reasoning about their object of research. According to Weiss, this method also narrowed the focus of research considerably. He makes plain that, as a result of using this instrument in observation practices, it was *structural aspects* of the cell that were designated as of special importance for further investigation (see ibid., 308). Weiss, however, shows that this focus was derived from an artefact of this observational practice due to a particular method of preparing the specimen presupposed by electron microscopy, namely the dehydration of the cell (see ibid., 307).

Due to this technique of preparation, the concept of a *cytoskeleton* became popular; that is, a network that consists of different types of filamentous protein polymers and, as such, is a part of cytoplasm. These structures were interpreted in analogy to the sustaining elements in human bones, thus leading to the idea of a cytoskeleton (see ibid., 308). Consequently, a *static* conception of the cell became the common point of view in the respective scientific community.

Nonetheless, as Weiss explains, this apparent essential structural feature of the human cell was nothing other than a mere artefact of the preparation process (see ibid., 309). This misconception, however, did not become obvious until a new method of microscopy was invented, namely electronic or computer-based microscopy, firstly by means of analogue, afterwards by means of digital cameras (see ibid., 309ff.). Video microscopy at this point marked the first step in this developmental process that allowed investigations of the living cell (see ibid., 311). This new observational method gave access to the *dynamic aspects* of the cell and made clear that the cytoskeleton was by no means so inflexible and inelastic as previous observations with the aid of electron microscopes had suggested (see ibid., 313). Concisely, then, only the moving image was able to correct the wrong conception of purely static structures in this context and return the dynamic aspect of the human cell to the scientists' understanding of their research object (see ibid., 311f.).

Furthermore, a variety of photographs can be used both as stills and as moving images. In astronomy, for instance, a series of photographs of a particular celestial region can be used in an animated version to detect planets, satellites, etc. via their movements in front of the more or less static background of fixed stars. As stills, the same pictures can be used to find out how many celestial bodies of a certain magnitude are visible in which part of the sky, that is, to produce a map of the celestial region.[16] Obviously, what matters here is the particular interest of research that determines whether a certain series of photographs is regarded in an animated version.

From an epistemological point of view, then, photography supports and enables scientific investigations of properties and entities not analysable by the naked human eye. Moreover, it allows the storage of visual data, making it available to a broader range of scientists – both on a synchronic and a diachronic level. Its ability to depict what has been detected makes this data cognitively accessible via vision – the primary human sense.

It has to be added, however, that in many instances it is *digital photography* that plays this role in science today. Even though one might wonder why this technological development should make a difference to what has been said about the role of photography in science so far, it has to be noted that the stance of some scholars towards this technology has indeed been influenced by this issue. In particular, the *evidential status* of photography is called into question by many scholars due to the advent of digital images and the increasing availability of easy-to-handle post-processing software. This discourse affects especially the trustworthiness of media reports making use of digital images (see e.g. Mitchell 1994; Nida-Rümelin, Steinbrenner and Lopes 2012; Savedoff 2008). Here again, this reasoning is guided by the assumption that photography allows for the automatic recording of events and thus for eliminating any interfering (and possibly distorting) human intervention. It is this ideal that now seems to be threatened by new technological developments.

However, and due to the technological nature of photography overlooked in this discourse in media studies and aesthetics, the digital development that is highlighted as a potential threat to the evidential status of photography above is often warmly welcomed in science, as Maynard points out.

> With amplification/suppression in mind, it is useful to compare how the advent of *digital* processing of photographs was greeted, as opposed to processes in medical and engineering imaging. Fears were immediately expressed about the potential manipulation of photo images for news, historical, forensic, political, or advertising purposes, without recourse to a stored original for comparison. By contrast, 'image processing' is the routine term for a generally observed *advantage* of electronic-optical imaging over the photochemical in the technical literature of engineering.
> (Maynard 2000, 143f., his italics)

One rationale for these different assessments of digital technology and its abilities is undoubtedly based on the fact that this development has contributed considerably to the broadening of photography's scope of application in science. It is digital technology that allows photography to play such an important role in astrophysics and related disciplines. Mounting a camera on a Mars rover[17] or connecting one to a spacecraft[18] or telescope such as Hubble only makes sense if the resulting images can be transferred safely back to Earth for analysis. Sending them via spacecraft would not only be quite an expensive solution, but also one entailing the considerable risk of losing the data altogether. Being able to transfer them via a computer downlink is therefore a highly appreciated alternative. Advantageous as these characteristics of digital photography are in science, the challenges of post-processing, mentioned above, still demand a more thoroughly epistemological analysis, to which I shall return.

The advent of digital images, however, has not only led to critical questions concerning the evidential status of photography, but has also affected other image producing processes that, at first sight, seem to be similar to photography. A particularly telling example in this regard is functional magnetic resonance imaging (fMRI), broadly used as a diagnostic tool in medicine. Even though fMR images – or *brain scans* as these images are often called – are produced in a completely different way to photographs, there are nonetheless certain commonalities and similarities that trigger a reasoning by analogy (to photography) concerning their evidential status. The discussion of this case in the next section will reveal why such an analogical reasoning constitutes epistemological challenges that have to be dealt with in the scientific context.

2.1.2 Imaging techniques

Contrary to the previous section and also in contrast to those to come, this section is labelled not in accordance with a category of visual representations, but explicitly invokes the processes of their production. The label of imaging techniques has become a familiar one in medical science, where it is used as an umbrella term to denote different technologies producing visual representations of (parts of) the human body. The category of 'imaging techniques' is thus orientated towards a shared object of research, but not so much towards commonalities in the respective processes of picture production. For instance, X-ray images as well as medical photographs are subsumed under this label, yet sonograms and functional magnetic resonance images (fMR images) also belong to this class of imaging techniques.

Are not all of these examples images nonetheless? Is it not therefore perfectly correct to subsume the respective technologies under this very label? Yes and no. My hesitant response is due to the fact that, although clearly faced with a broad mixture of imaging technologies here, their interpretation and, consequently, their status as diagnostic tools in medicine is, almost without exception, due to an analogy drawn between practices of interpreting those images in medical science and our familiar handling of photography. That is, we take it that a causal process holds between image and object so that the latter can be regarded as the cause of the former in the sense explained above. This is why we think that those images are informative about their causes and are thus useful as diagnostic tools. Furthermore, it is also this conception of imaging techniques that triggers attempts to transfer the particular evidential status of photography to technologies of this domain.

Yet although imaging techniques in medical science do demonstrate some important commonalities with photography, many of them seem to have more in common with the type of images that I will discuss in the following sections, namely *computer-generated images* such as graphs or diagrams. Most of them are digital images visualising measurement data. Those images visualise properties that are by their very nature not accessible by vision in any sense. They actually visualise the phenomenon at hand, i.e. they *produce* a visual appearance of something that can then be inspected by our perceptual means. Sonograms, for example, are, as the word says, produced from sound – but nonetheless *show* us their objects in a pictorial way. Why interpretative practices of such data images that draw on the evidential status of photography pose particular epistemological challenges, I will discuss in the following by analysing the case of fMR images in more detail. Here, I will take advantage of Kelly Ann Joyce's and Adina L. Roskies's investigations of the topic. However, a start can be made at this juncture by pointing out what kind of measurement data those images present and how.

> MRI technology was introduced to clinical practice in the 1980s. It was initially used to make pictures of the brain, the spine, and the area around the joints. Today, MRI technology is used to visualize the heart, the breasts, liver, arteries, and other soft tissues in the body.
>
> (Joyce 2008, 4)[19]

Interestingly, now, although MRI is regarded as an imaging technology, the way it functions does not imply the recording of any visible property of the human body. As Joyce states:

> MRI, despite its current construction as a visualizing technique, does not produce anatomical images in a straightforward fashion. It has no photographic lens, nor does it use x-ray techniques to create pictures of the internal body. Instead, MRI is used to numerically measure how hydrogen nuclei absorb and release energy in response to particular frequencies.
>
> (Joyce 2008, 25)

To put it briefly, then, MRI machines are scanner magnets whose measurement results are visualised. But what kind of data do they collect? Inside the scanner, the atomic nuclei of the human body are all aligned in the same direction by the magnet surrounding the tube. During the procedure, "[s]hort bursts of radio waves are then sent to certain areas of the body [which are in the focus of the respective diagnostic treatment, N.M.], knocking the protons out of alignment. When the radio waves are turned off, the protons realign [the time span to do so is called *relaxation interval* or *relaxation time*, N.M.]. This sends out radio signals, which are picked up by receivers", as the website of the National Health Service in England informs its readers (www.nhs.uk/Conditions/MRI-scan/Pages/Introduction.aspx, accessed March 31, 2016). It is these signals that constitute the measurement data of MRI. What kind of information they provide is explained in the same text as "information about the exact location of the protons in the body. They also help to distinguish between the various types of tissue in the body, because the protons in different types of tissue realign at different speeds and produce distinct signals" (ibid.). By recording those signals and later visually evaluating the results, fractures or cancer and other diseases can be detected by making use of these scans.

Joyce describes the procedure in the following way: "[t]he body is transformed into an array of numerical measurements that are then coded into images. The images are then interpreted by a physician, who is usually a radiologist, in a written report" (Joyce 2008, 14). One might wonder why the recorded signals are transformed into a visual representation at all. A crucial part of the answer is already present in Joyce's quotation by highlighting the radiologists' responsibility to evaluate the respective images. Why is it so important which professional group is in charge of the evaluation of results here?

As Klaus Hentschel explains, this technology was initially not developed for medical purposes (see Hentschel 2014, ch. 5.3), but as a measurement device in nuclear physics (see ibid., 188f.). Soon, it was also made applicable in chemical analyses (see ibid., 190). None of these initial applications made use of images. The question is, then, why did MR technology become an "imaging technique" when it entered medical science?

Joyce explains this interesting shift by pointing to the radiologists claiming responsibility for interpreting the results of those measurements in the clinical context. She draws our attention to a particular *visual culture* pre-configuring the practitioners' approach to this new diagnostic tool. "Radiologists are disciplined [...] to represent and interpret their area of focus [...] as black and white images. They do not work with the body in numerical form" (Joyce 2008, 38). This professional training then brought about two important changes in presenting MRI results. Firstly, although available and formerly accompanying the respective images, numerical data are now omitted. Secondly, the results are presented as greyscale images and no longer as colour images (see ibid., 39f.).

It can be argued that, indeed, making MRI results accessible to human vision for the evaluation of the data brings about certain advantages – in particular in the complex domain of diagnosis where a multitude of factors interact so that it seems rather unlikely that a machine or computer can suggest the relevant treatment by simply inferring it from numerical measurement data. However, the particular method of data presentation in the case of MR images – namely as a seemingly photographic picture of the inner body – also gives rise to certain epistemological challenges. The difficulties provoked by apparent analogies between these two representational means are indicated in Joyce's study when she states that "[...] the myth of photographic truth draws on the cultural belief that photographs reveal the physical world in an unmediated manner. With anatomical pictures as their final product, MRI examinations excite cultural ideas about photographs" (ibid., 151).

Why exactly such an analogical reasoning from photography to data images constitutes epistemological problems, is discussed in detail by Adina L. Roskies regarding functional magnetic resonance images (fMRI) – commonly also known as 'brain scans' (see Roskies 2007; 2008). Let us start with a brief note on what distinguishes fMRI from MR images.

As its name suggests, fMRI is based on the same kind of technology as MR images. Yet it is used for (slightly) different purposes in neuro- and cognitive science, namely, to put it bluntly, *to show the brain in action*. Contrary to MR images that show anatomical structures, fMR images show metabolic functions of the brain. The basis for fMR scans is what is called the *BOLD effect*, i.e. the *b*lood *o*xygenation *l*evel *d*ependent effect. What is measured here are changes in the oxygenation of the blood flow in certain areas of the human brain while the subject in the tube is performing specific tasks such as watching a movie or recalling items already learnt. A high rate of oxygen in the blood is related to neural activities in the sense that the latter increases the ratio of the former. This change in oxygenation can be measured by fMRI scans, as oxygen-rich blood has different magnetic properties to oxygen-poor blood. As a result, fMR images are supposed to show which brain areas are activated by different kinds of cognitive tasks and are thus interpreted as being the responsible parts of the human brain enabling these activities. Brain scans are therefore mainly used to map those activities of the brain to better understand the functioning of this organ.

Having pointed out what kind of information is processed to produce fMR images, let us now come back to the difficulties arising from drawing analogies between these images and photographs. Here, Roskies offers a detailed comparison between both types of visual representation. Her analysis is based on the background assumption that, normally, laymen regard brain scans not only in analogy to photographs, but even equate them with the latter. "Images are thought of as photographs of brain activity" (Roskies 2007, 861). Because of the particular epistemic status that photographs possess, such an analogy – if valid – might explain the habit of using fMR images as evidence in the context of medical diagnosis. If fMR images are like photographs, then the epistemic particularity of the latter might also hold for the former (see Roskies 2008, 21). But is this analogical reasoning justified?

The line of reasoning which Roskies presents is highly motivated by the causal and assumedly automatic method of picture production in the case of photography. Hence, a variety of objections could be raised with regard to the comparative results just mentioned if this conception of photography is not adhered to. Bearing this prerequisite in mind, however, agreement can be reached with Roskies that our *everyday conception* of photography is significantly informed by the ideal of a non-interventionist and automatic process of picture production. This becomes clear when taking into consideration the nature of our usual interest in photographs. Scruton describes our everyday attitude towards photography as one of curiosity concerning the object depicted. The background assumption thereby is that regarding x (i.e. the object of depiction) or regarding its photograph are interchangeable activities.

> In some sense, looking at a photograph is a substitute for looking at the thing itself.
>
> (Scruton 2008, 149)

> [t]ypically, therefore, our attitude toward photography will be one of curiosity [...]. The photograph addresses itself to our desire for knowledge of the world, knowledge of how things look or seem. The photograph is a means to the end of seeing its subject [...]
>
> (ibid., 152)

Such an attitude towards the visual representation at hand is also shown for MR images by Joyce. She highlights that often those images are presented and understood as being "interchangeable with the part of the body being scanned" (Joyce 2008, 49). Regarding her comparative project, Roskies focuses on the non-professional audience. That is, she refers to laymen with regard to neuroscience and its correlated theories and practices. It is here that laymen strongly tend to regard fMR images as analogous to photographs. But whereas the latter are *inferentially proximate,* as Roskies claims (see Roskies 2008, 22), the former are not. *Inferential distance* thereby refers to the effort of reasoning involved when inferring information about the object of depiction from the correlated image. Whereas there are only a few steps of reasoning involved in the case of photography, in the case of fMR images a lot more has to be taken into consideration to reach a conclusion about the depicted object.[20]

Although superficially similar, brain scans and photographs differ in some crucial respects which are also responsible for their difference in inferential distance. The first relevant aspect is about *visual properties.* Normally, photographs show only such features of the object which are accessible with our visual senses. Such pictures thus inform about the colour, the shape, the relative size of an object and the like. Brain scans, on the other hand, do not depict visual properties of the human brain.

> Functional MRI is not transparent in the sense that photographs are, for the information we are interested in and that the technique is sensitive to is not visual information at all. What fMRI allows us to visualize directly are magnetic properties of the water in the brain.
>
> (Roskies 2007, 863)

Thus, to understand what fMR images show, there is firstly a need to grasp the relevant background theory of the correlated instrument. To interpret the resulting images correctly, we have to know what the detecting device is sensitive to. Such background knowledge then restricts the set of possible interpretations of the respective image (see Perini 2012b, 166).

Yet the requirement of such background knowledge is exactly the point of concern in the case of fMR images. Here, Roskies highlights the fact that, contrary to photography, we do not "understand the causal and counterfactual relationships between the images and the data they represent" (Roskies 2007, 871). Of course, fMR images are causally dependent on a particular object. However, this is different to the case of photography in two ways. (1) Professional users of fMR images often lack an understanding of the technology involved; that is, they do not know, for example, when the measurement fails and only artefacts are produced. Thus, what the observer lacks in the case of brain scans is a *robust understanding of the imaging process.*[21] One might object that although a robust understanding of technological devices is seldom available to the user, one would not call into question their ability to make use of their devices. For instance, many people do not understand how their smartphones or computers work. Nevertheless, they are able to use them efficiently and effectively in a correct way. What is of relevance here is that at least some people do understand the technological functions correctly. By taking advantage of the *division of epistemic labour*, other users can then rely on expert knowledge.[22] However, such a rejoinder only partly undermines Roskies's critical evaluation of experts' competence in visual practices based on fMR images. There is another important dimension involved in a correct understanding and thus an interpretation of brain scans that also seems to be deficient here.

(2) This leads to the second deviating aspect with regard to the assumed analogy between photography and fMRI. Observers also lack a clear understanding of "the connection(s) between task and neural activity and the MR signal" (Roskies 2007, 865). As was pointed out above, brain scans are supposed to show activities of the human brain while the subject performs certain tasks in the tube. Two correlations are implied here that are not understood sufficiently by the scientists involved, as Roskies points out. On the one hand, there is the correlation between task and activity in the brain. On the other, there is the correlation between the activity in the brain and the oxygenation of the blood, that is, the measurement datum of fMRI. In this sense, a correct interpretation of fMR images presupposes, secondly, that we know what this visualisation might tell us about the human brain.[23] Explaining what brain scans show implies being able to identify different patterns of activity in the brain and to infer information from them; that is, being able to explain what these patterns might tell us about brain activities and how they might be related to mental activities such as remembering. Hypotheses about these correlations are entailed in neurobiological theories. Hence, background knowledge of those theories is also necessary to interpret and understand those visual representations correctly. Yet it is exactly here that many vagaries remain in the scientific evaluation of those images, as the relevant hypotheses have not yet been sufficiently worked out.

Comparing this practice of interpretation with the case of photography makes clear that especially the latter aspect is normally lacking in the interpretative processes of photographic pictures. Usually, we do not wonder what a photograph might show and how to interpret it correctly. We tacitly assume that it exhibits the properties of an object (being present in front of the camera during the exposure) which, in principle, we could also have observed with the naked eye.[24] There are no further theories involved here that are needed for the interpretation of the picture. Consequently, as the interpretation of fMR images presupposes background knowledge about a variety of additional theories, these images are, in Roskies's words, inferentially more distant from their object of depiction than photographs are. Concisely, the resulting difficulty is the following:

[o]n the one hand, the phenomena that neuroimaging studies investigate are inferentially far removed from the images produced; on the other hand, the image format is familiar and accessible. Like photographs,

brain images [...] appear to wear their content on their sleeves. A consequence is that images are liable to be mistakenly apprehended as inferentially proximate.

(Roskies 2008, 21)

Roskies's analysis highlights that, even though there is a referential relation between object and image, guaranteed by the causal connection between
object and measurement device and the correct sensitivity of the latter, this
relation can be more or less direct.

Furthermore, the interpretation of measurement data can be more
complicated if the relevant background theories – both with regard to the
instrument in use and the explanatory theory concerning the object of
research – are not understood well enough by scientists. This is exactly the
point that Roskies argues for in the case of brain scans. Perini broadens
the scope of this claim, when making us aware of the difficulty involved in
interpreting images in the realm of *basic research*. She points out that some
scientific images used in this context – for example, results of imaging technologies or photographs – can serve the function of detection, although not
all of them can be understood as depictions (see Perini 2012a, 158). Referring
back to Maynard's distinction between different functions that photography
can serve, Perini points out that detection simply means that a scientist can
infer that her instrument has recorded *something*. Moreover, she might also
read off her data that what has been detected shows certain characteristics.
However, regarding the results of such imaging techniques as depictions
and using them as representations of what has been observed in scientific
reasoning implies more than a mere glimpse at the resulting image. Here,
interpretative effort is needed. Regarding visual representations in science as
depictions presupposes that the observer has the relevant background knowledge to interpret the results as depictions of certain entities. In accordance
with Roskies's theses, Perini also mentions two kinds of epistemic resources
that are relevant to such an interpretive endeavour. The scientist has to have
background knowledge of the object depicted and the channels of information transmission, that is, a theory about the function of the imaging device
at her disposal. Both aspects can help the scientist to understand how the
image was construed as a depiction and, consequently, guide the hermeneutic
process. However, in quite a few instances, scientists lack this background
knowledge so that "[...] comprehending the image as a depiction can lead to
misrepresentation [...]" (ibid., 158).

In accordance with these considerations about the necessity of interpretation, we can state that fMR images are the results of *indirect measurements*
and presuppose a theoretical understanding of what is going on for their
correct interpretation. This is what Peter Kosso means when pointing out
that the amount of interpretation can vary enormously concerning different
objects of scientific observation (see Kosso 1988, 456). Sometimes we can
simply see what has been recorded, whereas on other occasions a considerable

number of background theories play a role before we are able to decipher the respective image.

For sure, this invocation of several theories relevant to a correct image interpretation might raise epistemological suspicions about the latter – and the more theories play a role, the greater these suspicions might become, which in turn raise such questions as whether there can be any certainty that the theories invoked are correct, whether it can be truly guaranteed that no errors occur in the process of interpretation, and whether all of the interim phenomena are sufficiently well understood so that scientists can rely on hypotheses about them in the interpretative process. It is here that James Bogen challenges the practice of using fMR images to gain knowledge about processes of the human brain. He points out that these images might blur the crucial distinction between *data production* or *detection* and *data interpretation* (see Bogen 2002, 144). As scientific images are normally produced along the lines of certain theoretical hypotheses, one might wonder whether scientists are tempted by these theoretical assumptions to omit details or emphasise others, although such considerations are sanctioned by theory alone and not by the data measured. Problems such as these are precisely the starting point for some people from STS and the history of science to look for a different account of what is happening in such cases (see e.g. Burri 2008; Carusi et al. 2014; Hennig 2011). They opt for a *constructivist* interpretation of scientific images, that is, they defend the thesis that, as those images are artefacts, they are not neutrally informative means about an independently existing entity in nature. With respect to MR images, Joyce puts this in the following way:

> [a]n MRI scan – like all representations – is a constructed artifact. Despite common narratives that position these exams outside of social relations, there are many sites in their production, interpretation, and use that transform them from conveyors of objective, authoritative knowledge into socially situated objects *that construct the body in complicated ways.*
>
> (Joyce 2008, 61, my italics)

Thus, imaging practices, such as the ones discussed by Roskies, trigger debates about the possibilities of basing claims to knowledge about a mind-independent world on visual representations as measurement results. Perhaps, as Joyce seems to suggest, the outcome of this consists merely of socially constructed artefacts. Yet scientists do not seem to give in to this premise – at least their epistemic practices of using visual representations seem to suggest that they can play the *evidential role* that is called into question by such constructivist claims.

In the following two sections, the tension mentioned between scientific practices and epistemological problems arising from them will become even more salient, as the types of visual representation analysed next are clearly even more remote from their object of depiction than the cases examined so far. We will proceed by investigating further the domain of data images or data visualisation.

2.1.3 Data visualisations

I will begin this section with an example, the discussion of which will make plain why certain problems related to the labelling of the section present themselves once again. My example will also help to point out which strategies might be chosen to face these difficulties and why instances of this type are often especially critically regarded when it comes to an epistemological assessment of their evidential role.

In 2012 the discovery of the so-called *Higgs boson* was announced by scientists working at CERN[25] in Geneva. To understand the huge relevance of this finding for contemporary physics, let me – very briefly – describe what this discovery was about. The so-called *Standard Model* in particle physics tells us which fundamental particles exist in the world and how they interact.[26] Part of this theoretical model is the Higgs boson, which is said to be the particle responsible for all others possessing mass. In this way, the Higgs plays a major role concerning the consistency of the Standard Model in physics. But even though its existence had been theoretically predicted by Peter Higgs and others in the 1960s (see https://physics.info/standard/, accessed January 9, 2018), its existence was not confirmed experimentally until 2012, when it was finally detected by the ATLAS[27] and CMS[28] experiments at the Large Hadron Collider (LHC) at CERN. The core of both experiments consists of general purpose detectors[29] located at LHC, the world's largest particle-collider. It accelerates two beams of protons in different directions in a circular collider. At certain points, that is, where the four main experiments are located, the two beams are brought to a collision. The protons are smashed at high energy levels, thereby producing other fundamental particles that can be registered by the detectors of ATLAS and CMS. In 2012, the detection of the Higgs boson was finally rendered experimentally possible in this way. An important part of the announcement of its discovery comprised various graphics visualising the measurement results within the instrument – the detecting devices of ATLAS and CMS at the LHC. Several of those images were circulated among the scientific community and the broader public as evidence of the successful experiments.

I will discuss one of them in more detail at this point (see Figure 2.4). It was part of the official announcement of the discovery in the respective CERN release in 2012 and also published in a related article in *Science* (see Abbaneo et al. 2012). What exactly does this image show and how was it related to the announced discovery? To answer this second question first, the image was published together with the respective measurement results and the statistical evaluation of data gathered between 2011 and 2012. That is, the announcement of the discovery was not based on one single detection, as many of the detected events can also be obtained due to other "known

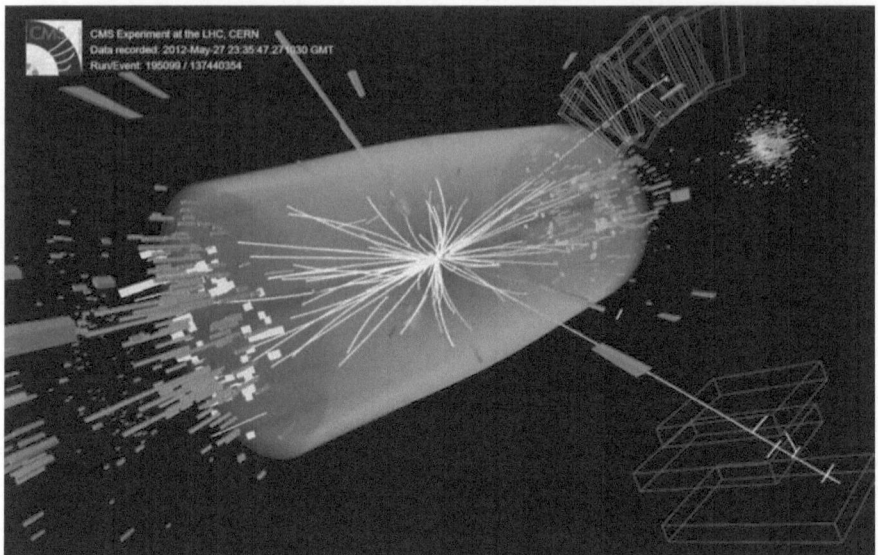

Figure 2.4 Higgs boson discovery at the LHC – CMS Higgs Search in 2011 and 2012
     data: candidate ZZ event (8 TeV) with two electrons and two muons.

Source: CERN, for the benefit of the CMS Collaboration. For a colour version of this image see
http://cds.cern.ch/record/1459462

SM [Standard Model, N.M.] background processes", as the authors of this
paper expressed it. Only by collecting data over a certain period of time, and
their subsequent statistical analysis, were scientists finally convinced that the
probability was high enough to postulate that their detective results showed
a Higgs boson. That is, physicists were finally convinced that their resulting
measurements showed a significant deviation in comparison to data that
would have been collected as the results of SM processes only (see ibid.,
1574f.).

Although Figure 2.4 does not then show *the* detection of the Higgs boson,
as it is only one recorded event amongst many others, it nonetheless is meant
to demonstrate what a typical event, taken into consideration when compiling
the data for their statistical evaluation, looked like. In this functional role, this
image also bears an *evidential role* in the context of the announcement. It is
evidence of what has been detected by the CMS experiment. Does this then
mean that it is also evidence of the discovery of the Higgs boson? Yes and
no. Two related aspects are responsible for this rather hesitant answer. On the
one hand, Figure 2.4 shows only one possible event of the data collected that
results in the relevant statistical significance supporting the announcement of
the discovery. On the other, the event shown in this image might also be due
to other SM processes. For a better understanding of this latter point, let us

turn to an investigation of what exactly this image shows. Does it depict the Higgs boson? Does it show its discovery? Or what do scientists mean when they talk about "events"?

In the colour version of this image we see an artificial tube and many orange lines with an apparently common centre. Moreover, there are two green and two red lines emerging from the centre at particular angles. But where is the Higgs? What do these coloured lines tell us about it? Does this image represent the Higgs boson at all? The caption to the image tells us:

> [e]vent recorded with the CMS detector in 2012 at a proton-proton center of mass energy of 8 TeV. The event shows characteristics expected from the decay of the SM Higgs boson to a pair of Z bosons, one of which subsequently decays to a pair of electrons (green lines in the tracker matched to green towers in the ECAL in the central region of the detector) and the other decays to a pair of muons (red lines). The event could also be due to known SM background processes.
>
> (Abbaneo et al. 2012, 1571, fig. 3)

Consequently, this is not an image of the Higgs boson at all. It is not a direct representation of this particle. It visualises the collision of subatomic particles – which are subsumed under the label of *theoretical entities* in the philosophy of science. Such entities are predicted by the theory at hand but are not observable with the unaided eye.[30] Furthermore, the Higgs boson is not a stable particle. It almost immediately decays into other particles when produced in a collision. There are at least five processes of decay possible, namely the decay of a Higgs to two photons, two Z bosons, two W bosons, two bottom quarks or two tau leptons – except for the photons, all of these particles will then decay further into particles detectable by the CMS experiment (see ibid., 1569). Thus, what is shown in Figure 2.4 is such a decay process of an assumed previously present Higgs boson. It is an *indirect measurement with the aid of indicators*. The theoretical assumption is that, if a Higgs boson decays into two Z bosons, the detectable radiation pattern will look like the one shown in Figure 2.4. Yet inferring statements from such indirect measurement data yields only results with a certain likelihood. They do not guarantee that the discovery actually obtained. This is what is expressed by the proviso that other SM processes might also show these characteristics. It is in this sense that it is *very likely* that Figure 2.4 shows the indirect measurement of a Higgs boson, but the image might also be produced otherwise.

Mario Bunge (2010) stresses the relevance of indirect measurements in the broader scientific context when pointing out that it is *indicators* of the phenomenon under investigation which are normally read off the respective instrument. "But most measuring instruments, whether old or new, do not show directly the value of a property: what they show instead is the value of an indicator, that is, an observable counterpart of the unperceivable item" (Bunge 2010, 86). Bunge makes use of this finding to formulate a harsh critique, namely that philosophers of science usually neglect the difference between *indicator* and *phenomenon* (see ibid.). However, it should be added that, although Bunge might be right in pointing out that the term 'indicator' is used rather seldom in philosophical discourse, this does not imply that its scholars are not aware of the distinction mentioned. Many of them use the term *evidence* in this context. Take, for example, Kosso's considerations on the topic of observability, discussed above. He makes us aware of the fact that, in some instances of data interpretation, more theoretically induced inferences are needed than in others (see Kosso 1988). It is these latter, more ambitious interpretive endeavours which are usually based on indicators, as Bunge calls them. Moreover, Kosso also shows that knowledge claims based on such inferences are not necessarily more epistemically challenging than those based on direct observations (see Kosso 2006). He makes clear that other aspects – such as obtaining corroborative results by using different measurement devices or methods, a tactic scientists at CERN followed when comparing the measurement results gained from the CMS and the ATLAS experiments relying on different detector technologies – might even turn them into a more reliable kind of evidence than direct observations can yield on other occasions (see ibid., 234f.).

Taking into account the way the Higgs image is derived in these processes of indirect measurements, one might be tempted to call it simply the product of another imaging technology like the fMR images discussed above. The Higgs image is produced from data visualised on a computer screen. It also shows phenomena unobservable to the unaided human eye. Yet there are apparently significant differences – beyond the divergent processes of measurement and image production. In particular, nobody would feel tempted to reason by analogy from the evidential status of photographs to the epistemic role played by images such as Figure 2.4. There is no similarity to photographs in any sense involved in this case that might motivate such reasoning. On the contrary, the Higgs image looks rather artificial.

I started this section by pointing out that it is difficult to name the type of image with precision. This difficulty arises as it has a variety of aspects in common with different image types, but also diverges from them in certain respects. Calling it the result of an imaging technique has already been dismissed, then, from the list of possible labels. Another option would be to call it a *computer graphic*, as it is undoubtedly produced with the aid of IT devices. However, such images are usually meant to depict something in a more or less naturalistic way, contrary to how the Higgs image represents the phenomena detected. Electrons are neither green, nor muons red. None of them leaves lines as traces. Moreover, what computer graphics depict are often material objects, that is, entities that have visible properties – such as, for instance, the cross-section of a machine in engineering science (see Schirra 2012, 335f.). In contrast to this, the Higgs image is so remote from the visible world that calling it a computer graphic seems to be inappropriate.

In his explanation of what computer graphics are, Jörg R. J. Schirra introduces another category that seems worth considering when looking for an image type to categorise the Higgs image, namely "information visualisation" (see ibid., 336). Here, algorithms are used to visualise data which is not pictorial, visually perceptible, or spatial. It seems advisable to subdivide this category further by making use of the last parameter, namely space. Tamara Munzner (2008) introduces the respective distinction when she explains that *information visualisations* imply the deliberate choice of spatial properties to visualise particular data, whereas the dimension of space is already inherent to the data which are used to produce what she calls *scientific visualisations* (see ibid., ch. 6). From the perspective of computer science, this latter expression seems then to be the correct term to apply to the Higgs image. Yet it apparently denotes a wide variety of very different phenomena, results obtained by imaging technologies included, which is why I chose "data visualisation" to label this section instead. This labelling captures two important characteristics of the Higgs image and similar visual representations. As should have already become clear, their production necessarily comprises (1) information-technology devices that visualise the results, and (2) data collected in measurement processes. These two aspects, then, are also the bone of contention in philosophical discussions about their potential epistemic capacities.

The two aspects mentioned are deeply intertwined in scientific practices. In particular, this is due to the data gathered in many scientific experiments by far exceeding the number of results that human individuals could cognitively process without any technological support. Many of today's sciences are thus called "big data sciences".[31] The experiments at CERN clearly belong to this category of science, as the following description of the data mining processes with regard to the discovery of the Higgs boson shows.

Although the LHC typically produces close to half a billion collisions in roughly 20 million bunch crossings per second, only a tiny fraction of these contain potentially interesting new phenomena, so it is neither necessary nor feasible to record all of the data from every single collision. CMS uses a two-level online trigger system to reduce the event rate from about 20 MHz to about 500 Hz, keeping only those events that are worthy of further investigation. The first level uses custom electronics close to the detector to analyze coarse information from the calorimeters and muon detectors to reduce the rate to 100 kHz or less. The second level uses a computing farm of 13,000 processor cores to analyze the full information from all subdetectors in order to make the final decision on whether to record an event. CMS has thus far selected several billion events, corresponding to more than 4 petabytes of stored event data. The recorded events are sent to computing centers at CERN and around the world to fully reconstruct the particles produced in each collision and allow subsequent analyses.

(Abbaneo et al. 2012, 1571)

In this quotation, the epistemological challenge of big data sciences becomes obvious. Due to the huge amount of data produced by those experiments, automatic data mining processes are implemented to single out the events that might be of interest to the scientists and are thus recorded and analysed further. That is, an automatic preselection process is installed to reduce the amount of data. Such a selection process is, however, theoretically guided by predictions and related expectations, by where to look for relevant phenomena within the flood of data. Moreover, data not meeting those criteria are not and, due to technological limitations, simply cannot be stored and later analysed. One might therefore critically argue that scientific investigations are highly *theory-laden*, which might narrow the scientists' analytical focus on the respective topic.

The problem of the *theory-ladenness of observation* lurking in the background of those experiments at CERN is a familiar one in the philosophy of science. Even though this poses certain epistemological challenges to practices of justification concerning knowledge claims based on such experimentally obtained data, their being theory-laden should nonetheless come as no surprise. Kosso makes clear that "*[a]ll* observation in science is influenced by theory" (Kosso 1993, 113, my italics). He summarises three aspects where theoretical assumptions are of significance: (1) theories tell us where to look for the relevant data which we need to verify or falsify our assumptions (see ibid., 114f.). There are plenty of observable facts in the world that are irrelevant to the theory under investigation. Theoretical assumptions can avoid wasting limited time and other resources by leading to the evidential data needed. This is what takes place when algorithms are installed to preselect relevant events at the detectors of the ATLAS and CMS experiments at CERN; (2) theories also influence the way measurement results are interpreted:

> [t]heories play a key role in the accountability of observations by supplying standards to evaluate the reliability of observational reports. The assessment of the viewing conditions, attesting that there are no distorting factors or correcting for any distortions that persist, is based largely on an understanding of the causal mechanisms of observation.
>
> (Kosso 1993, 115)

Thus, knowing how their detectors work will significantly increase the probability that scientists at the CMS and ATLAS experiments will interpret their results correctly. Finally (3), theories also provide us with the relevant vocabulary to describe what has been observed (see ibid., 117f.). With regard to our example above, theoretical particle physics tells scientists, for instance, that what their detector has recorded are 'decaying particles', namely 'Z bosons' that are indicators of the 'Higgs boson', etc.

However – and the problems enter at this point that arise from the challenge of theory-ladenness of observation – if scientific observation is thus heavily influenced by theoretical assumptions, how can it play the neutral evidential role it is supposed to fulfil? How can it be ascertained that the possibility of gaining falsifying evidence through experiments is not immediately precluded, i.e. that divergent data are not simply set aside? How can it be guaranteed that theoretical assumptions do not excessively narrow the focus of scientific investigations? To phrase this problem in the terminology of the example above: if scientists at CERN are looking for new physics, that is, physics beyond the Standard Model as they usually claim, how can they pursue such an investigation true-heartedly in the context of experiments that are so heavily influenced by theoretical assumptions derived from the very same model? The situation seems to be even worse when it is taken into account that alternative data, that is, data that do not fit the criteria of the preselecting algorithms, are simply not available, as they are dismissed right from the start and cannot be used as a point of comparison in later steps of the research process.

Thus, one of the main epistemological concerns about data visualisations such as the Higgs image is their theory-ladenness, a difficulty that seems to be intensified by the apparently automatic method of data processing that precedes image production. Do such processes still yield data images that can play an evidential role, *what or whom we can believe concerning what we see*, as Kosso describes the way that evidence plays its part in science?[32] Or does the whole experimental set-up, i.e. the preselection of data based on algorithms and the visualisation of the data gathered, entail *unauthorised evaluations* or *interpretations* that undermine the suitability of these data images as evidence?

Undoubtedly, the Higgs image poses challenges to anyone using it as evidence of the discovery of the Higgs boson. Yet, as pointed out at the beginning of this section, it is not this image alone on which the announcement of the discovery was based in 2012. *Diagrams* were also of importance in this context. This type of visual representation is therefore to be examined next to see if it fares better in epistemological terms.

2.1.4 Diagrams

The last paradigmatic example of visual representation in science to be discussed here are diagrams. They are widely used in the scientific context and exhibit a diversity of appearances. This section begins by characterising the phenomenon at hand more precisely in order to get a better understanding of how and why diagrams are employed within scientific processes.

Although diagrams clearly belong to the category of visual representations, as analysed in this study, they nonetheless diverge from the representational style found in photographs and adjacent kinds of images. They are not realistic depictions, whereas the former often are. Nonetheless, and despite their more artificial style of representation, diagrams can and often do show real properties of the object under investigation. Yet, similar to other kinds of data images, they might only visualise indicators as results of indirect measurements.

Several attempts have been put forward to conceptualise this somewhat complex relation between *diagrams* and *pictures*. Let us take a brief look at some of these. To draw a distinction one might, for example, point to the fact that diagrams are essentially a *hybrid* kind of representation, i.e. composed to varying degrees of visual, numerical and linguistic expressions, whereas pictures are only composed of visual representations. Figure 2.5 is a good example of this. Similar to the Higgs image above (see Figure 2.4), this diagram also featured in the announcement of the discovery of the Higgs boson in 2012. Actually, the CERN announcement was accompanied by several diagrams (see Abbaneo et al. 2012). The one discussed here plots the data collected between 2011 and 2012 to show the statistical significance of their measurement (see Figure 2.5). As pointed out above, the detection of different products of decay (in this case, two photons) could also be due to other Standard Model processes registered by CMS detectors, and not just due to the decay of a Higgs boson. Yet the data plotted in this diagram show a significant deviation from these so-called background processes. Moreover, "[t]he observed excess is consistent in shape and size with that expected for diphoton decays of SM Higgs bosons" (Abbaneo et al. 2012, 1572f.). The diagram therefore presents strong evidence for the successful discovery of the Higgs boson.

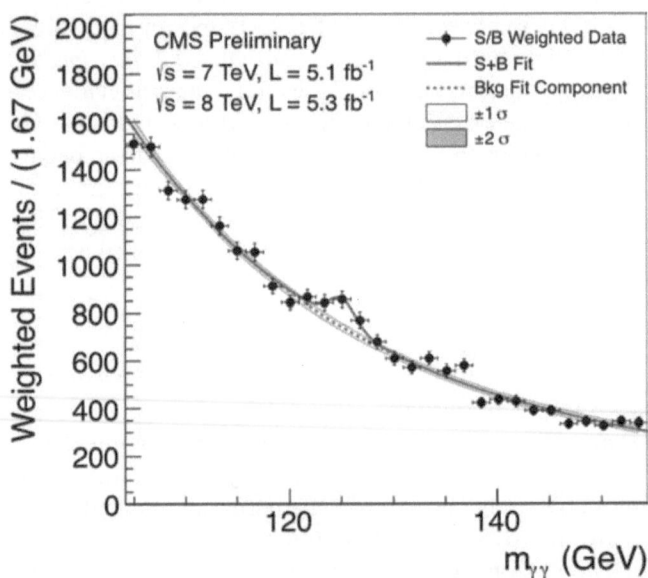

Figure 2.5 Higgs boson diagram: "di-photon (γ,γ) invariant mass distribution for the CMS data of 2011 and 2012 (black points with error bars). The data are weighted by the signal to background ratio for each sub-category of events. The solid red line shows the fit result for signal plus background; the dashed red line shows only the background." For a colour version of this diagram see http://cds.cern.ch/record/1459463

Source: CERN, for the benefit of the CMS Collaboration.

Figure 2.5 nicely illustrates the hybrid status of diagrammatic representations. On the one hand, there are the title and the captions of the ordinate and abscissa which are linguistic in kind. On the other, the labelling of both axes is numerical. And finally, there is the curve representing the relation between both values, which is pictorial in kind. Thus, the example illustrates the functional interaction of all different kinds of representation to transmit the relevant information.

However, the comparative characterisation stressing the hybrid status of diagrams rather quickly turns out to be problematic, as pictures can, of course, also contain non-pictorial elements. A telling example in this respect is presented by Janet Vertesi. She discusses how scientists working on the *Mars Exploration Rover* project enrich photographic pictures, obtained by their rover's cameras on Mars, via digital post-processing. Not only do those scientists guide what the viewers see by colouring certain regions, etc., depicted on the photographs, they also often add explanatory symbols, linguistic expressions and numerical values (see Vertesi 2015, ch. 4).[33] All of this information is merged in the resulting image so that the hybrid style of representation, pointed out as characteristic of diagrams above, seems also to fit rather well as a description of those images. A simple instance of such a kind of enriched image is the earlier photograph showing Pluto and its satellites (see Figure 2.1). Another interesting case concerns scanned text pages. Depending on the mode of scanning, the result of the process can be a digital image, although the information depicted might be exclusively linguistic in kind. Thus, the hybrid status of representational means can occur even if no diagrammatic style is used. Is there then an alternative criterion available to mark the distinction between pictorial and diagrammatic representations?

An influential proposal for such a distinction was put forward by Goodman (see Goodman 1976, 170ff., 229f.). Because of his complex terminology, however, the reader has to have at least a basic understanding of Goodman's theory of depiction first, before she can adequately understand his distinction between diagrams and pictures. Hence, I will discuss his ideas in detail in section 2.2.2 of this book, but here I will only briefly and in a simplified manner describe his criterion of demarcation and then focus on a successor to his approach in more detail, in order to point out the epistemological particularities of diagrams in scientific processes.

The easiest way to understand how to discern between pictures and diagrams in Goodman's sense is by analysing a visual example. In explaining the difference between *pictorial symbol systems* and *diagrammatic symbol systems*, he refers to an illustration similar to Figure 2.6. Of course, the two images in Figure 2.6 present different contents. The first one is a photograph of Mount Fuji, depicting the landscape from a certain angle, while the second is a diagram showing the relation between two sets of numerical values.[34] But this obvious difference in semantics is not the point. What is of interest is the question why it is the case

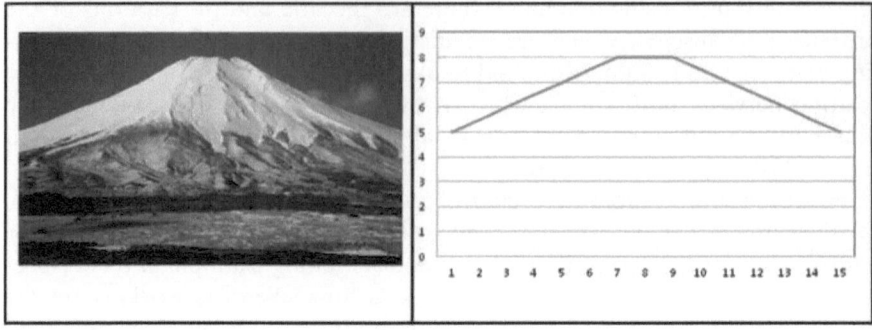

Figure 2.6 Mount Fuji and diagram.

Source: own image, photograph taken from http://commons.wikimedia.org/wiki/File:Mount_Fuji_from_Hotel_Mt_Fuji_1995-2–7.jpg

that, although the two exemplary images appear superficially similar in shape, they belong to two different symbol systems.

Goodman regards visual representations as parts of *symbol systems* which can be analysed on syntactic, semantic and pragmatic levels. Within this framework, he thinks that the answer to the previous question is to be found at the *syntactical level*. Changes at this level affect the meaning transmitted by a given image. Neither a change in the colour nor the thickness of the curve in the diagram of Figure 2.6 influences its meaning. Thus, these characteristics do not belong to the syntax of diagrammatic symbol systems. Varying the units of the abscissa and ordinate, however, would affect the information transmitted in the same way as replacing the solid line by a dotted one. Hence, this visual information belongs, from Goodman's point of view, to the essential syntactic elements of diagrams. With respect to the photograph of Figure 2.6, however, colour and thickness of lines are indeed of importance to transmit the respective information. For example, changing the brightness of the colour of the sky might indicate that the picture shows the mountain at a different time of the day. Consequently, Goodman defends the thesis that many more elements a picture is composed of are relevant to its meaning than would be the case with diagrams. This feature he calls "relative repleteness" (ibid., 230). And it is here that he thinks that pictorial symbol systems diverge significantly in comparison to diagrammatic symbol systems.[35]

Of course, Goodman's distinction has not remained uncontested in the philosophical debate. On the one hand, one might, for instance, call into question that, according to his approach, pictures have to be rather detailed, yet there are clearly many counter-examples to this claim. Does any detail really matter? Does, for example, blurring the outline shape of Mount Fuji in this depiction really change the meaning of the picture? Or does this merely demonstrate that some pictures are more and others less accurate as depictions?[36] On the other, it can be questioned whether the representational power of diagrams is actually restricted to just a few visual characteristics. Add, for example, a second curve to the diagram in Figure 2.6 and its colour can indeed be of significance. A famous critique along these lines is presented by Christopher Peacocke. By adding an increasing amount of information presented in the diagram and transmitted to the viewer via visual features of the graph, he shows that a diagram can finally reveal the same characteristics that Goodman suggests as the relevant features of pictures alone. Diagrams can exhibit *syntactic density* and *relative repleteness* without being pictures (see Peacocke 1987, 405ff.).

Despite these critical remarks, Goodman's suggestion has nevertheless served as an influential framework to analyse diagrams in philosophy. Perini, for example, adopts his account to discuss the role of diagrams in biological explanations (see Perini 2005a) and there will be a return to her epistemological considerations at the end of this section. Meanwhile, however, I would like to introduce another follow-up project that is loosely related to Goodman's idea. The detailed discussion of this approach will offer the stepping stone to investigate some crucial epistemological aspects related to the use of diagrams in science.

Valeria Giardino and Gabriel Greenberg make use of Goodman's suggestions only in a very rough sense. They accept the thesis that different kinds of representation can best be understood as elements of different symbol systems (see Giardino and Greenberg 2015, 4). Moreover, they also agree with Goodman that pictures and diagrams belong to different types of systems in this respect, yet they do not follow his proposal of how to mark the distinction between these two types. Contrary to Goodman's criteria, they suggest taking into account that pictorial representations necessarily entail a certain point of view, whereas diagrammatic representations do not.

> Among the class of PICTURES we include the likes of perspectival drawings, photographs, paintings, and film clips. Among the class of DIAGRAMS we include the likes of graphs, charts, timelines, and Venn diagrams. The distinction is certainly not a sharp one, with the likes of maps and pictographs occupying intermediate positions; it may instead reflect polar ends of a spectrum of kinds [...] pictures are *perspectival* representations, while diagrams are not. That is, pictures alone necessarily express content which describes the world relative to some spatial viewpoint of perspective.
>
> (Giardino and Greenberg 2015, 2f.)

Although initially this criterion of perspectivity seems to be a plausible one, it becomes problematic depending on which of the instances mentioned above are regarded as typical examples of diagrams. Here, Giardino and Greenberg focus on *Euler* and *Venn diagrams*, respectively on Peirce's refined approach to Venn diagrams (see ibid., 9ff.).[37] As they note, these diagrams "exploit *spatial* relationships between shapes to represent relationships in some other domain – in this case, *logical* relationships between classes of objects" (ibid., 9). Regarding these examples, they are clearly correct in claiming that perspectival considerations do not make a difference to such visual representations. However, a difference emerges if diagrams that present information in more than two dimensions are examined, for example three-dimensional graphs depicting a development in the two dimensions of space and time. It seems questionable whether a particular perspective is also irrelevant to those depictions. Similarly, we could call into question whether pictures necessarily require a perspective. The more abstract pictures are, the less plausible their criterion seems to be. Thus, what turns out to be of major significance in Giardino and Greenberg's account is how we define the polar ends of the spectrum that their criterion of distinction delineates.

Despite these difficulties in pointing out the criteria involved in drawing a distinction between pictorial and diagrammatic types of representational systems, Giardino and Greenberg mention some quite interesting aspects of the latter with regard to their cognitive processing and thus their epistemic capacities. In particular, they point to an apparent tension between the "naturalness" and the "expressive power" of those diagrams (see ibid., 13). Both features apply not only to Euler and Venn diagrams, as they are used in logic and mathematics,[38] but also to diagrams in general, and are therefore worth considering in detail here. What is meant by those characteristics is most easily understood when regarding a concrete visual example (see Figure 2.7). The *expressive power* of diagrams is related to the precision of their content-transmission. Both Euler and Venn diagrams can, for example, be used to express general statements. The statement 'All Xs are Ys' is expressed in both diagrams in Figure 2.7. But even though both types of diagram clearly deal well with the task of expressing this general statement, the Euler diagram does so only at the expense of simultaneously expressing the statement 'There are Ys that are not Xs'. Contrary to this, the Venn diagram is more precise in its content, as it only transmits the former expression (see also ibid., 12).

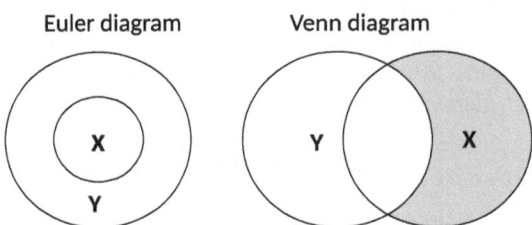

Euler diagram   Venn diagram

Figure 2.7 Euler and Venn diagrams: general statement (All Xs are Ys) expressed by Euler diagram and Venn diagram.

Source: own image.

Venn diagrams are apparently more precise in their content transmission and are also suitable means by which to communicate existential claims; that is, a set of statements which Euler diagrams cannot express. Yet although Venn diagrams thus reveal a greater expressive power than their competitor, they do so only at the expense of their *naturalness*, as Giardino and Greenberg convincingly show. This feature is explained by pointing to the system's allowance content to be transmitted "in a way that, for human cognition, is particularly easy to grasp, especially as compared to familiar symbolic systems" (ibid., 14). That is, in regard to the example above, it is relatively easy to understand the general statement expressed by the Euler diagram. Apparently, we only have to *look* at the diagram to grasp its content. In contrast to this, the recipient needs some further information about the representational system to decode the same content expressed by the Venn diagram. In particular, she has to know that *shading* signifies an empty set. This conventionally established meaning of the visual feature of shading has to be learnt in advance in order to decipher the respective diagram in Figure 2.7 correctly. Consequently, the greater expressive power of Venn diagrams is gained only at the expense of reducing their naturalness.

The feature, termed "naturalness" by Giardino and Greenberg, is also discussed by the cognitive scientists Jill H. Larkin and Herbert A. Simon in their analysis of diagrams as tools in reasoning processes (see Larkin and Simon 1987). They suggest the following three reasons as a rationale to explain why diagrams can exhibit particular virtues with respect to the cognitive processing of the information presented by them.

1 Diagrams can group together all information that is used together, thus avoiding large amounts of search for the elements needed to make a problem-solving inference.
2 Diagrams typically use location to group information about a single element, avoiding the need to match symbolic labels.
3 Diagrams automatically support a large number of perceptual inferences, which are extremely easy for humans (Larkin and Simon 1987, 98).

The last point refers back to Giardino and Greenberg's assumptions about the naturalness of diagrammatic representations. Recipients can rely on their eyes, so to speak, to gather and comprehend the information transmitted by diagrams. However, as the example of Euler and Venn diagrams shows, this thesis about the naturalness of information representation is a matter of degree. The more conventionally elements are introduced in a given diagrammatic style, the more complex the interpretive task becomes to understand them correctly and, consequently, the more they lose of their characteristic of naturalness.

This interpretative challenge becomes clear in the example of the Higgs diagram (see Figure 2.5). Not only does one have to know how the representational style used in plotting the data works, but also what kind of data is plotted, and to which theory it refers. Kosso's thesis can be introduced at this juncture, that we can gain observational data even about entities that are "unperceivable in fact", that is to say, unobservable without the use of instruments (see Kosso 1988, 455). The Higgs boson illustrates this case nicely, as it is only by detecting the traces of its products of decay, which appear after the collision of particles artificially brought about within the LHC, that scientists could claim its detection. With respect to such observational results, Kosso lists four "dimensions of observability" to make clear where exactly difficulties can obtain within the process of observation and the interpretation of observational data. These problems might then trigger the claim that certain observational results are more reliable than others. Those dimensions are (1) immediacy, (2) directness, (3) amount of interpretation and (4) independence of interpretation (see ibid., 454ff.). A brief examination of them is required at this point to find out in what sense the interpretation of diagrams is affected by them.

Kosso's first dimension is captured by the distinction between entities that can be observed with the naked eye, ones that are detectable via instruments – such as the Higgs boson – and entities that are in principle unobservable such as, for example, causality or the 'Big Bang' (see ibid., 454f.). The second dimension of directness is related to a quantitative question, namely how many intermediate steps between the cognitive processing of the incoming information by the epistemic subject and the object under investigation have to be taken into account when analysing the observational process, how many instruments are involved, and how many procedures are used in order to prepare the object for observational purposes. In the case of the Higgs boson, for example, at the very least the particle accelerator – the LHC – as well as the different detector mechanisms at the CMS experiment and the information-technology devices that evaluate, select and plot the data would have to be considered. The next issue then, relating to the third dimension of Kosso's concept of observability, is the question of how many different theories related to each instrument used are involved in the interpretational process. Kosso explains the difference between (3) and (2) in the following way: "[t]his dimension [the amount of interpretation, N.M.] is different from the previous one, directness. Something which is observable via many interactions is not necessarily dependent on commensurately many laws to account for the information" (ibid., 456). It might be that, to observe a particular entity, the same kind of instrument has to be used a number of times. In such a case, only one theory about how the instrument works and derives its data has to be taken into account. With regard to the Higgs diagram this is plainly different, as quite a few technologies are in charge of the respective experimental set-up. Kosso's final dimension, "independence of interpretation", tackles the question of whether the observational result that is supposed to serve as evidence is dependent on any of the theories considered necessary to build and operate the instruments that produce these data. It is here that the *problem of theory-ladenness of observation* begins.

Regarding the interpretation of diagrams, Kosso's assumptions permit for now a more precise location of the difficulties obtaining. The results plotted in the Higgs diagram (see Figure 2.5) are less direct than results gained, for example, by photographic means – despite the fact that basically both processes rely on causal mechanisms. Additionally, the Higgs data also require a greater amount of interpretation to be understood correctly. Admittedly, the data visualised in the Higgs image (see Figure 2.4) do not seem to fare much better in this respect. However, the aspect introduced above as *naturalness* suggests that the more conventionally defined elements that particular diagrams contain, the more background knowledge has to be presupposed on the recipient's part in order to work sufficiently well as a cognitive tool. Here, Larkin and Simon point out that this kind of background knowledge is not only relevant to the interpreting recipient, but also affects the producer of the diagram, who has to invent a suitable design to transmit the information she intends to transfer.

> The advantages of diagrams, in our view, are computational. That is diagrams can be better representations not because they contain more information, but because the indexing of this information can support extremely useful and efficient computational processes. But this means that diagrams are useful only to those who know the appropriate computational processes for taking advantage of them. Furthermore, a problem solver often also needs the knowledge of how to construct a 'good' diagram [...]
>
> (Larkin and Simon 1987, 99)

Moreover, as becomes clear from the above quotation, the design process can address at least two different audiences. On the one hand, the scientist developing the diagram might intend to transmit her results to her fellow researchers by this means. On the other, the *problem solver* to be addressed might simply be the scientist herself, in which case, by pondering about diverse design possibilities, she simultaneously muses over different possibilities of interpreting her data and thus thinks about different explanatory strategies concerning a given phenomenon.

It is here that, as a concluding remark, I want to draw attention to Perini's discussion of the role of diagrams in scientific explanations (see Perini 2005a). She points out that, in biology, *functional explanations* are quite common, that is, approaches "in which some capacity of a system is explained by the capacities of components of the system" (ibid., 264). By analysing two case studies from the biological sciences, Perini shows that diagrams serve the purpose of revealing such functional explanations particularly well.

She points out that the two-dimensional format of diagrams allows two kinds of information that are essential to such explanations in biology, namely relations among components of a system and the identity of those components revealing their most crucial capacities within the relations

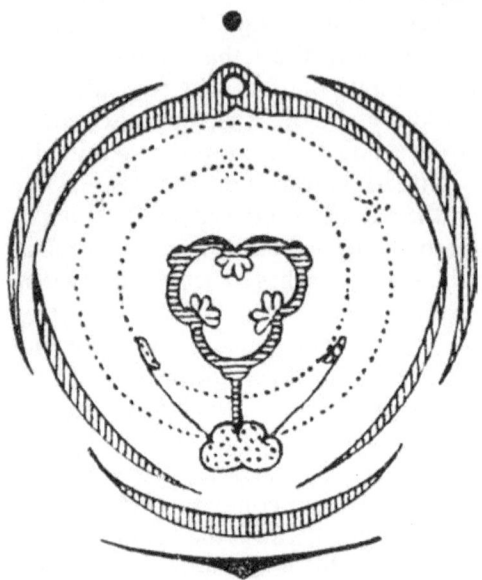

Figure 2.8 Flower diagram of orchis.

Source: Strasburger, Noll; Schenck, Schimper (1900). Lehrbuch der Botanik für Hochschulen. 4th ed., Gustav Fischer: Jena, p. 438, provided online by Wikimedia Commons, https://commons. wikimedia.org/wiki/File:Orchis_flowerdiagram.png

shown (see ibid., 265). Figure 2.8 serves as a simplified example. It shows the morphology of an orchis flower. This diagram shows how the petals, the labellum and the column are related to each other. Moreover the diagram is supposed to demonstrate that the characteristics of the flower, which is predominantly pollinated by insects, have evolved their shaped accordingly. Thus, the flower diagram is not only supposed to show the relations among elements of the blossom, but also to reveal functional roles played by these parts within the system.

Here, we can connect the results of Perini's analysis of the role of diagrams in functional explanations in biology with Larkin and Simon's findings about why diagrams often serve as cognitive tools in science. Points (1) and (2) of their list, namely the grouping and localisation of relevant information, definitely match Perini's ideas about why diagrams can play the explanatory role they are clearly supposed to fulfil.

Beyond that, Perini makes plain that, although diagrams often exhibit less naturalness than other visual representations in science, cognitively accessing diagrammatic information via perceptual means can still turn out to be epistemically advantageous. Comparing diagrams to linguistic expressions, she states that:

[...] the visual format can function differently in scientific communication from linguistic representations. The human visual system works by integrating sensory cues into a perceptual whole. The result of this process is a perception in which the parts are related to the whole. It is not feasible for a person to infer equivalent content from linguistically encoded information.

(Perini 2005a, 267)

Here, she makes it clear that the visual format of information presentation can support *scientific understanding* of phenomena and their functions in a way that linguistic expressions cannot. Visually processing such information might allow scientists to realise connections and thus infer explanations that they would not otherwise be able to produce by relying on the interpretation of their data alone.

Bearing this important epistemic advantage in mind, I now wish to take stock of what this discussion of paradigmatic instances of visual representations in science has revealed concerning their epistemic virtues and vices more generally.

2.1.5 Interim results: what can be learnt from paradigmatic instances?

The above discussion of four paradigmatic kinds of visual representations in science, namely photographs, imaging techniques, data visualisations and diagrams, has made clear that scientists take them both as important sources of information about their object of research and as tools for communicating that information to other scientists and the broader public. In particular, the causal relation, between image and object, between measurement device and the object under investigation (see Harré 2010, 31) – understood according to Woodward's interventionist account of causation – seems to support their assumption, namely that visual representations are sufficiently informative in the relevant sense about their object of depiction to fulfil the purposes just described in an appropriate manner.

Although scientists seem to rely heavily on visualisations in their epistemic practices and apparently do so on the assumption that their reliance is epistemically justified, a couple of difficulties emerge in the above discussion that seem to undermine this very assumption. Thus it will be one of the main tasks of the following investigation (see Chapter 3 of this book) to critically analyse from an epistemological point of view the reasons to make use of and to rely on visual representations in science.

A first step in this direction consists in delineating some of the main difficulties that have already shown up in the discussion above. Two of them need summarising in more detail, namely problems related to (1) *objects* of visualisations and (2) their *contexts*.

The first aspect (1) is related to many visual representations in science literally *visualising* their object of depiction. That is, they render entities observable that cannot be seen with the naked eye. Many of the examples discussed earlier are related to this aspect. They make clear that *observability* is a matter of degree. Pluto's satellites in Figure 2.1 are not observable without the aid of a powerful telescope. But even though the Higgs image (see Figure 2.4) also visualises an unobservable entity, it nonetheless seems to be of another evidential quality than the photograph of Pluto's satellite. The hypotheses inferred from the Higgs image are only based on indicators, whereas the latter image apparently allows direct inferences about the existence and nature of its depicted object.

The Higgs image illustrates that what is often visualised in science are *theoretical entities*, which are basically not accessible to the unaided human senses, hence the need for the aid of *instruments*, such as particle accelerators, microscopes, telescopes, photography, etc. Additionally, today's observations and experiments in science normally involve *information technology* to plot the data received. Such devices are not only employed to make the unobservable accessible to the human eye, but also as a heuristic means to deal with big data. Many scientific experiments – such as the ones at the LHC in Geneva – produce huge amounts of data that no human observer could evaluate without technical support. Computer programmes, for example, preselect which events are worthy of further analysis, since storing all of this data for later investigations is technically impossible today. In this sense, information technology serves indispensable purposes.

Yet all of these technological intermediaries can be regarded as potential sources of error. Moreover, the need to make use of an agglomeration of devices to produce a visible trace of what is supposed to be investigated, for example the Higgs boson, increases the susceptibility of those images and raises questions about whether they can really be regarded as reliable evidence. If those entities are not accessible without making use of those instruments, how can it ever be proven that the properties they reveal are proper ones and not just artefacts, as pointed out in Weiss's example of the cytoskeleton in microbiology? And if no certainty can exist about properties indirectly observed and measured, then what guarantees are there that the entity itself exists?

The question of whether objects only visible with the aid of instruments are real or mere artefacts is not new. It lies at the heart of the debate between *scientific realists* and *anti-realists*. Realists such as Richard Boyd describe their position as follows:

> [s]cientific realists hold that the characteristic product of successful scientific research is knowledge of largely theory-independent phenomena and that such knowledge is possible (indeed actual) even in those cases in which the relevant phenomena are not, in any non-question-begging sense, observable.

> (Boyd 2010)

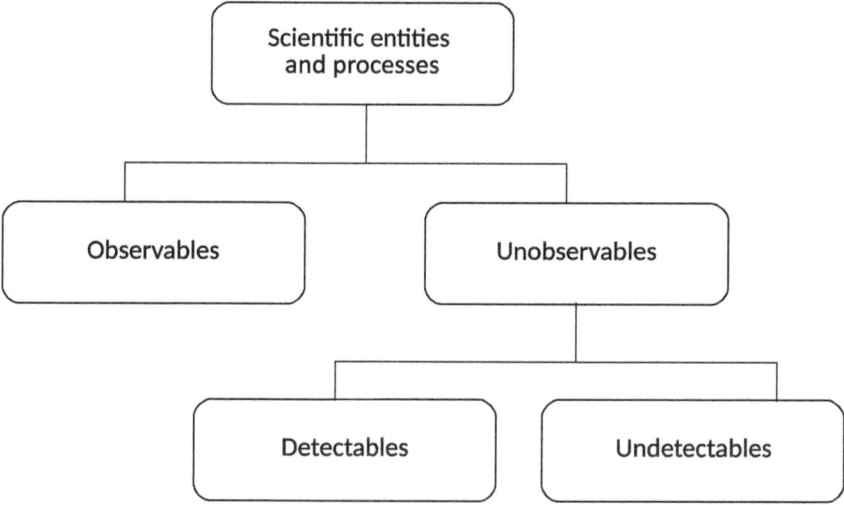

Figure 2.9 Observables and unobservables in science.
Source: image adapted from Chakravartty 2007, 15.

Here, it becomes clear that the dispute between scientific realists and anti-realists is triggered by the unobservable element in scientific investigations (see Chakravartty 2013, sect. 1.1), but what exactly do they mean when they talk about 'unobservables'? Anjan Chakravartty highlights an important distinction in this regard in Figure 2.9, namely by pointing out that 'unobservables' can nonetheless be 'detectables' due to the use of instruments. Such nomenclature also implies the belief that our instruments are reliable, at least in this minimal sense, that is, in detecting the entities, although they might not be able to tell us more about their specifics. Furthermore, Chakravartty makes clear that the distinction between observables and unobservables in science is vague and has indeed often been criticised within the debate (see ibid.).

However, this distinction, despite its vagueness, plays a crucial role in the debate between scientific realists and anti-realists. This can best be understood when considering the questions discussed in this context in more detail. Chakravartty points out that there are three different levels of inquiry involved, namely an ontological, a semantic and an epistemological one.

Ontologically, scientific realism is committed to the existence of a mind-independent world or reality. A realist semantics implies that theoretical claims about this reality have truth values, and should be construed literally, whether true or false. [...] Finally, the epistemological commitment is to the idea that these theoretical claims give us knowledge of the world.

(Chakravartty 2007, 9)

Moreover, these three levels can also be combined in different ways. One can, for example, be a realist with regard to the ontological level, but at the same time not with respect to the epistemological. That is, a possible claim can be that there is an external, mind-independent world, but we do not possess the cognitive capabilities to acquire knowledge of it.[39] Consequently, a scientific realist who looks for support to back her thesis that a particular unobservable entity E exists (or has the characteristics x, y, z) can refer to visual representations produced with the aid of instruments in the process of E's detection. Such representations will then be regarded as fallible evidence of E's existence (or characteristics) and thus as evidence of a mind-independent reality at the level of unobservable entities. Now, in what sense do scientific anti-realists differ from this perspective?

Scientific anti-realists, such as Bas C. van Fraassen, are sceptical about theoretical terms and unobservable entities presumably connected to them. It seems easiest to explain their resentments by focusing on van Fraassen's influential approach. He claims that all a scientist *can and does* long for is an *empirically adequate theory*. His account, labelled as "constructive empiricism", reads as follows: "[s]cience aims to give us theories which are empirically adequate; and acceptance of a theory involves as belief only that it is empirically adequate" (van Fraassen 1980, 12). 'Empirically adequate' thereby means that those theories offer true descriptions and explanations about the observable part of the world, but do not attempt to account for any unobservable phenomena. As it cannot be discovered whether the theoretical components that are not testable via direct observation are correct or not, one no longer speaks of approximately true theories. Consequently, theoretical entities are set aside by scientific anti-realists, as they are of no further epistemic value – let alone thought of as real entities in the outside world. On this account, visualisations of unobservable phenomena do not provide evidence of their objects' existence.[40] As epistemic access to the domain of the unobservables is therefore impossible, one simply has to remain agnostic about their existence and their characteristics. Van Fraassen's anti-realistic approach is particularly interesting in that he explicitly discusses the topic of scientific images. Not only does he call into question the status and the explanatory power of depictions showing observable entities, he is also sceptical about visual representations within the domain of observable phenomena. From his point of view, *usage* is the key to our understanding of how representation works. He states that "[t]here is no representation except in the sense that some things are used, made, or taken, to represent some things as thus or so" (van Fraassen 2010, 23). In this context, resemblance does not provide an answer to questions about the characteristics of depiction, but it may yield a rationale for why certain representations are successful (see ibid., 33). He denies, however, that a resemblance relation between picture and object can serve as evidence of the object's existence. "That an image trades on resemblance, on any level, does not imply that it resembles what it represents, *nor that there is something that it resembles, nor even that there exists something that it represents*" (ibid., 35, my italics). In the next section we will come to see that van Fraassen puts a wide-ranging challenge to the fore here, as quite a few scholars traditionally point to resemblance relations as the essential characteristic of depiction. However, as resemblance is a relational property, accounts of this type are also committed to the claim that next to the image *there is* an object that the former can resemble. Apparently, this is a strong commitment that proponents of resemblance-based theories have to accept and their possible rejoinder to van Fraassen's challenge will be discussed in section 2.2.1 of this book in more detail.

Meanwhile, van Fraassen provides a reminder that there is an important difference between *appearance* and *phenomena.*

> 'Appearance' I reserve strictly for the contents of (possible) measurement outcomes. Phenomena are observable, but their appearance, that is to say, *what they look like in given measurement or observation set-ups*, is to be distinguished from them as much as any person's appearance is to be distinguished from that person.
>
> (van Fraassen 2010, 284f., his italics)

And summing up these points, he comes to the conclusion that "[...] the measurement outcome shows not how the phenomena *are* but how they *look*" (ibid., 290, his italics). Thus, van Fraassen raises a clearly sceptical voice with regard to visual representations in science. In particular, he calls into question their evidential status and also their suitability to serve as a more generally informative means concerning the object of research.

It would be wrong, however, to think that scientific realists have yielded to van Fraassen's arguments in particular and to the scientific anti-realists' critique in general.[41] On the contrary, they have worked out various strategies of defence. Chakravartty, for example, discusses a variety of arguments supporting the scientific realist point of view in Chakravartty (2013), for example, discusses a variety of arguments supporting the scientific realist point of view.

A famous line of reasoning in favour of scientific realism is, for instance, the "no miracle argument" developed by Hilary Putnam. To put it briefly, it raises the question of how we can explain the success of our scientific theories in predicting new phenomena if we do not want to grant that they, including their theoretical statements concerning the unobservable, are – at least approximately – true. Scientific realism is then the only approach that does not turn the success of science into a miracle (see Chakravartty 2013, sect. 2.1). This assumption can be backed by referring to scientific practices making broad use of *inferences to the best explanation* (see Suhm 2005, ch. 5.2). Scientific realism seems to be our *best explanation* of why scientific disciplines are thus successful in their endeavours. Taking into account that the inference to the best explanation is a legitimate method to defend the likelihood of the conclusion in question, it seems warranted to accept scientific realism as (approximately) true. In this sense, scientific realism can be regarded as an empirical thesis that can be tested by taking into consideration historical developments in science. If it can be shown that there is a continuity in this development in the sense that phenomena predicted by scientific theories once are observable later – here we make use of Kosso's detailed concept of observability – it seems justified to regard these developmental processes as an approximation to true descriptions of the world and scientific realism, as the meta-theory accounting for this developmental process, as thus also approximately true (see Suhm 2004, 177f.). Connecting this with the foregoing discussion about visual representations in science, the possibility of defending scientific realism also means not having to yield to the anti-realist position on the evidential status and informativeness of scientific images. Of course, this is not to say that there cannot be errors or fraud. But neither does it mean that we have to take a thoroughly sceptical attitude towards visual representations in scientific epistemic processes. These preliminaries might then suffice to frame our subsequent discussion of the evidential status of scientific images, which will be analysed in more detail in Chapter 3 of this book.

The second aspect (2), then, which proved to be a source of potential difficulties with regard to the epistemic capacities of visual representations in science, is related to their contexts of usage. As our above discussion made clear, such contexts seem to play a major role in ascribing a particular meaning to an image at hand. The Higgs image (see Figure 2.4) exemplifies this. Originally, this image shows an event detected by the CMS experiment at the LHC. It was evaluated by experts as the data output of the detection process. Later on, the very same image became part of a variety of articles produced for interested laypeople. Here, the image was used for public-relations reasons and not for research. Hence, we are sometimes confronted with the same visualisation serving different purposes in different contexts.

Making general statements about visual representations in science and, in so doing, muddling the requirements and effects affecting scientific images in those contexts seems to be one of the main reasons for the confusion amongst philosophers concerning the epistemic status of visual representations in science. For example, post-processing might play quite a different role in the two contexts indicated by the Higgs image example. Such techniques might be used to improve the image in such a way that it attracts the attention of laymen, and yet post-processing might also be an activity rendered necessary in order to enhance the readability of the plot to ensure that the detection fits the relevant parameters. Consequently, whereas the former procedure might lead to a simplified and thus more superficial depiction, the latter is meant to enhance the precision of the visual representation. Therefore, it seems reasonable to keep contexts apart in order to be more precise in the analysis of scientific images. This suggestion will guide the analysis set out in Chapter 3 of this book.

Another consequence of this observed difficulty is that the meaning of a particular visual representation can only be grasped fully if we take into account its context of usage. This suggestion is defended in detail by the semiotician Søren Kjørup. He thinks that it is the *communicative situation* as a whole that we have to consider in order to understand what is meant by a particular image:

> [a]nd the point is that pictures should not be discussed as such. Any discussion of pictures as symbols must take its point of departure in a communicational situation where some person (or some group of persons or legal body or person) is using one ore more pictures to 'say' (in a very general sense of 'saying') something to somebody.
>
> (Kjørup 1978, 57)

In particular, he thinks that, because of the ambiguity of pictorial representations, they are always in need of linguistic fixing. In accordance with Roland Barthes, he thus points out that captions and the like are always necessary to disambiguate pictorial representations (see Kjørup 1989). Consequently, these linguistic auxiliaries help to clarify what the referential parts of a picture are and what exactly they refer to (see ibid., 310f.) – at least, they are meant to do so. In this sense, Kjørup makes us aware of the relevance of considering the intertwining of different kinds of representation.[42] Setting aside for the moment the question of whether one is dealing with 'symbols' in this context, I agree with Kjørup about this basic idea. It is often the context of usage that permits jettisoning or, contrariwise, creates certain problems when it comes to clarifying the particular meaning of a visual representation. Moreover, by considering the broader context, we will note the role played by information presented by other representational means when analysing the meaning of scientific images. The above investigation has shown that often there is a functional interaction between the different kinds of representation (visual, linguistic and numerical). This is particularly true in the case of diagrams, but Vertesi's discussion of imaging processes, adding various layers of information (pictorial, numerical and linguistic in kind) to photographs taken by the cameras of the *Mars Exploration Rover* project, makes clear that this intertwining of different representational means also affects other types of images and can be regarded as a common practice in science.

An approach, based on Kjørup's ideas, that explains the meaning of pictorial representations more generally is Scholz's proposal of *image games* (see Scholz 2011). This concept is modelled along the lines of Ludwig Wittgenstein's concept of *language games*,[43] which is meant to highlight the fact that 'speaking a language' is part of an *action* or a lifestyle (see Wittgenstein 1995, § 7). This idea connects Wittgenstein's term with Scholz's concept of image games. Scholz points out that knowledge about the concrete context of usage is necessary to correctly understand the meaning of a picture from a historical point of view (see Scholz 2011, 377). The term *image games* often refers to the different acts of communication that a person performs with the aid of images (see ibid., 373f.). Scholz's examples thereby refer to ordinary situations of image usage, such as showing someone a photograph of a friend to inform her about the latter's physical appearance or by nailing a sign to a fence showing a furious dog to warn strangers of the watchdog, etc. Yet the very same functions can also be fulfilled by scientific images. By showing, for example, the photograph of the Plutonian system (see Figure 2.1) to a colleague, I can inform her about the physical ordering of the system and what it looks like in the night sky. Although I do not want to claim that image games exhaust the meaning-constitutive part of visual representations in science, I agree with Scholz that, in many instances, background knowledge about them might contribute a lot to the correct understanding of images in this context. We will come back to this interesting point in more detail in Chapter 4 of this book. In a nutshell, then, both considerations about (1) the object of depiction and (2) the contexts of image usage give rise to certain challenges when arguing for the epistemic capacities of visual representations in the information processes of science. These challenges result from our analysis of paradigmatic instances of scientific images and are thus related to the *extension*, that is, the scope, of the term. Now, the next question to consider is whether anything can be discovered about an assumed *intension* of the concept. The function of representation in the visual domain and what particularities are related to this process will therefore be examined next. Following this purpose, it seems natural to turn to *picture theory*, that is, the theoretical branch of philosophy whose proponents attempt clarification of what the essence might be of 'pictorial representation' in general, which is what will be focused on next.

2.2 The nature of depiction

The above considerations about paradigmatic kinds of visual representations in science demonstrate the wide variety of different phenomena subsumed under this label. Despite this diversity of appearances as well as methods of production and treatment, it seems that those phenomena have nevertheless something in common, namely their *visuality*. However, visuality is not an end in itself, but serves the purpose of information transmission. For instance, scientists use images as a means to distribute their research results amongst colleagues, or to communicate them to the public in the explanatory context, or to analyse them as the output of measurement processes in the exploratory context. All these practices presuppose the assumption that images are suitable means to contain and transmit certain information, i.e. that they are able to *represent* certain notions that recipients can extract.

However, there are other items – such as numbers or words – that can also be used for representational purposes. The capacity to represent alone, therefore, does not allow us to discern between these phenomena and to single out an assumed common core of the concept of visualisations. What is needed in addition is the *special mode* of representation involved, namely its *visuality*. For certain, words and numbers can also be taken to represent their information visually when they are written down. In such cases, we also decipher them with the aid of our eyes. Yet there seems to be something special about the kind of visual representation realised by images. This particularity becomes clear when discussing the main advantages and disadvantages that are connected with the visual mode of representation. They will be discussed in the following.

Obviously, one major advantage of visual representations consists in their ability to transmit *vast amounts of information* without much effort. This is pointed out, for example, by John Kulvicki (2010a) and by Philip Kitcher and Achille Varzi (2000). A whole story can be told in just one picture. Take for example Figure 2.10. This *plaque* was attached to the *Pioneer 10* and *11* spacecraft in the 1970s. They were meant to transmit greetings from planet Earth in case some extraterrestrial being might stumble across them. Most of all, the plaque contained information about the planet where the spacecraft came from and about the beings that had sent them into space.

First of all, let us take a closer look at the depiction at the bottom of the plaque. It shows the starting point of the spacecraft in our solar system. The following information can be extracted from this graphic: the solar system consists of one sun, depicted on the left-hand side, and nine planets.[44] Moreover, the graphic not only shows the sequence of the planets, but also their relative size in comparison to the sun. The little digits next to each planet are meant to tell the viewer the relative distance of each planet to the sun. And, finally, the designers tried to make clear that the Pioneer spacecraft started from the third of these planets – Earth. All of this information is encoded in this single image in Figure 2.10. Compared with a verbal version of the same, the graphic offers much more information using fewer means. Moreover, the relations of the individual pieces of information (for example, the number of the planets and their relations to each other and to the sun) are also much easier to access cognitively than a linguistic explanation could provide. Consider also the line drawings of the two human beings depicted on the plaque. They are meant to show what the species looks like that created the spacecraft. How many words would be needed to describe the two of them accurately? And more specifically, what would the description be like, presupposing that the recipient had never seen human beings before?

Information transmission via visual representations seems to happen immediately and not stepwise as in the case of linguistic presentations. Apparently the designers of the plaque assumed that an extraterrestrial recipient can simply *see* that the sun is located centrally to these 14 pulsars and where they are situated relative to our parent star. Thus, Kulvicki speaks about the *"immediacy"* of information transmission (Kulvicki 2010a, 296). Kitcher and Varzi point this out in a similar fashion with regard to maps. "Our point is not that a picture is worth so much because it is not a linguistic entity but rather because a map says a lot of things at once" (Kitcher and Varzi 2000, 379).

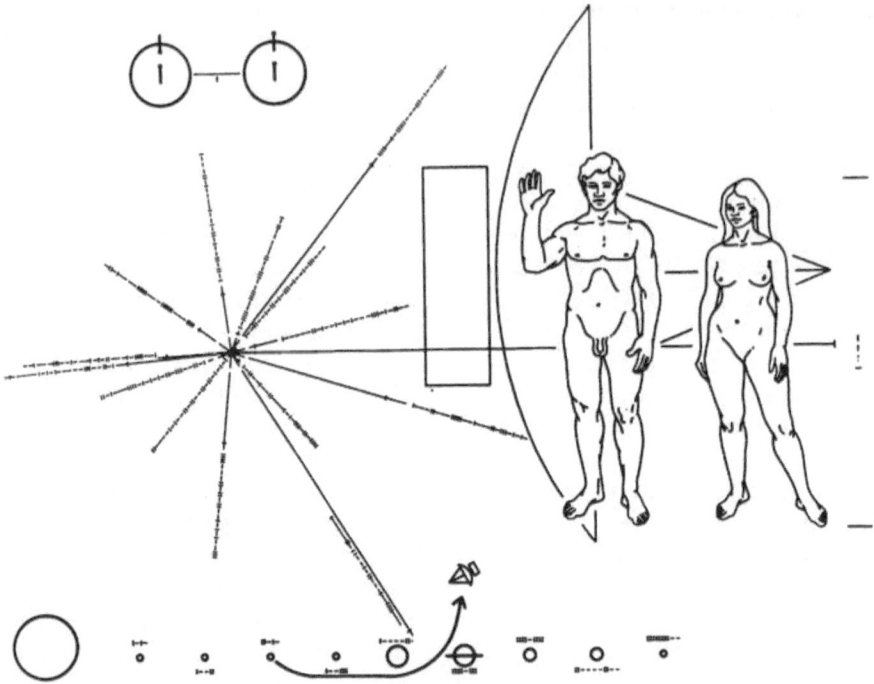

Figure 2.10 Pioneer plaque: "this image shows the *Pioneer 10* and *11*'s famed plaque,
          which features a design engraved into a gold-anodized aluminium plate,
          152 by 229 mm (6 by 9 inches), and was attached to the spacecrafts'
          antenna support struts to help shield it from erosion by interstellar dust."
          Copyright: NASA.

Source: https://commons.wikimedia.org/wiki/File:Pioneer_plaque.svg

Furthermore, a common claim with regard to pictures concerns their
apparent *universal understandability*. That is, it is assumed that everyone can
understand images just by looking at them without the necessity to learn how
to interpret them in the first place. Of course, this claim is not uncontroversial and it will be analysed critically in due course. Setting these problems
aside for the moment, however, the initial assumption seems to support the
thesis that pictures might be used as a universal language, a *lingua franca* so
to speak. Of course, this was also a reason behind the design of the *Pioneer
plaque*. It cannot be rationally assumed that some unknown extraterrestrial
being could understand any of the languages spoken on Earth – though, obviously, scientists believe that such beings could extract information from visual
representations.[45]

To illustrate this point more clearly, let us take a look at another more fully
developed proposal of a visually orientated system of communication that
was suggested by Otto Neurath, a famous member of the *Vienna Circle* in the

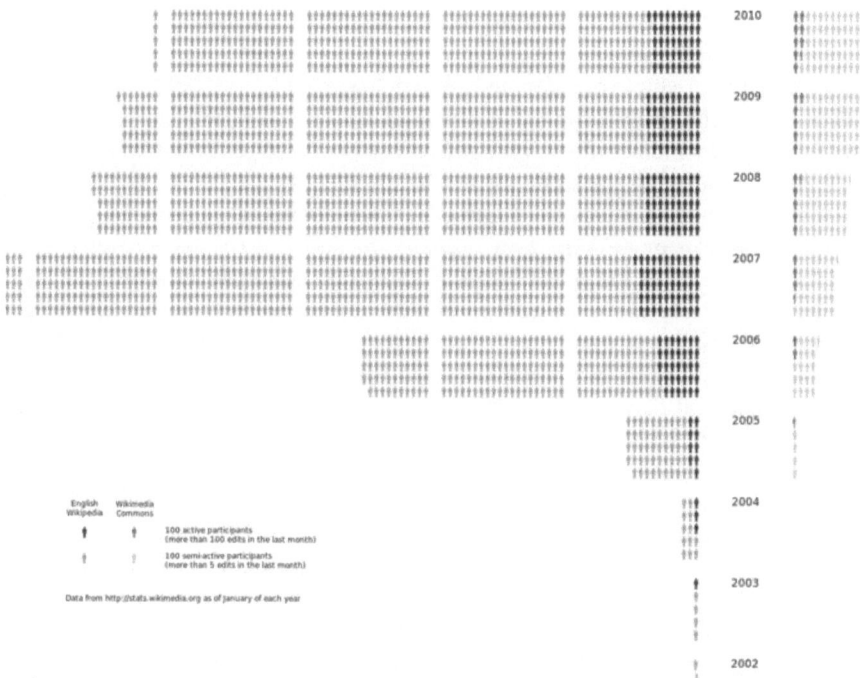

Figure 2.11 ISOTYPE: Isotype-style diagram showing the temporal evolution of the number of participants (and their level of activity) on the English Wikipedia and Wikimedia Commons. Data is based on stats.wikimedia. org as of January of each year. Each minature figure represents 100 participants; people are grouped by 100. Copyright: Guillaume Paumier, guillaumepaumier.com, CC-BY. For a colour version of this image see http://commons.wikimedia.org/wiki/File:Demographics_English_ Wikipedia_and_Commons.svg.

1930s. He developed a pedagogical concept to communicate in a visual format statistical data of common concern to uneducated people (see Neurath 1991). The key to his account is "ISOTYPE" (*I*nternational *S*ystem of *TY*pographic *Picture Education*). Figure 2.11 shows a modern adaptation of Neurath's concept of visualisation and highlights some of its main features. The diagram depicts the steady development of the English version of *Wikipedia* and *Wikimedia Commons* between 2002 and 2010. The same symbol (a human being) appears both in the left column and the right column of the diagram, only coloured differently and thereby demarcating the two different online projects, namely *Wikipedia* and *Wikimedia Commons*. The diagram allows us to infer, without great effort, the rapid growth of both projects and also to notice the difference in their quantitative developments. The recipients can thus learn a lot about the developmental history of these two projects by comparing the different columns.

The core of Neurath's idea was to communicate difficult statistical data about economic or social concerns to the masses of the working class in a way that would enable them to understand the relevant aspects, although they may not be educated readers or informed laymen (see e.g. Neurath 1927; 1929; 1931). Accordingly, the visual display of the data had to be rather simple, symbols had to be used which could be recognised easily in different contexts but nonetheless to be carrying the same meaning (see Neurath 2010, ch. 5).[46] The fully coloured symbol of a man in Figure 2.11 represents 100 active participants in the *Wikipedia* or *Wikimedia Commons* project, as the legend tells us. In this sense, Neurath's ISOTYPE does not entirely dispense with ordinary language. The symbol refers to human beings in general so that a legend has to tell us what *exactly* it stands for – participants of an online encyclopedia, or employees, or citizens, or whatever. Nonetheless, as Elisabeth Nemeth points out, the pictorial form of the symbols allows recipients to remember better what they have learnt. Remembering a bar graph or a pie chart does not tell the learner what the respective content was about; remembering symbolised people might at least give them another mnemonic clue to recall what the diagram was about (see Nemeth 2011, 69). Depicting statistical data in such a way also means omitting certain details. For example, the values depicted in Figure 2.11 are rounded up in such a way that only developmental steps including around 50 people can be shown. Thus, if there were an addition of, say, 338 participants or 321, this numerical difference could not be represented accurately. Although Neurath was aware of this patent flaw in the ISOTYPE system, he nevertheless thought that it would be better for his target group to have a rough idea of what was occurring in their environment than to have no knowledge at all thanks to the usual awkward ways of presenting such data (see Neurath 1929, 185).

Is ISOTYPE then an example of a *visual language* – a *picture language* – allowing universal understanding as announced above? Nemeth thinks that this is not the case. She points out that "Isotype pictures are constructed discursively. We cannot grasp what they mean through immediate 'intuition' [...], but only by 'going back and forth' and gradually uncovering what the comparison is about" (Nemeth 2011, 71). Her description of the process of understanding an ISOTYPE chart contradicts the above thesis of an apparently universal picture language that can be understood *immediately*. Nemeth explicitly denies the assumption of immediacy with regard to ISOTYPE (see ibid., 70). For her, Neurath's conception has more in common with an "intellectual tool" to stimulate one's imagination (see ibid., 72, 73). She highlights the fact that Neurath did not want his ISOTYPE charts to provide *the* ultimate answer to a question, but to provoke curiosity among the recipients and thus to pass on a "scientific attitude" to them (ibid., 76f.). The charts are supposed to motivate the recipients to find out themselves what the information presented is about. Learning is conceptualised as an *activity* here and not as a passive absorbance of 'convenience food', so to speak. This seems to hold true concerning our own experience when regarding Figure 2.11.

However, activating the participant to enhance the process of learning does not preclude the possibility of visually transmitting information – a feature that would constitute the core of a visual language.

What might pose a bigger problem, if we were to defend the thesis that Neurath's ISOTYPE constitutes an example of a picture language, are the legends attached to his charts. As mentioned above, without their aid, the charts cannot be understood appropriately. Without paying attention to the attached legend, we could not claim to know what exactly the increase depicted on the chart in the example above is about. Only by reading the legend do we come to know that the charts refer to the two internet services and that the growth depicted concerns the number of edits and not, for instance, downloads. To be concise, although Neurath made use of visualisation's representational advantages in order to transmit particular kinds of information to his target group, he did not create an autonomous visual language.

But even though visual representations seem particularly good at transmitting certain information, as the above example suggests, a variety of disadvantages connected with this type of representational means becomes evident when it is employed for communicational purposes. Criticisms in this respect are, for example, expressed by Umberto Eco (see e.g. Eco 1994, ch. 7). He highlights different shortcomings of images in comparison with linguistic representations when it comes to the transmission of *precise* meaning.

First of all, he makes us aware of the fact that an *apparent universality* of picture languages presupposes that the resemblance relation between image[47] and depicted entity is constant and consistent (see ibid., 178), otherwise the user would not be able to recognise what the sign is supposed to represent. It is this recognisability of particular similarities that seemingly allows us to assume a universal comprehensibility. However, as Eco points out, there are different resemblance relations which a recipient might regard as being relevant. For instance, some might hold that a circle, as shown on the *Pioneer plaque* above, resembles the sun, whereas others will point out that this is not the relevant property to pick out a star from other (more or less) circular entities such as planets. Hence, the required one-to-one correspondence (here: circle means sun) cannot be established because the relevant resemblance relations are neither fixed nor defined.

Eco discusses mnemonics and sign language as examples (see ibid., 179ff.). As a consequence of their polysemy, he infers the necessity to anchor the particular meaning of an image with the aid of the conventional rules of a verbal language (see ibid., 183).[48]

Furthermore, Eco points out that it is difficult to express actions, tenses, adverbs, or prepositions by using visual means alone (see ibid.). How would a particular action be depicted as taking place in the past or as only thought of? How is something to be depicted that is not the case? And although Eco thinks that it might be possible to find a way to express such contents visually, he claims that, in the end, those visual strategies would nonetheless be parasitic on the semantics of linguistic expressions. In this context, he refers to Goodman's insight that pictures can represent different meanings at the same time (see ibid.). How could such contents be depicted in a way that no misunderstanding occurs? Eco seems to think that no positive answer can be given. He ironically illustrates this negative point of view by describing how to solve the problem of nuclear waste dumps, namely by establishing a new priesthood that will retain the relevant knowledge by creating myths that will prevent us from digging into dangerous ground some 10,000 years in the future (see ibid., 186f.). No sign – neither pictorial nor linguistic in kind – can do this job, as people will not remember its meaning after such a long period of time.

Less ironically, but equally critically, Scholz discusses the assumption that resemblance relations allow an understanding of pictures without presupposing any kind of learning. In sharp contrast to this, he thinks that more is required to understand pictures than just looking at them. Thus, he challenges the thesis of immediacy, as Nemeth does, by discussing what cognitive performances are involved when we speak about 'understanding a picture'. He shows that various aspects play a role. Actually, he makes a distinction between *nine different levels of understanding* (see Scholz 2009b, ch. 5.4.3). The following questions illustrate the recipient's complex cognitive performances. *Does she understand that what she sees is a picture? Does she understand what is depicted? Does she understand what the picture refers to? Does she understand what the painter wants to communicate via the image?*

These questions seemingly support Scholz's thesis that one has to 'read' an image nearly in the same way as one reads sentences in a written language (see ibid., 41). The point is that to understand a picture correctly, background knowledge is needed to decipher the visual code. One important aspect with regard to both artistic and to scientific images, concerns *conventions* of depiction.[49] In this context, Kjørup highlights the diversity of different kinds of conventions that have been developed in art (see Kjørup 2009, 50f.). He makes clear that unfamiliarity with the convention employed in a particular picture might prevent the recipient understanding the image correctly (see ibid., 51). As these styles depend on cultural, historical and stylistic backgrounds, it comes as no surprise that misunderstandings can occur if the recipient does not share the relevant background knowledge. In this sense, Kjørup also calls into question the thesis of immediacy with regard to pictorial understanding.

To summarise the features of visual representations mentioned so far: images show the capacity to transmit *vast amounts of information*. I agree with this thesis because it is a quite moderate claim, it being about capacities only; that is, not all visual representations have to fully realise their respective capacity. The discussion concerning the second and the third characteristics of visual representation is more controversial, namely the *immediacy of information transmission* and their apparent *universal comprehensibility*. They do not apply equally well to all kinds of visual representations. Both seem to be more gradual in manner, depending on the context considered.

Nonetheless, there seems to be a difference with regard to these aspects in comparison to linguistic representations. As Scholz points out, although one has to admit that pictures are not universally understandable and are, therefore, not an appropriate means for constructing a *lingua franca*, they nevertheless show a particularity concerning how we learn to 'read' them. He states that "inductive learning" plays a much bigger role in this context than in the acquisition of verbal languages (see Scholz 2009b, 46f.). Contrary to linguistic representations, visualisations do not have to be learnt gradually, like the vocabulary of a foreign language. Visual representations can often be understood by learning only some examples of the respective system of depiction, after which the learner can then extrapolate to new instances without much effort.

The residual question, then, is whether we can say more about this special kind of *visual representation* being responsible for the virtues and vices mentioned above. Actually, there is an intensive debate about this topic in philosophy, namely in *picture theory* – a theoretical branch of aesthetics whose proponents also take into account attempts from further domains such as cognitive science, theories of perception, semiotics, the philosophy of mind and the like.[50]

Proponents of picture theory try to answer the question *what is a picture?*, that is, what the definition of the term 'picture' is supposed to be. Accordingly, they are looking for the necessary and sufficient conditions of being a picture. What then are the essential characteristics of pictures? As a first approximation, they can be regarded as entities that *depict* one or more objects or states of affairs.[51] Thus, another question of picture theory concerns the notion of *depiction*, which, roughly speaking, can be understood as *visual or pictorial representation* (see e.g. Newall 2011, 2).

Picture theorists, however, do not only aim at defining what a 'picture' is and how to conceive of 'depiction'. Michael Newall, for example, suggests that, in addition, such a theory should also explain *how depiction works*. "That is, it should tell us, in general terms, how surfaces marked in particular ways come to be understood as depicting particular things" (Newall 2011, 52). Knowing what *a picture depicting its object* means might help us to understand how scientific visualisations represent information. A similar point is made by Mark Rollins, who notes that a theory of depiction should – beyond the task of definition – also explain why pictures have a particular meaning (see Rollins 2008, 383). Consequently, a theory of depiction should also tell us something about why it is that we *understand* pictures in a certain way. Some of the cognitive particularities – such as *immediacy* or *universality* – have been discussed above. To be concise, theories of depiction are supposed to tell us both something about the object involved, that is, what it is to be a picture, and about the cognitive process that it engenders in the recipient, that is, how and why we understand visual representations.

The nature of depiction has been discussed in philosophy since Plato. As a result, there are a lot of different theoretical approaches to this topic nowadays.

It would go well beyond the scope of this investigation to try to examine all of them. Luckily, some philosophers have already worked out a synoptic view by trying to categorise the different theories. In the following, I will adopt Newall's suggestion. He makes a distinction between five types of explanatory theories, namely *resemblance theories, conventionalism, experience-based theories, recognition theories and mixed theories.*

In the next sections, these categories will be considered and individual theories examined to show the developments within these classes.[52] One focus will be on objections raised to the different theoretical approaches, as they show the motivation of development within this debate. The guiding question will be which of these theoretical approaches might suit best to our purpose of developing an explanatory account of visual representations in science.

2.2.1 Resemblance theories

Resemblance theories belong to the oldest attempts of philosophy to explain what 'depiction' means. As Newall puts it: "[r]esemblance theories hold that pictures depict in virtue of resembling their subject matter. Resemblance theories have a long tradition extending well beyond the modern scholarship on depiction; Plato and [Charles Sanders] Peirce are among the most notable of their proponents" (Newall 2011, 2). Here, 'depiction' is understood as implying that a picture (x) resembles its object (y) in a certain way. However, what does *to resemble y in a certain way* mean? Obviously, more precision is required regarding the notion of resemblance implied here.

The simplest account would be to claim that the picture x resembles its object y iff x and y share certain properties. However, the question remains which and how many properties the picture and its object have to have in common in a case of depiction: all, some, or the essential ones? And if the essential features are chosen, then what does 'essential' mean?

This apparently vague concept of resemblance leads to a variety of more or less obvious objections.[53] A complete discussion of all of them would be too wide-ranging a task for my current purpose, yet an examination of the most famous arguments will be helpful to better understand what resemblance-based accounts are about. Mark Rollins mentions the following two.

The first one can be traced back to Plato who pointed out that "everything resembles everything else in *some respect*" (quoted from Rollins 2008, 384, my italics).[54] In this sense, the property of *having a head*, for example, relates me to the rest of mankind and to a good deal of the animal kingdom as well. And what holds true in this broad respect also affects our more finely tuned aim of discovering the essential difference between pictures and other kinds of representation. If everything resembles everything else in certain respects, and this is all that we are able to claim to know about the characteristics of pictures, then this suggestion does not constitute a helpful demarcation criterion. We have just stated a *triviality* about general relationships between all kinds of things whatsoever in our world.

In addition to that, Rollins mentions a second crucial objection to resemblance theories that Nelson Goodman puts forward and which can also be found in Plato's writings,[55] namely:

> that resemblance does not distinguish a representation from what it represents. Resemblance is symmetrical and reflexive: a picture and its object each resemble the other and a picture resembles itself more than anything else, but objects do not represent pictures nor do pictures represent themselves.
>
> (Rollins 2008, 384)

Goodman takes this to support the statement that "resemblance in any degree is no sufficient condition for representation" (Goodman 1976, 4). To see the point, the case of twins should be considered. They may resemble one another perfectly, but we would not suggest that one depicts the other (see Scholz 2009b, 25ff.).[56]

Furthermore, Goodman not only defends the claim that resemblance is *not a sufficient condition* of depiction, he also thinks that it is *not even a necessary one*. His argument is based on assumptions concerning the reference of pictures. Representation normally implies that we refer to something – the represented object or subject matter. But, as Goodman shows, a relation of resemblance is not even necessary for such a referential relation. "Nor is resemblance necessary for reference; almost anything may stand for almost anything else" (Goodman 1976, 5). Presenting an apple in my bare hand and declaring '*This represents the Earth*' may allow me to establish a referential relation and, thereby, also a representational relation between the apple in my hand and our planet. It simply does not matter whether there is really any kind of resemblance involved between fruits and planets. What is of importance is just my declaration and your acknowledgement of it. Such arbitrary referential relationships are often established with the aid of *conventions*, i.e. rules accepted within a particular community.[57] These rules will then define which symbol refers to what entity. Regarding pictures as symbols in this way is Goodman's account of depiction, stating that pictures are symbols whose meaning is arbitrarily ascribed to them, which will be discussed in due course.

Goodman's assumption thereby undermines Peirce's distinction between *symbols*, *icons* and *indices*. Peirce has pointed out that representational relations of signs can have three different foundations, namely conventions, resemblance and causal relations. But whereas he thinks that images are based on resemblance relations and thus should be regarded as icons, Goodman characterises them as symbols that are said to bear their meaning via conventions in the Peircian system.[58]

Another challenge to be met by proponents of a resemblance theory of depiction are *failed images*. Scholz claims that some images show only minor or even no similarities to the object of depiction, like those, for instance, produced by incompetent painters. Nevertheless we would still call them pictures (see Scholz 2009b, 29, 85), thus giving rise to an explanatory gap.

Newall discusses two different strategies to cope with these difficulties. Firstly, proponents of resemblance theories try to save their preferred account by *specifying the kind of resemblance* they dwell on. In this context, Scholz highlights the fact that 'resemblance' can be understood in a variety of different ways. He presents an elaborated synoptic view of the diverse accounts and analyses their respective difficulties (see Scholz 2009b, ch. 2.6, 2.7). Secondly, a resemblance relation could be taken as constituting only a *necessary condition* of depiction (see Newall 2011, 68f.) so that we have to look for further conditions to complete the analysis of 'depiction'. Many philosophers choose this second strategy and thereby accept Goodman's point that resemblance alone cannot be a sufficient condition for depiction. Yet, despite Goodman's further critique, they maintain that resemblance can still be a necessary condition. How such a line of reasoning might work with respect to visual representations in science is examined as follows. In this context, we may admit that a picture of y can be used by a recipient to represent x, as pointed out above, if there exists a respective conventional relation between x and y. However, such a *convention-based usage* does not preclude the possibility that other referential relations are more basic to determining a kind of *genuine meaning*. And it is here that resemblance relations can fulfil the role of a necessary condition. Conventions of the kind just mentioned only seem to bring to our attention the meaning that a representation gains during a particular communicative act.

I want to put forward such a mixed nature of what constitutes the meaning of a particular visual representation in regard to scientific visualisations. Here, too, conventions play a role, but only in addition to further meaning-constitutive relations, especially, though not exclusively, causal connections. Many scientific images are technically produced by measurement devices, instruments, or imaging technologies. It was mentioned above that one relevant aspect of defining what the correlated visual representation shows is constituted by the causal connection between instrument and datum. The image produced by a detecting device can only show what has been in front of the detector, that is, what the instrument is sensitive to. Thus, what (partly)[59] constitutes the meaning of these images in the first place is this causal connection between the detecting instrument and the object under investigation. The sensitivity of the instrument to register certain parameters in addition to this causal connection between the world and the detecting device allows the instrument to show and sometimes also to register this incoming information. Such a basic recording of information does not rely, then, on conventions, but this does not mean that conventions do not necessarily play a role when extracting and thereby constituting the meaning of those signs, nor are resemblance relations entirely irrelevant in this regard.

For example, Figure 2.12, which shows *Olympus Mons* on Mars, the highest volcano in our planetary system, is a *false colour photograph*. This means

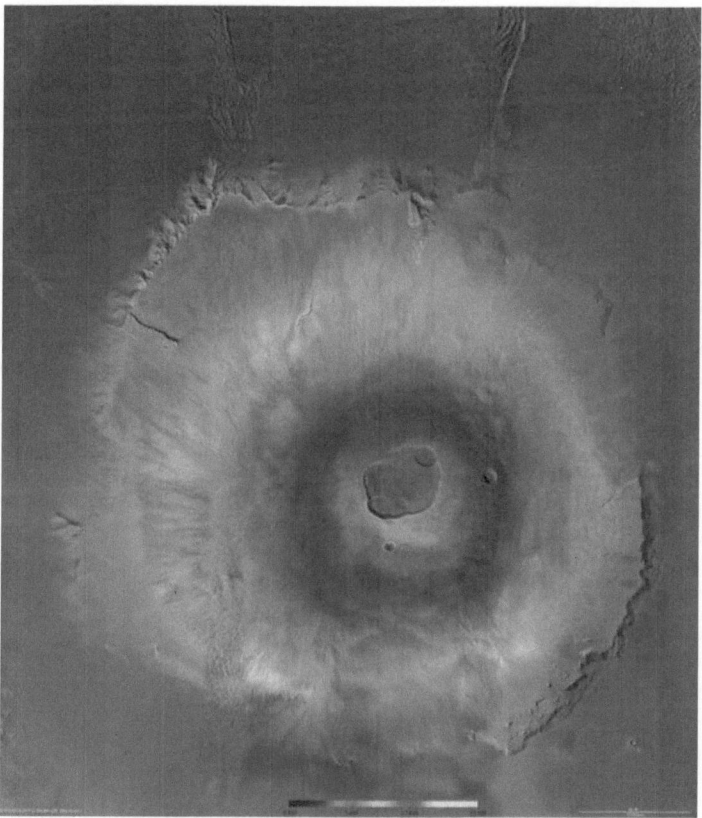

Figure 2.12 Olympus Mons in false colours; Copyright: ESA/DLR/FU Berlin, CC BY-SA 3.0 IGO. (Note to the reader: for the genuine colour version of this image see www.dlr.de/mars/en/desktopdefault.aspx/tabid-4677/7747_read-11947/gallery-1/gallery_read-Image.8.5155/, accessed August 4, 2017. Converting the colour image to a printable black and white version was not possible without a loss of information. The legend of the image shows that the same grey scale is used for two different colours of the genuine image)

that, on the one hand, the camera produced an image because the respective volcano was in front of its lens. On the other, the colours were added later and represent the different altitudes of Olympus Mons. A correlated legend informs how to read them. It indicates, for instance, that a reddish or whitish region means a region of high altitude, whereas blue stands for low altitude regions.

Consequently, the information about the outline shape of the volcano is transmitted causally, whereas the information about its altitude is transferred to us via convention. In this sense, conventions are secondary without losing their relevance in information transmission. In this sense, I think that proponents of a conventionalist account are right in pointing out that conventions are *often* necessary to *understand* images correctly. However, this does not imply that the depicting relation has to be constituted by conventions alone. The need of knowledge about conventions to understand a picture does not preclude that resemblance or causal relations (or both) constitute the basic pictorial meaning, that is, a marked surface.

However, regarding resemblance as a necessary condition, as pointed out in the above line of reasoning, is not free of difficulties either. As Scholz makes plain, proponents of such an approach owe an explanation of how to conceive of *pictures without a referent*, pictures of fictional entities such as unicorns, for example (see Scholz 2009b, 30). In such instances, there exists no object in the actual world that could be mimicked by the image. Nonetheless, we would still call such representations 'pictures'. In what sense could a resemblance relation play an even necessary role in constituting depiction here? Leaving this open for further discussions in picture theory, I will now turn to another important point about the concept of resemblance provided by Dominic Lopes and Michael Newall.

Starting with the same question, namely what kind of resemblance proponents of such an account seek, they draw our attention to another conceptual particularity here. Although neither thinks that resemblance relations are the key to understanding what depiction is, they analyse the concept involved quite thoroughly. One important result of their respective analyses is to point out that there are two different kinds of resemblance relations involved that should be kept apart, namely a *subjective* and an *objective* concept of this term. The upshot of their investigations, then, is that proponents of resemblance-based theories would have to base their accounts on the objective concept, but fail to do so, which is why resemblance-based accounts cannot fulfil their supposed task to explain depiction. Although aiming at the same distinction, Lopes and Newall put a slightly different focus on these terms in their analyses. It therefore makes sense to discuss their approaches separately. I shall begin with Lopes's conception first. He makes a distinction between a "representation-dependent" and a "representation-independent" kind of resemblance (Lopes 2006, 16f.). This differentiation is made along the lines of the viewer's background knowledge to detect the respective resemblance relation.

- *Representation-independent resemblance* does not presuppose any background knowledge. It can be noticed just by looking at the picture – so Lopes's thesis. The object and its depiction share some properties which are easily visually identifiable by the observer. For example, a picture of a house may show a roof, some windows and a door, exactly as the object itself does. In this sense, anybody should be able to identify the shared properties just by comparing picture and object. This can then be conceived of as the objective kind of resemblance, mentioned above.
- A *representation-dependent resemblance relation*, on the other hand, presupposes a certain amount of background knowledge on the part

of the viewer. To detect the similarity, the viewer has to know what is depicted and why it is depicted in a certain way. As an example, Lopes reminds us of streaked lines depicting the motion of animals or people in comic strips, etc. If we know what they are meant to show, we can detect a similarity to the blurred background of moving beings (see ibid., 16f.). But if we do not know this, it might be difficult to find out what exactly has been depicted here and, as background knowledge varies among viewers, this can then be regarded as the subjective sense of resemblance in depictions.

It has to be inserted here that the difficulty involved in producing clear examples of either case makes the distinction at least questionable. Admittedly, Lopes does not hesitate to discuss problems connected with this distinction either, yet he nevertheless presupposes that the above distinction can really be drawn along the lines of background knowledge. Although I think that, contrary to Lopes's assumption, Scholz has a point in stating that *all pictures have to be read* (see Scholz 2009b, 41), this also implies that no resemblance relation is completely representation-independent: that some background knowledge has to be presupposed in any case – at least a certain familiarity with the depictive style employed to detect the relevant similarity.[60] This finding might be regarded as supporting Lopes's assumption, as will be seen in due course, but it might also be used as evidence that Lopes finally fights a straw man, as no case for a fully representation-independent resemblance relation can ever be made.

Lopes points out that theorists who want to emphasise the important role of resemblance relations for the nature of depiction have to promote the representation-independent variant. What they need for their theory to be viable is a relation between image and object which is detectable for any viewer. If they need some background knowledge beforehand, it would not be resemblance but the cognitive resource which constitutes the depicting relation. Concerning the consideration above, however, Lopes's demanding a completely representation-independent resemblance relation appears too strict a requirement at this point because, if Scholz's thesis is correct, at least each pictorial style has to be individually learnt in order to correctly understand the pictures of the respective style.

Taking the distinction between the two kinds of resemblance seriously, Lopes then raises a 'chicken-and-egg' problem with regard to the relation of resemblance and depiction. "Resemblance is inextricably connected to depiction, but it remains to be seen whether we understand pictures by noticing resemblances or notice resemblances as a result of understanding pictures" (Lopes 2006, 17). In this sense, Lopes himself seems to be puzzled by the possibility of a representation-independent resemblance relation. However, his question about dependencies should not be exaggerated, as there seem to be clear instances where observers can detect similarities, even if the latter are not connected to pictures at all. Similarities between twins are a common example here. Newall's suggestions of how to draw a distinction between subjective and objective resemblance relations in depiction is perhaps more successful (see Newall 2011, 81ff.). He starts by admitting that, in a variety of cases, there might be a kind of resemblance relation involved in depiction. Nonetheless, he thinks that these cannot play the crucial role that some philosophers ascribe to them in this context and, with regard to this problem, he highlights the difference between what he calls *viewer-dependent* and *viewer-independent* resemblance relations, pointing out that the former can be regarded as a subjective instance of resemblance while the latter constitutes an objective one. Likes Lopes, Newall stresses the point that the proponents of a resemblance theory of depiction would want to defend the stronger claim. From his point of view, they are committed to the thesis that the picture and the object depicted share certain real properties, that is, they resemble one another in an objective sense.

In this context, Newall offers some counter-examples (see Newall 2011, 91f.) – for instance, optical illusions and figures exploiting similar phenomena of human visual perception such as Figure 2.13. These examples are meant to undermine both classical as well as refined approaches of resemblance theories of depiction. They convincingly show that the resemblance relations the viewer apparently notices in the pictures are not evoked by objectively shared properties, for example, sameness of colour or shape. For instance, railway tracks do not converge like the lines in the geometrical form in Figure 2.13.

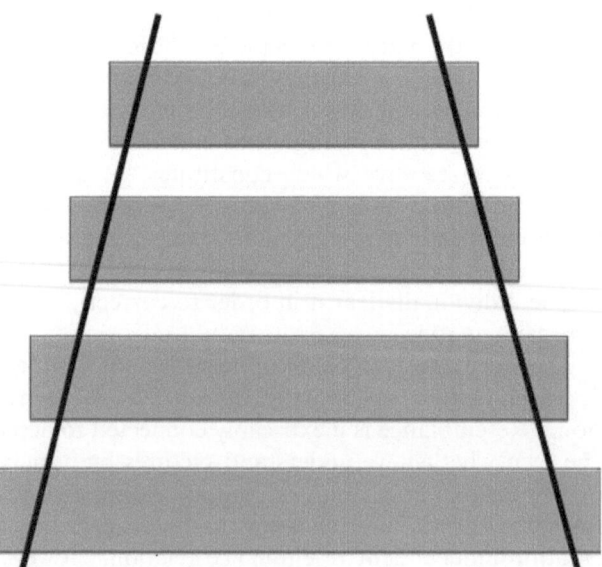

Figure 2.13 Railroad image.
Source: own image.

They do not have gaps. The impression of resemblance that appears here between image and object is only the result of the viewer's visual experience of the picture; that is, the resemblance in play is viewer-dependent and not an independent relational feature of depiction. In the above example, the picture thus evokes an impression of resemblance in shape although – in fact – there is none.[61]

It might be objected that, like Lopes's requirements of an objective resemblance relation, Newall's counter-examples also demand too strict a condition of demarcation. With respect to the above example, we can make this problem more precise by pointing out that even a photograph of railway tracks apparently shows converging lines, although the real object does not possess this property. Moreover, comparing the visual experience of looking at real railway tracks and regarding a picture of them shows that the visual impression in both instances is similar, namely one of converging lines.[62] This is probably the foundation of the claim that the invention of perspective drawing in the Renaissance era contributed so enormously to a greater realism in painting: it echos the everyday visual experiences as human observers.

Refined approaches of resemblance theories are guided by the insight that similarity lies in the visual experience of the observer rather than in the objective properties of the entity depicted. A popular account in this sense has been put forward by Christopher Peacocke[63] who defends the thesis that the place to look for the relevant similarity is in the *visual field* of the observer (see Peacocke 1987, 385). The visual field is a mental space which can be understood best by imagining it as a drawing grid between object and observer like those introduced by Albrecht Dürer. It is here – in this mental space – that the perceptual properties of a particular object match the visual experience provoked by its image, but only when the former is regarded under the intended observational conditions. This last insertion makes plain that this approach to similarity is only one part of Peacocke's explanation of how depiction should best be understood. Further requirements must also be met (see ibid., 388).[64]

Peacocke's approach highlights what seems to have been neglected in Lopes's and Newall's criticism above, namely that it is *visually noticeable similarity* which is relevant with regard to depiction, and not similarities in unobservable essences. Moreover, visually detectable similarities of course partly depend on the characteristics of human sense perception, yet this does not imply their subjectivity in comparison with an allegedly objective resemblance relation via shared essential properties. The development of perspective drawing seems at least to suggest that an *inter-subjective standard* can be established with respect to properties of visual experience. In the context of depiction, this seems to me all that can be expected when demanding an objective criterion of resemblance. Consequently, with regard to both criticisms, namely Lopes's and Newall's critique described above, I would object that their respective requirements are too strict and also have one looking in the wrong place for relevant characteristics of comparison when discussing resemblance-based accounts of depiction. Although there is a variety of other objections with respect to resemblance theories of depiction, I will conclude by discussing one last critical point put forward by Scholz (2009a) and frame his critique in the investigation of scientific visualisations. Obviously, resemblance theories can deal well with cases where an existing object is depicted, but what happens if images show something that does not already exist in nature? There are all kinds of drafts, blueprints, sketches and the like in the engineering sciences – and not just there. Some scientists might use visual representations to come to terms with their own ideas; they might scribble certain things to see correlations that they merely assume. In such instances, it would be wrong to say that there is a resemblance relation involved, as the entity depicted does not exist prior to the imaging process. The *direction of fit*, as Scholz expresses it, differs from the usual instances with which resemblance theorists deal (see Scholz 2009b, 38f.). Resemblance theorists normally dwell on a world-to-image direction of fit; that is, the object exists first and only afterwards is the image produced. Drafts and the like are of a reverse direction, that is, image-to-world: an image is produced first and afterwards the machine is built in accordance with this model.

It might be argued that, in the case of drafts, they still resemble something, namely the ideas of the scientist or draughtsman, but what exactly does 'resemblance' then amount to? Obviously, the properties of a mental image or state are totally different from the ones of material objects. Consequently, drafts do not seem to follow the same requirements of depiction as *ordinary* pictures do. Moreover, they show that resemblance alone does not fully explain how depiction works. What else might then be put forward in this context?

In this section, the topic of conventions and their possible contributions to the process of depiction have already been mentioned. Correlated approaches whose proponents focus on conventions as an explanation of depiction now need to be more closely examined.

2.2.2 *Conventionalism*

A different explanatory approach to the topic is taken by conventionalists. This theoretical stance was prominently developed and defended by Nelson Goodman (see Rollins 2008, 384), also in response to the above-mentioned problems of resemblance theories of depiction. "Conventionalism, as developed by Goodman, holds that depiction shares with language a basis in conventional rules, but is set aside from language by a distinctive structure" (Newall 2011, 2). This short characterisation already strikes a nerve with regard to this account. What exactly constitutes the element of *conventionality* in Goodman's theory? Is it, as Newall suggests, a result of rules? And, if so, what stance does Goodman take with regard to the latter? Or can conventionality be rationalised otherwise? Clarifying what is meant by 'convention' in Goodman's theory seems to be necessary, as a good deal of criticism is based on the assumed thesis of conventionality. A start will thus be made by discussing two approaches to Goodman's theory, namely Newall's and Lopes's interpretative accounts of how to understand the term 'convention' in Goodman's theses.

The first approach to consider is Newall's detailed analysis of Goodman's theory of depiction. A good deal of Newall's criticism is thereby built on the concept of convention involved. He begins his critique with the assumption that conventions should be understood as *solutions to co-ordination problems*. He thereby adopts David Lewis's account on the topic. What does this mean in the context of visual representations? Newall states that:

> [...] the solution to a co-ordination problem involves the agreement of the community to co-ordinate their collective behaviour in one way or another. [...] Words and symbols are conventional representations, for any configuration of syllable may be used as a word and any inscription may be used as a symbol, provided that a community of users agrees on the particular use.
>
> (Newall 2011, 9f.)

Thus, Newall suggests understanding conventions concerning images to be the result of negotiations within a group of people on how to visually represent particular entities. A consequence of this account is that Newall defends the claim that not all methods of depiction are governed by conventions, as not all of them are subject to co-ordination problems. For example, he states that no convention is necessary to determine how to depict light and dark areas in a picture (see ibid., 11, 13), as there seems to be no alternative way to using light and dark colour hues. Although *inverted pictures* can be mentioned as a counter-example to this rather simplified assumption concerning the potential scope of conventions in depiction, Newall's conception of convention can be granted here briefly in order to examine his critique of Goodman's conventionalism.

Newall thinks that conventionalism as a theory of depiction, understood this way, has mainly been dismissed by philosophers, although it still has adherents in art history (see ibid., 3). To support this thesis, he presents two arguments against conventionalism that he takes to be the decisive ones. "Both criticize conventionalism for its failure to explain particular facts about depiction. The first of these is a feature usually called 'natural generativity' [...]. The second is the apparent ability of members of 'pictorially innocent' cultures to interpret pictures" (ibid., 11f.).

The latter argument clearly rests on empirical considerations. Last but not least, it is an empirical question whether pictorially innocent cultures, that is, people who are not familiar with the use of pictures, are not actually able to make sense of depictions and, hence, to understand pictures correctly. Consequently, Newall cites ethnological studies doubting the passed-on assumption that such people misconceive depictions (see ibid., 13f.). If the results presented in those studies are correct, it cannot be a matter of initially learnt conventions (alone) which explains why people understand pictures correctly.

However, besides scientific (that is, ethnological) studies not being conclusive about this subject matter, as Newall points out, the argument itself seems to confuse the following three questions, namely *what constitutes depiction, what is essential to understanding pictorial representations*, and *what does it mean to understand a picture correctly*.[65] Without doubt, many items of background knowledge are often required *to understand correctly* what a particular picture is about. Think of Roskies's discussion of the difficulties that even experts have to face when interpreting fMR images as an example. Or take Figure 1.1 as an illustration of this point. Maybe the viewer simply does not know what an aurochs looks like. In this case, she will have considerable difficulties in figuring out what exactly this image is supposed to show. Yet nonetheless the viewer will be able to identify the physical object, presented to her in this book as Figure 1.1, *as a picture* of whatever content. Thus, knowing what a picture shows and knowing that she faces a picture represent two different levels of understanding.

Now, do misunderstanding or ignorance with respect to pictorial representation tell us something about the notion of depiction? In particular, do they show that conventions are inessential to the latter, as Newall suggests? His counter-example would undermine the conventionalist's approach if, and only if, understanding a picture as a picture essentially depends on grasping the conventions it is based on (this does not necessarily imply understanding what it depicts, as shown above), namely the depictive styles used by the artist. Yet, as the distinction between different levels of understanding is not taken into account in the argument about pictorially innocent cultures, it remains a rather vague objection. It does not exclude, for example, the possibility that pictorially innocent people might not understand the picture as a picture because they are not familiar with a *causal* method of picture production. Thus, it is by no means clear that divergent conventions play the constitutive role of depiction at this point if people of particular cultures do not understand what depiction is about. Setting this argument aside and turning instead to the first objection, mentioned by Newall, it can be noted that his argument also stresses the aspect of learning. Here the claim is that conventionalists are not able to explain the phenomenon of *natural generativity*. That is, having mastered a minimal pictorial competence, we are usually able to grasp the meaning of a broad variety of images. Scholz also indicates this special and apparently simple way of *inductive learning* as a particular characteristic of pictures in comparison with other kinds of representation (see Scholz 2009b, 46, 150). If depiction rests on conventions – and provided that *knowing how to interpret* images can account for *knowing how to produce* images – this feature of pictorial understanding seems to speak against a convention-based account of depiction in the following way. Conventions of depiction would have to be learnt by anyone who wants to be regarded as a competent member of the depiction-using community, and we would have to learn a lot if conventions were really essential in this context. As Newall states, "[b]ecause conventions are arbitrary rules, knowing one cannot shed light on any others" (Newall 2011, 12). Yet it seems to be an empirical fact that we are able to deal with new pictures – i.e. depictions of objects we have not seen as depictions before, or new types of depictive styles – almost effortlessly and at once, whereas we have to learn vocabulary of a new language item by item (see ibid., 12f.). Thus, so the argument goes, it seems as if there is a real difference between learning a pictorial symbol system and a linguistic symbol system, which would be a curious finding if both were governed by conventions in the same way.

As a rejoinder, we might point out that this distinction is somehow exaggerated. Although it can be admitted that learning a language implies (explicitly or implicitly) learning rules of grammar, the range of this task is much more restricted than Newall suggests in his argument. Apparently, speakers are able to produce an infinite set of words and sentences from a finite set of characters and rules for their application. Thus, language acquisition does not imply learning rules of grammar by rote. If asked, many native speakers are not even able to name them correctly. Obviously, then, we do not have to learn convention after convention to be able to express ourselves correctly. As a result, the distinction between learning a linguistic symbol system and learning a pictorial one would only be a matter of degree, not a difference in principle.

This finding also does justice to the conventionalists' rejoinder to the objection based on natural generativity, cited but immediately dismissed by Newall. Here it is suggested that there are *only some* conventions that have to be learnt in early life and which allow for the feature of natural generativity later on – also with regard to depiction (see ibid., 12f.). If we grant such a rejoinder for the mechanisms of acquiring and using our mother tongue, why not also apply it to depiction?

This is then the point to bring in Lopes's discussion of Goodman's conventionalism, as Lopes puts forward a similar critique with respect to Newall's objection concerning natural generativity, as mentioned above (see Lopes 2006, 70ff.). He explains why he thinks that Goodman is able to deal with this topic without much effort in his theory of depiction. The relevant point that Lopes puts forward is that pictorial systems and verbal systems are both *symbol systems* from Goodman's point of view. Starting from this common basis, it does not seem odd to suggest that ways of learning and mastering those systems are similar. If one has succeeded in interpreting and understanding one instance of a particular system, one will be able to grasp other items belonging to the same system in exactly the intuitive way the generativity claim suggests. Furthermore, he thinks that such an intuitive way of understanding new items of a particular system is enabled by some basic rules. Accordingly, Lopes states that:

> the claim that pictures are arbitrary symbols is not in conflict with the facts about pictorial competence, so long as pictorial symbol systems are rule-governed. Our mastery of a system of depiction might consist in a grasp of rules that govern it, enabling us to understand novel pictures in the system in a way that adds to our knowledge of their subjects.
>
> (Lopes 2006, 73)

Here, the issue about what is meant by 'convention' in Goodman's theory of depiction demands more significant clarification. Apparently, Lopes's own strategy to defend Goodman's claims against the objection based on natural generativity comes into conflict with his statements about how to conceive of conventions in the latter's theory.[66] The above quotation apparently suggests that conventions are supposed to be understood as *rules* that govern different symbol systems. Yet it is exactly this rule-following that Lopes calls into question as an appropriate interpretation of Goodman's claims. As a consequence, Lopes doubts that *conventionalism* can be regarded as an appropriate label for Goodman's theory at all. He states that:

> [t]he principles of correlation underlying a system are entirely a matter of the habits and practices of the system's users. When principles of correlation defining a system enter into the practices of picture-users in a context, the system becomes 'entrenched' in that context. However, Goodman's symbol theory is not a convention theory of depiction. Conventions are rules, and Goodman is sceptical about pictorial practices (or any symbolic practices) being rule-governed.
>
> (Lopes 2006, 65)

Indeed, Goodman introduces the term 'habit' to explain why people regard certain methods of depiction as more *realistic* or *naturalistic* than others (see Goodman 1976, 34ff., 38, 231). That is, realism or naturalism in depiction are always relative to a certain symbol system and its standards of depiction. The more a particular depiction is in conformity to those standards, the more realistic it will be judged by a viewer familiar with those standards. In this sense, Goodman explains evaluations of depictive styles as being guided by the viewer's habits. Thus far, Lopes is right to claim that Goodman avoids conventions as an explanation of the viewer's assessment of depictive styles. However, this does not exclude the possibility that conventions might play a role on another level when dealing with pictures, as Scholz points out (see Scholz 2009b, 148). We will come back to this point in due course, when taking a look at Scholz's suggestion of how to complete Goodman's theory of depiction.

Beforehand, however, it can still be asked whether referring to habits relevant to assessing realism coincides with what constitutes depiction in the first place. Answering this question leads directly to the heart of Goodman's theory. He states that "[t]o represent, a picture must function as a pictorial symbol; that is, function in a system such that what is denoted depends solely upon the pictorial properties of the symbol" (Goodman 1976, 41f.). According to this, Goodman takes pictures as being symbols within systems exhibiting particular properties. As symbols, pictures fulfil the very same functional roles as linguistic expressions, namely denotation (that is, reference) and predication, as Lopes states regarding Goodman's theory (see Lopes 2006, 68). The question then is what the difference consists of between linguistic and pictorial symbols.

The answer is to be found at the *formal level*, in the way *pictorial symbol systems* are characterised in contrast to other systems of representation.[67] The pictorial system's relevant features then are *analogicity* and *relative repleteness* (see ibid.). Certainly these are only necessary conditions leaving room for further qualification, although, without constituting a complete list either, the two conditions mentioned already demand clarification. What does Goodman mean by those terms?

Goodman analyses syntactical and semantical relations amongst symbols within particular symbol systems. As he regards all kinds of what I have called 'representations' as symbols, he introduces a lot of terminology to be able to talk in a neutral way about them – a way that allows comparisons between those different kinds of symbols not already preconfigured by previously held stereotypes. On the contrary, he builds up a theory of symbolic signs starting with morphological considerations, running through syntax and up to semantics. In the following, I will confine myself to discussing only those terms that are directly connected to the demarcation problem mentioned above.[68]

The first concept, then, is 'analogicity'. Goodman makes a distinction between *analogue* and *digital symbol systems* (see Goodman 1976, 159ff.). In digital systems, a clear difference between particular symbols can be recognised both at the syntactic and the semantic level. Digital thermometers are of this kind, as well as, for example, digital scales or watches. In each of these cases, the recipient will notice a clear distinction between the different symbols – for instance, numerical values – shown on the display of the digital scales. Reading off the data poses no problem; *either* it shows, for example, 25 *or* 26 grams, but not a vague mixture. In the same way, the numerical value indicated by the scales refers to the weight of the object under investigation.[69] Thus, neither on the syntactic level – with respect to the design of the signs – nor on the semantic level – with respect to the notion of the signs – do overlaps and thus vagueness or other insecurities with respect to interpretation arise (see ibid., 161). Goodman thinks that verbal languages, that is, linguistic symbol systems, can be characterised in this way (see ibid., 140, 225f.).

Now, from his point of view, pictures obviously differ from the symbols of such systems. No such clear-cut distinctions can be made amongst the respective symbols. Consequently, he suggests that pictures belong to *analogue systems* which he characterises as follows:

> [a]nalog systems are thus both syntactically and semantically undifferentiated in the extreme: for every character there are infinitely many others such that for some mark, we cannot possibly determine that the mark does not belong to all, and such that for some object we cannot possibly determine that the object does not comply with all.
>
> (Goodman 1976, 160)

With this in mind, Goodman also describes these symbol systems as "dense" (see ibid.). On the syntactic level this means that "if it [a schema, N.M.] provides for infinitely many characters so ordered that between each two there is a third" (ibid., 136). Consequently, the symbols of such systems cannot be demarcated in the same way as those of digital systems. On the semantic level, this leads to ambiguities and vagueness.

Figure 1.1 can be regarded as an example here – the photograph of a cave painting showing parts of different animals and rock formations in the background. Obviously, it is not easy to decipher the image and to tell what *exactly* is shown in this photograph, namely which parts are painted on the walls, which ones belong to the rock in the background or whether the colours belong to the different paintings or to the background. Apparently, the photograph alone does not convey this information. It can be assumed that a verbal description of this photograph, for instance, the comment of an archaeologist explaining this image, might be less ambiguous in this respect – though we have to admit that language itself can also be vague.

For the sake of argument, Goodman's point may for the moment be assumed that pictorial and linguistic symbol systems can be demarcated in this way. However, this is only a first step in clarification. The pictorial system is still underdetermined because the question remains how its elements diverge from those of *other analogue symbol systems*. Goodman discusses the case of diagrams in this respect. He states that it depends on how we read a diagram as to whether it belongs to an analogue or to a digital system (see Goodman 1976, 170). If such different ways of reading are, however, possible, then analysis is needed to ascertain whether diagrams also belong to pictorial symbol systems or to a system of their own. Since Goodman suggests the latter, he also has to explain the difference between them, and it is for this purpose that he introduces the second term mentioned above, namely "relative repleteness".

The difference between pictorial symbol systems and diagrammatic symbol systems is only of a gradual kind (see ibid., 230).[70] He explains the distinction by an example similar to Figure 2.6 (see ibid., 229). Here, he points out that aspects constitutive of the picture are only contingent to the diagram. To read the diagram correctly, the observer has to take into account only the values indicated by the graph with respect to the abscissa and the ordinate. Other aspects, such as the thickness or the colour of the line, are completely irrelevant to the meaning of the graph. Concerning the picture this is different. Here, any change in appearance may also imply a change in meaning – this is what is meant by the criterion of relative repleteness, which allows the demarcation of pictorial symbol systems from other kinds of analogue symbol systems such as the diagrammatic.

Goodman's account, although without doubt highly influential in the philosophical debate about depiction, has not remained uncontested. Not only have philosophers critically discussed the role and status of conventions as already indicated above, but also Goodman's criteria of demarcation discussed in the previous paragraphs.

Peacocke, for example, poses a serious challenge to Goodman's account (see Peacocke 1987, 404ff.). He offers a counter-example to show that Goodman's criteria of analogicity and relative repleteness are neither sufficient nor necessary for depiction (see ibid., 405). To show this, he tells his readers to imagine a diagram whose graph is stepwise enriched by adding further information. The important point is that this information is encoded by using hue, brightness and saturation of the graph's colour (see ibid.). As a consequence of this procedure, Peacocke states that in the end "[t]his would be a dense and relatively replete scheme: the hue, brightness and saturation at every point in the diagram matter. But it is not a picture [...]" (ibid.). Thus, the thought experiment shows that Goodman does not succeed in establishing the relevant distinction among analogue symbol systems by suggesting the criterion of relative repleteness.

Beyond this, it can also be asked whether the criteria mentioned by Goodman are properties of the respective system or not. He claims that "[a] system is representational only insofar as it is dense; and a symbol is a representation only if it belongs to a system dense throughout or to a dense part of a partially dense system" (Goodman 1976, 226). This sounds as if the criteria of the system are significant to decide whether a particular symbol is a picture or not. Yet the above explanations make clear that these are not criteria of the system, but relational properties of its respective elements, that is, of its symbols. Goodman writes that "[t]he articulation that distinguishes descriptions from representations is not, I must insist, a matter of their internal structure. [...] The significant difference lies in the relation of a symbol to others in a denotative system" (ibid., 228). This vagueness in characterisation, however, provokes another difficulty that becomes especially apparent in Scholz's attempt to complete Goodman's theory by adding theses concerning the *usage of pictures* (see especially Scholz 1993; 2009b; 2011). He reminds us of the fact that 'being a picture' does not depend on any intrinsic features of pictorial symbols but on being an element of a symbol system showing the relevant characteristics. Moreover, both Goodman (see Goodman 1976, 226, 231) and Scholz (see Scholz 2009b, 121f.)

claim that the very same symbol can be an element of different systems. At this point another difficulty for Goodman's account arises. Either it depends on properties of the symbols which allow their classification (a possibility that Goodman denies) or it depends solely on features of the system, which is not possible as depiction is constituted by relational properties of respective elements. This vagueness in Goodman's theory allows Scholz to claim that it is up to the user to decide in what sense the symbol at hand should be understood: as a picture or as a symbol of another system. In this sense, Scholz suggests replacing the question "what is a picture?" by "when is a picture?" (see Scholz 1993).

The term 'user' refers to three different groups of people: the producer of the picture, its recipients and people using the image in different ways – in advertisements for example (see Scholz 2009b, 140). Bolstering Goodman's theory in this way produces an interesting side-effect on our discussion of the aspect of conventionality in his account of depiction. Scholz points out that conventions come into play regarding all three parties involved. To start with, the producer (1): here social conventions determine her authority to tell her audience what she intends her picture to show, that is, what it depicts (see ibid., 147). Social conventions regulate that the producer is in a kind of privileged position to tell what a correct interpretation of her picture would look like.[71] Furthermore, the different manners of use (2) are also governed by conventions. Scholz's suggestion to take "image games" as essential to the constitution of meaning of pictorial symbols has already been observed (see Scholz 2009b; 154ff.; 2011). Within those contexts, images are parts of communicative acts, to warn someone of something, say, or to tell someone what she is supposed to buy, etc. These communicative acts and the parts that pictures can fulfil within them are governed by conventions (see Scholz 2009b, 161). Finally (3), the recipient's understanding of the content depicted also presupposes knowledge about particular conventions of depiction (see ibid., 176). These conventions refer to different styles of depiction and thus it is only this last claim concerning conventions that Goodman's thesis on habituation deals with. Contrary to Lopes's claim, then, the symbol theory of depiction as developed by Goodman and completed by Scholz nevertheless seems to be classified correctly as *conventionalism*.

Helpful as Scholz's theoretical approach to the problem of depiction might be, it can still be asked whether his additional theses are in accordance with Goodman's initial account. By looking for an answer to this question, another imprecision in Goodman's theory can be noted. Discussing symbol systems and beginning his analysis at the morphological level of such systems, he nevertheless fails to explain what exactly the *elements of a pictorial symbol system* are. Concerning linguistic systems, it is not difficult to follow Goodman when he claims that such systems contain letters that constitute words, sentences and texts, by using conventional rules which are also part of these systems, but he leaves unanswered what exactly corresponding elements in pictorial systems are. On the one hand, by his suggesting criteria such as syntactic and semantic density, as explained above, it seems as if Goodman intends similar small units of forms in regard to pictorial systems. This would, for instance, explain claims such as "[t]he significant difference lies in the relation of a symbol to others in a denotative system" (Goodman 1976, 228), but what exactly these units are remains unclear.

A second way of reading his thesis comes to mind, namely that pictorial systems are constituted by fully finished pictures. Such a method of interpretation seems, for example, to inform Lopes's understanding of Goodman's account. This becomes apparent, for instance, in his discussion of the "natural generativity" objection (see Lopes 2006, 70ff.). In this context, he suggests "that depiction is only partially generative between pictorial systems. [...] Depiction is generative, however, within pictorial systems" (ibid., 71). Lopes here divides pictorial systems along the lines of the depictive style used, for example, in Albertian depiction. Thus, his claim amounts to the thesis that knowing one picture produced with the aid of a particular style usually allows us to understand innumerable others of the same system. In this sense, he suggests taking both (1) pictures of particular depictive styles and (2) conventions determining how to produce such images as the relevant units of Goodman's pictorial symbol systems. Understood this way, Goodman's criteria of syntactic and semantic density no longer seem to fit. They no longer describe the units of the system correctly, because they are not applicable to depictive styles. Furthermore, symbol systems can be invented that violate Goodman's initial conditions, as their depictive styles allow the production of images that are neither dense in the syntactic nor in the semantic way.

Despite these difficulties, Goodman's theory has been highly influential not only within the context of picture theory, but also with respect to theoretical accounts trying to explain the nature of scientific images. Laura Perini's approach, for example, is guided to an important degree by Goodman's ideas. This becomes especially apparent in her early papers on scientific visualisation (see Perini 2005b; 2005c). Goodman's analytic approach of taking pictures as symbols comparable to other symbols such as linguistic expressions meets Perini's attempt to argue for an equivalent status of visual representation to linguistic representation in scientific arguments. Her ideas will be examined more closely in Chapter 4 of this book.

In a similar fashion, I briefly want to mention Zachary C. Irving's approach to scientific visualisations (see Irving 2011). He also connects his ideas to Goodman's theory of depiction and to Perini's approach to scientific images. Most of all, he tackles the problem of a presumed arbitrariness of conventions within both accounts. He tries to defend the thesis that visual conventions are not completely arbitrary, as they are in accordance with human cognitive abilities. Consequently, he calls his own approach "nonarbitrary conventionalism" (see ibid., 779). Of course, this combination raises many questions, but again I will postpone a detailed discussion until section 4.3.3 of this book.

Beyond those approaches to make use of Goodman's theory, other philosophers have suggested that a combination of perceptual (that is, resemblance theories) and conventionalist theories of depiction might also be possible (see e.g. Lopes 2006, 57; Sachs-Hombach 2006). This seems to be a promising way to avoid difficulties that are connected to each of them in isolation, and such mixed approaches will be examined in due course. Before analysing combinations of different theories, however, two more approaches to depiction as independent accounts should be discussed.

2.2.3 Experience-based theories

Obviously, both convention- and resemblance-based theories face serious difficulties with regard to their explanations of depiction. Some alternative accounts are thus based on *visual experience* instead. As Newall states: "[e]xperience-based theories, such as [Ernst H.] Gombrich's 'illusion' theory, and [Richard] Wollheim's 'seeing-in' account, claim that pictures depict in virtue of occasioning a particular kind of visual experience" (Newall 2011, 2). Some proponents of experience-based theories have already been mentioned in the two previous sections. Peacocke's account draws our attention to the *visual field* of the viewer. He thinks that it is this mental space where visual experiences of regarding the object x and the picture of x will be compared, and where similarities between both can therefore be detected (see Peacocke 1987, 386).

Two more people, already mentioned, namely Walton (see Walton 1990) and, following his approach, Maynard (see Maynard 2005; 2011), both defend an experience-based account of depiction. They maintain the thesis that a depiction of x occasions the imagination of seeing x directly. It is in this sense that they hold a particular visual experience as being typical and also explanatory of depiction.

Obviously, most *phenomenological approaches* to depiction can also be listed in this category. Examples are Lambert Wiesing's account (see Wiesing 2006) and Gottfried Boehm's approach (see Boehm 2008). However, as this work is part of the analytic tradition, I shall not discuss these approaches in detail, but suggest instead as further reading Antje Kapust's overview on phenomenological approaches (see Kapust 2009) and Aud Sissel Hoel's work. Hoel makes use of phenomenological accounts (for example of Maurice Merleau-Ponty) to explain the nature of scientific visualisations (see Carusi et al. 2014).

In the following, I will discuss two theoretical approaches within the context of experience-based theories whose proponents are more akin to the analytic tradition of philosophy, namely Ernst H. Gombrich and Richard Wollheim. Taking a look at their theses will reveal some insights into the developments of this category of theories, as they present successive accounts of depiction. I shall begin with Gombrich's so-called "illusion theory" of depiction.

Gombrich offers his account as an alternative to resemblance-based theories. He regards the latter as being unsuitable to explain depiction. Combining art history and the psychology of perception in his analysis, he thinks that there are two main challenges for such theories.

Firstly, Gombrich argues that *artists' techniques* set constraints on the degree of resemblance that can be achieved in a picture (see Gombrich 2004, ch. 1.1). He claims that artists translate rather than copy what they see (see ibid., 30, 42). He discusses this point especially in relation to colours and their hues. Gombrich highlights the fact that depictive styles vary with regard to what they allow their users to show. Moreover, he points out that the phenomenology of colours is complex, depending on lighting (see ibid., 33) as well as on adjacent colours (see ibid., 40ff.).[72] Within this context, Gombrich describes the development of colour scales by artists to cope with these difficulties. Moreover, he also points out that viewers get accustomed to certain conventions of particular styles and this adaptation raises correlated expectations about visual appearances presented in pictures (see ibid., 47ff.). Thus, it is not only the artist who is affected by the restrictions of her techniques of depiction but also her audience. Since such variations in depiction are possible, Gombrich questions what the demand of a "faithful reproduction" is supposed to amount to (see ibid., 40).

Secondly, Gombrich argues that not only techniques of depiction restrict what a particular artist can show in her image, but also *schemata*, i.e. learnt ways of how to depict certain objects. On the psychological level, they influence and set constraints to the creative process of depiction (see ibid., ch. 1.2). Artists are taught how to draw certain entities and in this sense become equipped with specific schemata they can use as starting points to depict even unfamiliar objects. Gombrich compares this educational equipment of the artist's mind with the acquisition of a particular vocabulary that is needed to create works of art – poetry as well as painting (see ibid., 75). As a consequence of this, Gombrich suggests keeping in mind that the way someone draws an object does not imply that this is also the way she sees that object (see ibid., 64). From his point of view, drawing means choosing an appropriate schema from the artist's learnt repertoire and adapting it to the current purpose so that each new picture will start from known methods of depiction (see ibid., 72). Obviously, Gombrich's allowance of modifications to those schemata suggests that he thinks that improvements of realism in depictions are possible despite the artist's guidance by what she has previously learnt. But even though Gombrich admits this possibility of improvements, he defends the claim that there is still a limit to faithful copying. He thinks that a depiction of x would be a faithful copy of x if it conveyed an identical amount of information in comparison to the one our eye would receive from regarding x directly (see ibid., 78). Yet, as the information of the former is too complex, no single depiction can be called a faithful reproduction in this sense (see ibid.).

Both constraints on faithful copies as ideal pictures undermine the plausibility of resemblance as a sufficient condition of depiction. As an alternative, Gombrich proposes his "illusion theory", that Newall explains in the following way: "[t]his 'illusion theory' claims that a picture depicts its subject matter because it generates, in the viewer, a visual experience that under the right conditions is apt to deceive the viewer into believing the subject matter is actually present" (Newall 2011, 24).

It has to be added here that Gombrich would be reluctant to describe the viewer's visual experience as *deception*. Deceiving the recipient would imply that she is misled into believing that she is not regarding a picture of x, but x itself. Yet Gombrich points out that only the viewer's eye is tricked by the picture, but not her mind. "Are we not always aware of sitting or standing in front of a picture? Of course we are. The illusions caused by this operation are again not delusions" (Gombrich 1969, 65). He uses the term 'illusion' instead of 'deception' in order to emphasise this distinction.

Moreover, Gombrich is aware that picture perception means both seeing the object of depiction and the picture as a picture, that is, its design and surface. It is also this dual perceptual information gained by regarding an image that motivates his distinction, stated above, between illusion and deception. From his point of view, this dual perception does not occur simultaneously (see ibid.). On the contrary, he defends the claim that these two perceptual experiences are comparable to what happens when we take a look at picture puzzles such as Wittgenstein's famous duck-rabbit (see Gombrich 2004, 4f.). *Either* we see the drawing as a rabbit *or* as a duck, but we can never see both figures simultaneously. In the same way, we cannot enjoy the illusion provoked by a painting and examine its design to find out how it works.

From Gombrich's point of view, both the artist and the viewer play an important role in achieving such an illusory effect evoked by a picture. We already mentioned Gombrich's emphasis on tradition and education in the arts. This way of approaching the topic of depiction guides his description of the artistic process along the lines of Karl Popper's dictum that the history of science and of scientific progress can best be understood as a steady process of *trial and error*. Popper's thesis thereby constitutes the model for what Gombrich holds to be the common way artists produce pictures. Being equipped with certain schemata by their instructors and, more generally, by the tradition they get accustomed to during their training, they will use these basics as a starting point for a picture. Furthermore, the artists will modify those schemata as long as they are satisfied with the result in the sense of meeting their expectation of what they want to show (see ibid., 64, 126). It is along these lines that Gombrich picks up Popper's central idea of trial and error in his account of depiction.

Beyond these efforts of the artist, Gombrich thinks that the part played by the viewer is no less important. He calls the viewer's contribution to the illusion-creating process the "Beholder's Share" (Gombrich 1969, 40) and states that "[...] no two-dimensional image can be interpreted as a spatial arrangement without such a constructive contribution of our spatial imagination" (see ibid., 40f.). This constructive contribution is the result of the psychological effect known as *projection* (see Gombrich 2004, 89ff.) and it is through this that the viewer meets the artist's efforts half-way when interpreting the picture. In particular, Gombrich points out that it is the viewer's *attitude* that leads to certain expectations, finding their expression in particular projections, and in this way influences what she sees or otherwise perceives (see ibid., 157).

Gombrich also thinks that this participation of the viewer is not only necessary to constitute the meaning of the picture, but might also explain our fascination with pictorial representations. He points out that viewers experience a particular kind of joy when being involved in the creative process of constituting meaning (see ibid., 235). This he makes especially clear when discussing the innovations in artistic style developed by proponents of Impressionism and defending the claim that it is this method of painting that brings the *principle of guided projection* to a climax (see ibid., 169). Furthermore, it is this principle that he holds to some degree basic to all kinds of depiction (see ibid., 170).

Moreover, Gombrich uses Popper's *principle of trial and error* not only with regard to the artist's efforts to adapt a learnt schema to what she wants to show via her image, but also with respect to the recipient's interpretative endeavour (see ibid., 188, 198ff., 231).[73] According to this approach, *understanding a picture* means being able to categorise it as a whole in a way that our experience has taught us. Arising contradictions to such trials of categorisation reveal that we are mistaken in our interpretation and we have to start anew. It is in this sense that our experience constitutes a test case for our interpretation of pictorial representations (see ibid., 198). Obviously, these considerations are based on the assumption that all pictures are ambiguous (see ibid., 211) and thus allow for multiple interpretations (see ibid., 330).

However, it remains to be clarified what exactly the recipient's attitude towards a given picture constitutes. It is important to note at this point that Gombrich thinks that there is no "innocent eye" (see ibid., 251). This he makes plain in contrast to George Berkeley's theory of perception. Contrary to Berkeley, Gombrich defends the claim that we do not perceive sense data, but immediately interpret, classify and categorise what we perceive. And it is our experience with the real world that accounts for our attitudes and resulting expectations as interpreters of everyday perceptions as well as of pictures (see ibid., 265).

Finally, Gombrich comes back to his criticism of resemblance-based theories of depiction. Regarding the history of art as a whole, he points out that with respect to the different aims pursued by the artists,[74] what is of importance is the *equivalence* between the real object and the depicted one. This can, but does not necessarily, imply a resemblance relation between both. It is, however, essentially based on the *ability to provoke similar reactions* by the recipient (see ibid., 292). This can be considered the essential idea in Gombrich's illusion theory of depiction.

A critical discussion of Gombrich's claims is offered by Lopes (see Lopes 2006, ch. 2.1, 2.2). Firstly, he objects that the illusion theory of depiction does not cope with the diversity of pictorial phenomena (see ibid., 38f.). Lopes admits that some pictures might lead to the illusion of seeing their object of depiction; *trompe-l'oeil* pictures, especially, are of this kind. They are meant to trick the eye of the viewer. However, this is but one kind of the broad variety of depictive styles and most of the remainder are neither meant nor able to deceive the observer's eye in the illusory way that Gombrich requires,[75] for example, data visualisations such as the Higgs image (see Figure 2.4). If the illusion theory were right, calling them *pictures* would then be incorrect, because they would not lead to an illusory impression of their subject matter. Yet being forced to exclude them from the scope of a general theory of depiction would be a very unpleasant consequence. As Lopes points out in his "diversity constraint" of such theories (ibid., 32), a picture theory worthy of its name should be able to permit an explanation of all, rather than carefully selected, pictorial phenomena.

Secondly, Lopes calls into question whether Gombrich's illusion theory of depiction is correct concerning the phenomenology of picture experience (see ibid., 41ff.). Positing that a central feature of pictures is that they evoke illusions of seeing their subject matter directly seems to imply that our perception of pictures is always and exclusively like seeing the real object. Yet, in Lopes's opinion, this is not what our normal perception of images is like. Beyond seeing the object of depiction, we are also *simultaneously aware of the design of the picture*, that is, its canvas, brushwork, etc. However, if these latter impressions are an essential part of our common perception of pictures, then images cannot cause the illusion that Gombrich suggests. The viewer would be constantly aware of the fact that she is regarding an image and not the real entity.

This is then the point at which to introduce the second philosopher of an experience-based theory of depiction, namely Richard Wollheim, who takes this tension between the acknowledgement of our awareness of the materiality of pictures and their assumed illusory effect on the viewer's picture perception developed in Gombrich's theory as the stepping stone to introduce his own theory of depiction.

Similarly to Gombrich, Wollheim stresses the relevance of the viewer's visual experience of depictions. He states that:

> [w]hat is distinctive of seeing-in, and thus of my theory of representation, is the phenomenology of the experiences in which it manifests itself. Looking at a suitably marked surface, we are visually aware at once of the marked surface and of something in front of or behind something else. I call this feature of phenomenology 'twofoldness'.
>
> (Wollheim 1998, 221)

The feature of twofoldness is not only the phenomenological particularity of picture perception which Wollheim thinks is essential to pictorial representation, it is also the relevant modification that Wollheim introduces in his theoretical approach in comparison with Gombrich's account.

As has been noted above, Gombrich points out that two different visual experiences are involved when regarding a picture, namely one of its surface and the other of the object depicted. Although contrary to Wollheim, Gombrich speaks of "seeing-as" when referring to this phenomenology of picture perception. He thinks of this phenomenon in analogy to our perception of picture puzzles such as the duck-rabbit. The observer switches between two different perceptual impressions, between two aspects of the same figure: either she sees a rabbit or she sees a duck. In the same way, the viewer is either aware of the picture's surface or of the object depicted, according to Gombrich. Wollheim contests exactly this description. He is convinced that the phenomenology of picture perception should not be regarded as two separate visual experiences, as is stated in Gombrich's seeing-as thesis, but as "a single experience with two aspects" (Wollheim 1998, 221). This phenomenon he terms *seeing-in*. Thus, the *twofoldness* of picture perception is one of two *simultaneously* visual experiences, namely (a) the perception of the medium and its materiality, i.e. the picture surface with its markings made by brushstrokes, etc. and (b) the perception of the picture's subject matter.

Lopes agrees with Wollheim in this regard. He also rejects Gombrich's thesis of seeing-as and adds the following rationale:

> [...] switches in attention between two contents are not analogous to switches in attention between design and content. That the duck and the rabbit cannot be seen simultaneously does not show that the duck cannot be experienced together with the picture's design [...]
>
> (Lopes 2006, 41)

Thus, Lopes denies the validity of the analogical reasoning used by Gombrich to defend his thesis of seeing-as. Aligning himself with Wollheim, he suggests that twofoldness is an important feature of (many) pictorial experiences. He thinks that any appropriate account of depiction should be able to deal with this particular phenomenology of picture perception if it occurs;[76] this is what he calls his "twofoldness constraint" (Lopes 2006, 42, 50f.). From Lopes's point of view, not being able to meet this standard discards Gombrich's illusion theory as a suitable explanatory candidate.

Beyond the twofoldness of picture experience, the phenomenon of seeing-in is described as a "perceptual skill" of the viewer by Wollheim (Wollheim 1998, 221). Newall adds "[w]hen we understand a picture, he [Wollheim, N.M.] held, we *see* the picture's subject *in* the picture; hence, 'seeing-in'" (Newall 2011, 25, his italics). Yet it can be objected that we are able to see many different things in a particular picture. In this sense, it is still underdetermined how exactly seeing-in works. Wollheim's approach suggests that there are correct and incorrect ways of seeing-in. But what is a relevant criterion of comparison, a *standard of correctness* here?

Wollheim is aware that seeing-in allows for ambiguous perceptual impressions of a given picture and therefore proposes that "a *suitable spectator* looks at the picture" (Wollheim 1998, 217, my italics). He defines this somewhat cryptic term as denoting a person who "is suitably sensitive, suitably informed, and, if necessary, suitably prompted" (ibid.). The first and the last conditions are physiological and psychological in nature. That is, the viewer has to have an average ability to visually regard the picture. She should not suffer from colour-blindness, for example. Moreover, the viewer should also be aware of regarding a picture, that is, not merely glancing at it or otherwise neglecting it. Sensitivity and promptness, however, still do not explain why a particular seeing-in impression is correct whereas an alternative is not. In this sense, one is left with Wollheim's suggestion that the viewer is also supposed to be "suitably informed", but what this means and what the viewer is supposed to be informed about is left unanswered and thus can only lead to the assumption that it is background knowledge about the *artist's intention* that constitutes the standard of correctness one is seeking at this point. Wollheim mentions it as another important condition of depiction, when he states that:

> [r]epresentational meaning, indeed pictorial meaning in general, is, on my view, dependent, not on intention as such, but on fulfilled intention. And intention is fulfilled when the picture can cause, in a suitable spectator, an experience that tallies with the intention [of the picture producer, N.M.].
>
> (Wollheim 1998, 226)

Thus, what constitutes depiction in his sense is the phenomenology of seeing-in in combination with the artist's intentions.

Wollheim's approach to a theory of depiction is critically discussed by Lopes (see Lopes 2006, ch. 2.3, 2.4). Firstly, Lopes complains about a missing explanation of the nature of seeing-in in Wollheim's theory. Even worse, as Lopes points out, Wollheim apparently believes that such an explanation *cannot* be given (see ibid., 43f.). The vague conception of seeing-in that Wollheim offers, however, does not suffice to make plausible the distinctiveness of pictorial experience.[77] Wollheim states that seeing-in is essential to pictures and depiction, but is reluctant to explain how it works. As Lopes puts it:

> [w]ithout an account of the perceptual basis of seeing-in, what is there to rule out the possibility that seeing-in is not a perceptual process[78] at all? [...] Nevertheless, Wollheim's scepticism about the possibility of giving an

account of these mechanisms [that is, our process of picture perception, N.M.] means that the only content his theory is left with lies in the claim that seeing-in is a distinctive kind of seeing.

(Lopes 2006, 44)

To fill this obvious gap, Bence Nanay suggests an explanation based on results from the cognitive sciences concerning human vision (see Nanay 2011). He proposes that the phenomenon of twofoldness in seeing-in can be explained by referring to two different visual subsystems in the human central nervous system, namely the *ventral* and the *dorsal* stream. Although these two subsystems usually work together in processing visually gained information, empirical studies tell us that they can nevertheless be anatomically and functionally differentiated. "To put it very simply, the ventral stream is responsible for identification and recognition, whereas the function of the dorsal stream is visual control of our motor actions" (ibid., 464). Nanay now suggests that our twofold perception of pictures is the result of the processing and representing of information about the picture surface or design and about its object of depiction in these two different cognitive subsystems. "[...] [I]t is constitutive of our experience of seeing things in pictures that the depicted scene is represented by our ventral vision, whereas the surface of the picture is represented by our dorsal vision" (ibid., 466). According to the above description of the system's function, the correlation between our perception of the object depicted and its representation in the ventral system seems plausible. A more challenging vindication has to be afforded for the second pairing, that is, the design of the picture as represented in the dorsal system, because it then has to be explained what properties of the picture's surface might have in common with a cognitive subsystem visually co-ordinating our motor actions. Nanay defends both claims in detail by arguing that object perception is represented ventrally, but not dorsally, and that surface perception is represented dorsally, but not necessarily ventrally (see ibid., 467ff.). He supports the more controversial second thesis with the findings of empirical studies analysing the compensation capabilities of our visual system. Clearly, we do not perceive pictorial subjects as distorted when regarding them from oblique angles, and only when pictures are too far away does our cognitive system no longer compensate for these perceptual effects (see ibid., 471). According to Nanay, both phenomena can be explained by our representing the orientation of the picture surface dorsally. Unfortunately, this does not answer the above question satisfactorily. The point might be further improved by adding that an undistorted perception of the subject depicted is possibly relevant for guiding potential actions concerning pictures, such as reaching out to touch the strokes on the canvas forming the figure perceived. However, the impression remains that there is still something missing in Nanay's explanation, although his strategy to employ the cognitive sciences to clarify the phenomenon of twofoldness seems promising – if only because, as he states, it turns the original claim into a "testable and, therefore, falsifiable" hypothesis (ibid., 477).

Lopes objects that Wollheim's theory not only is too vague to explain the phenomenon of depiction correctly, but is devoid of any significant content because it does not offer an explanation of seeing-in at all (see Lopes 2006, 50). To support this claim, Lopes discusses Wollheim's comparison between *seeing-in* and *seeing-as* in some detail (see ibid., 44ff.). Wollheim mentions three differences here, namely regarding content, localisation and twofoldness, of which only the last does Lopes consider as crucial.

Since Lopes successfully shows that the former two criteria of demarcation are not convincing, attention should focus on the last criterion, namely the claim that the crucial difference between seeing-in and seeing-as consists in the fact that only the former involves twofoldness in picture perception:

> Wollheim observes that one cannot at one and the same time see x *as* y and be aware of the features of x which sustain seeing y – it is necessary to switch attention between x and y. Yet seeing x *in* y means simultaneously attending the features of x which sustain an experience of y in it. Pictorial seeing is discontinuous with other varieties of seeing, because it alone provides for the possibility of twofold experience.
>
> (Lopes 2006, 47)

The question which is then put forward by Lopes is whether this characterisation of picture experience is correct or not.

To start with, Lopes points out that Wollheim holds twofoldness to be an *essential* feature of picture perception. Lopes calls this assumption "strong twofoldness" (see ibid.), but does not think that Wollheim can show convincingly that twofoldness really is such an essential feature of picture perception. In order to demonstrate this, he discusses three arguments which Wollheim presents in defence of his view (see ibid., 48ff.):

1 Our normative standards for aesthetic judgements of pictures make essential use of twofoldness: we appreciate pictures most because of their artistic design which allows us to see the subject matter in them.
2 Twofoldness is called for to explain perceptual constancy: distortions in pictures, due to perspective depiction, are not experienced as such because of twofoldness.
3 The distinctive phenomenological feature of seeing-in is twofoldness: it demarcates the latter from normal visual perceptions.

Lopes finds none of these arguments convincing. For example, he accuses both Wollheim's first and third argument of not accounting for the diversity of pictorial representation. In his first argument, Wollheim suggests that design features are of significance for the viewer's aesthetic appreciation of the picture. Yet Lopes objects that not all pictures are subjects of aesthetic judgements. That is why he criticises Wollheim for a non-permitted extrapolation here – from the case of artistic pictures to all other kinds of pictures. With regard to the second argument above, Lopes objects that "perceptual constancy is not essential to depiction" (ibid., 49).

Another unpleasant consequence of a strong twofoldness requirement is, as has often been discussed with respect to Wollheim's account, that it excludes *trompe-l'oeil* paintings from the scope of depiction. *Trompe-l'oeil* pictures are successful when the viewer gets tricked by the painting in the sense of not recognising it as a picture. In such instances, the viewer is solely aware of its subject matter, but not of its design. Wollheim acknowledges this and accepts the consequence that *trompe-l'oeil* paintings are not pictorial representations in his sense. This admission, however, has provoked further responses in the debate.

Susan L. Feagin, for example, argues that *trompe-l'oeil* paintings are presentational rather than representational (see Feagin 1998). Thereby, she tries to show that the difference, resulting from Wollheim's theory of depiction, does in fact make sense. Others, such as Jerrold Levinson, think that this is a completely counterintuitive consequence and try to argue for the contrary. Levinson states that "[o]nce you grasp that something is a trompe l'oeil you can attend to its surface, and in its visual aspect, even though you cannot by hypothesis see the surface as such" (Levinson 1998, 228). Thus, from his point of view, after the initial illusion has faded and the viewer understood what she is looking at, she can finally direct her attention to the picture's design.

Levinson's suggestion then paves the way to questioning the twofoldness condition in its strong version. He thinks that *pictorial seeing* – which he contrasts with Wollheim's seeing-in (see ibid., 229) – may include twofoldness in Wollheim's sense, but may also, on other occasions, entail a "switching back and forth between awarenesses or focusings of attention of those two kinds, seeing sometimes only pure pattern, sometimes only pure object" (ibid., 230).

In a similar fashion and despite his critique of Wollheim's thesis, Lopes does not want to deny that twofoldness plays a significant role in picture perception. He thinks, however, that it would be wrong to regard this as an "all or nothing" criterion (see Lopes 2006, 51). Contrary to this, he suggests regarding twofoldness as a gradual feature. Some pictures – like van Gogh's – are more subject to this phenomenon than others. And, some – like *trompe-l'oeil* – may even show no relation to twofoldness at all. Consequently, Lopes claims that a picture theory worthy of its name should be able to deal with all the different degrees of twofoldness along this spectrum (see ibid.).

A last critical point concerning experience-based theories of depiction in general that I want to mention here is put forward by Newall. He mentions the psychological phenomenon of *blindsight* as a counter-example to stating that visual experience is a necessary condition of understanding depictions (see Newall 2011, 39f.). 'Blindsight' means that people affected by certain diseases or damage to the brain can have cognitive access to the content of pictures without being aware of having visual experiences of them.[79] Medical case studies of this kind show, according to Newall, that visual experience – of whatever kind – cannot be a necessary condition to understanding pictures.

As a rejoinder, I want to highlight again the distinction between the different levels of understanding elaborated by Scholz (see Scholz 1993; 2009b; ch. 5.4.3). Picking out and observing just two of these more closely makes clear that the phenomenon of blindsight does not perform the subversive function that Newall suggests. Obviously, blindsighted people can understand what a picture shows; that is, what it refers to – the object of depiction – and how it is characterised. However, they do not understand the picture as a picture, as they do not have the corresponding experience. A theory of depiction should offer an explanation for both. If the mental disorder of blindsight shows that people affected by it cannot perceive pictures as pictures despite having cognitive access to the visually presented content, this seems to me – contrary to Newall's suggestion – to indicate that they lack a relevant resource to understand pictures and thus depiction correctly. A particular experience therefore seems to be characteristic of depiction. Yet the problems discussed above with respect to theories of this kind suggest that more research has to be invested in this proposal to discover its particularities. Moreover, gaining this insight by reflecting on the different levels of understanding involved also makes clear that visual experience can only be part of the explanation.

2.2.4 Recognition theories

An alternative perceptual capacity of human observers is put in charge as an essential condition of depiction by proponents of recognition-based theories. As Newall states, "[r]ecognition theories of the sort suggested by Flint Schier and developed by Dominic Lopes explain depiction in terms of a picture's capacity to engage appropriate visual recognitional abilities as being essential to pictures" (Newall 2011, 2). This first characterisation of recognition-based theories seems to suggest that they could be regarded as a subcategory of experience-based theories of depiction. It could be argued that visual recognition is a particular kind of visual experience. Nevertheless, defending Newall's suggestion that these approaches be discussed separately, I think it depends on which component is being emphasised by proponents of this theoretical branch – the *visual experience* or a particular *capability of human vision*. As the latter seems to be the case – of course, without denying that a particular visual experience plays a role in picture perception – listing these approaches as a separate category seems reasonable.

Recognition-based theories are developed to avoid difficulties mentioned in correlation to resemblance- and convention-based theories. Thus, the question emerges in what sense their proponents consider them to be superior to the former. To answer this query, one of these theoretical approaches merits closer and more detailed examination, namely Dominic Lopes's "aspect-recognition theory" (Lopes 2006, 111).

Lopes understands 'depiction', i.e. pictorial representation, in the following sense: "[a] picture represents by conveying aspectual information from its subject, on the basis of which viewers can recognize its source" (ibid., 136). To fully comprehend what Lopes has in mind about how pictures represent their subjects, at least the following three questions have to be answered: how exactly do pictures *convey information*? What is *aspectual information*? And, finally, what exactly is the role of the *viewers' recognition* within this theoretical framework?

A first hint to a better understanding of Lopes's account consists in acknowledging that, similar to Goodman, he analyses *systems of representation* and not isolated images. However, it is not to Goodman's account that Lopes relates his theoretical approach, but Gareth Evans's theory of reference. According to Lopes, Evans suggests that, within "information systems", epistemic subjects can identify the sources of certain pieces of information, that is, discover what their referents are (see ibid., 102ff.). Lopes takes this as a model of pictorial reference. Accordingly, he states that:

> [o]n this model, pictures are part of an information system, individual pictures conveying perceptual information from their subjects. A picture represents an object only if it conveys information from it on the basis of which it can be identified. To understand pictures, viewers must employ a specifically pictorial mode (or modes) of identification which single out, on the basis of their contents, the pictures' sources.
>
> (Lopes 2006, 107)

Thus, two ways of information transmission have to be analysed further with respect to this model, as illustrated in Figure 2.14. On the one hand, how is the information *encoded* in the image? That is, how are the source and its image related so that the latter can convey information about the former? On the other, how does a viewer *decode* the information transmitted by the image? How is it that she understands the picture?

Explaining these two ways of information transmission, Lopes states that his account "bears the marks of a hybrid theory" (ibid.), as both the original source and the content of the picture contribute to the meaning of the image. To find an answer to the above questions, how information encoding and decoding work with respect to pictorial representations and to find out what components Lopes's hybrid theory consists of, it makes sense to examine more closely how he distinguishes his own approach from resemblance- and convention-based theories of depiction.

First of all, Lopes thinks that his account is compatible with both kinds of theories of depiction (see ibid., 111). Even though he sees virtues in both theoretical approaches that should be retained, he is convinced that neither of them presents an appropriate account of depiction. Beyond a variety

Information decoding

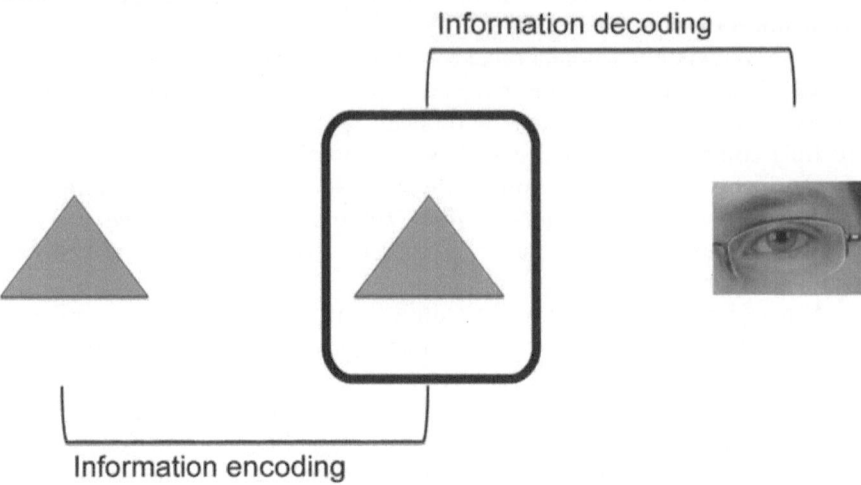

Information encoding

Figure 2.14 Ways of information transmission between object (source), image and
          viewer.
Source: own image.

of difficulties that Lopes discusses separately with respect to each of these
approaches (some of his objections have been mentioned above), he puts
forward the following deficiency that both sets of theories have in common.
From his point of view, an adequate theory of depiction should not only be
able to answer the question, "what distinguishes pictorial systems from other
systems of representation?", but also, "what distinguishes pictorial systems
from each other?" (ibid., 127). Furthermore, he claims that proponents of
resemblance- as well as of convention-based theories have devoted their time
solely to answering the first question and offer no solution to the second.
Even worse, Lopes is convinced that their approaches *cannot* offer a suitable explanation with respect to this second difficulty, "[...] for the features
that these theories claim distinguish depiction from other representational
systems – resemblance, analogicity, and the like – are unvarying features of all
pictorial systems" (ibid.). Thus, a different approach is needed that is able to
deal with both questions appropriately.

As an alternative approach able to handle this task, Lopes suggests his
"aspect-recognition theory" (ibid., 111). This theoretical approach implies
that "pictures present recognizable aspects of objects" (ibid.). Thus, we have
to clarify what is meant by "aspects" and also how the recognition of aspects
operates. Both components then help to clarify in what sense pictorial systems
are to be distinguished from other representational systems, that is, how depiction operates according to Lopes.

Following Gombrich, Lopes defends the claim that all images are essentially selective (see ibid.). This means that pictures can represent only a small
amount of the visually available information of real-world entities.

As an illustration, compare the diversity of colour hues perceptible by the human eye. Especially in the context of digital photography, this poses a challenge. Today, digital cameras are usually able to record a wide range of colour hues. However, recording is often only a means to a different end, namely e.g. for printing these pictures, or regarding them on a display. Technological devices in these contexts, however, are on many occasions not sensitive to colour in the same way as the genuine recording device was. Thus, photographers are advised to work with different colour ranges or "gamuts" to get better results, that is, images that bear more similarities to what the human eye perceives (see Freeman 2009, 462ff.).

Yet this pictorial selectivity does not imply that what a picture can show has to be in accordance with (so to speak, a reduced form of) normal visual perception (see Lopes 2006, 121, 124). Pictures can also depict entities which are not visually accessible in such a way. They may let us see radiations of a source to which the human eye is not sensitive, such as ultraviolet and infrared (see Figure 2.2). Another example is Figure 2.4 depicting subatomic particles, imperceptible of course to the unaided eye. In this way, the feature that, as Lopes points out, has to be made clear in order to understand a picture's selectivity is already under investigation, namely what exactly is meant by the "aspects" presented by a picture.

He explains that pictures are both committal and non-committal concerning the particular properties of entities they depict (see ibid., 119). Briefly, an image is committal about the property x if it presents its object of depiction as showing property x, and it is non-committal about y if it presents its object such that it is indeterminate whether the object actually possesses property y or not. Hence, the depiction is precise with respect to x and imprecise or indeterminate with respect to y. An illustration of this is provided by Figure 2.12, the false colour photograph of *Olympus Mons*. This picture informs its viewer about the outline shape of the volcano with respect to its different altitudes (property x), albeit at the expense of information about its true colours (property y), which a human observer would notice when observing the volcano with a common telescope suitable for the task.

However, as Lopes claims, the presentation of aspects understood in this way is not essentially pictorial. Verbal descriptions could do this job as well. Thus, more precision is required about *what kinds of aspects* pictorial representations are dealing with contrary to verbal descriptions and other systems of representation. The argument is returned here to the first question of demarcation, posed above. The solution to this difficulty is to be found at the level of *spatial properties*.

> Pictures are selective because, in order to represent some spatial features of their subjects, they are precluded from representing others. The reason is simply that not all spatial relations between objects in three-dimensional space can be represented on a two-dimensional surface. Selecting to represent some spatial relations makes other relations unrepresentable.
>
> (Lopes 2006, 125)

In a nutshell, then, pictures present *selected spatial information* of the entities that they depict.[80] Beyond this suggestion of what kind of information is characteristic of depictions, Lopes also points out how recipients glean information from visual representations. With respect to Figure 2.14, this is then the second way of information transmission: the viewers' decoding of the visually presented information in the picture. Here the recipients' mental faculties of recognition play the essential part.

Lopes thinks of *recognition* as a "perceptual skill that enables us to think of objects, and so potentially to refer to them" (ibid., 140). Thus, this issue takes us back to the framework of his theory, namely Evans's account of reference in information systems (see ibid., 102ff.). Evans states that information transmission takes place within systems so that the recipient can identify the source via the mentioned conveyance of information. Presupposed the process takes place successfully, it eventually enables the recipient to select the correct referent of information gathered from it. Now, Lopes thinks that understanding pictures functions in the same way via the process of recognition. Viewers recognise the depicted object x as the picture of x which makes mentally available information formerly transmitted by and, thus, gathered from the real-world entity (x) (see ibid., 137ff.). In this sense, Lopes conceives of the viewer's recognition abilities as the relevant reference-constituting mechanism in picture perception. It is in this way that the identification of the picture's source is possible (see ibid., 136).

Concerning picture perception, Lopes makes a distinction between two kinds of recognition tasks operating in the recipient's decoding of the visually presented information, namely "content-recognition" and "subject-recognition" (ibid., 145). The former task is about recognising a particular content within the design of the picture. The latter consists in recognising a particular subject in this content. Thus, a viewer can recognise the subject as it is presented via certain aspects and she can recognise content that is presented via a particular design. Following Lopes's conception, gleaning information from an image then is a *dual process*. Figure 1.1 is an illustration of this. Content recognition here means that one is able to recognise that the dark spot to the left in the design of the painting is meant to represent the eye of an animal, the curved line to its right an ear and, all together, the head and horns of an aurochs. Subject recognition, on the other hand, means that the recipient is able to recognise "pictures' contents as of their subjects" (ibid.). For example, a particular human being depicted in an advertisement is recognisable as the famous actor xyz. Thus, not only does one recognise that the depiction – the design of the image – shows a human being (content recognition), but also that it shows xyz (subject recognition).

That much said about depiction and what therefore distinguishes pictorial systems from other representational systems, the second question can now be answered, namely how to make a distinction between different systems of depictions. Here, Lopes suggests that these systems can be discerned by the aspects of the world they make visible. "It is the potential for making different commitments, resulting in different kinds of spatial aspects, that I shall suggest accounts for the diversity of systems of depiction" (ibid., 127). This means that pictures of system A are committal with respect to property x and non-committal concerning property y of a real-world entity. It is the reverse case with pictures of system B; they are committal about y and non-committal about x. Thus, two pictures can show the same entity, but depict it radically differently depending on the system of representation to which they belong. Lopes explains this phenomenon by stating that:

> [...] two pictures embody distinct 'aspects' of an object if and only if there is at least one property with regard to which one is committal and the other is not, or one is explicitly non-committal and the other is inexplicitly non-committal.

(Lopes 2006, 119)

As an illustration of the above, Figure 2.12, the false colour photograph of *Olympus Mons*, informs its viewer about the outline shape of the volcano with respect to its different altitudes, whereas Figure 2.15 shows a completely different aspect of the same volcano. It depicts its surface structure and texture and, in particular, solidified lava flows on its flanks. Comparing both depictions makes clear what Lopes meant by his criterion cited above. Figure 2.12 is committal about the volcano's property of altitude and non-committal about its surface structure, whereas Figure 2.15 is committal about the latter and non-committal about the former. In this sense, Lopes states that distinct pictorial systems are committal about different types of aspects they present an entity as showing (see ibid., 128).

Moreover, it is not only the diversity of pictorial systems that can be explained by this theory, but also the phenomenon of *generativity*, as Lopes claims (see ibid., 148f.). He defends the view that our competence to grasp relatively quickly how depiction works is *system-relative*: if we know one example we can understand other pictures of the same system quite easily. Yet understanding one kind of images does not imply that we can understand pictures in general. Knowing how false colour photographs work does not imply being able to understand cubist paintings, for example.

The explanation offered as to why Lopes's account accommodates the viewer's competence to understand pictures of the same system relatively easily, in contrast to pictures in general, draws on the recipient's recognitional abilities. Lopes points out that one particularity of this ability consists in the fact that people can recognise entities although they may appear differently in varying contexts. For example, we are usually good at recognising faces, although the visual appearances of people might change considerably over time. Thus, there is a certain range of visual appearances of one and the same object recognisable by a particular observer.[81] Now, Lopes calls "the kinds of aspects with respect to which objects can vary but remain recognizable

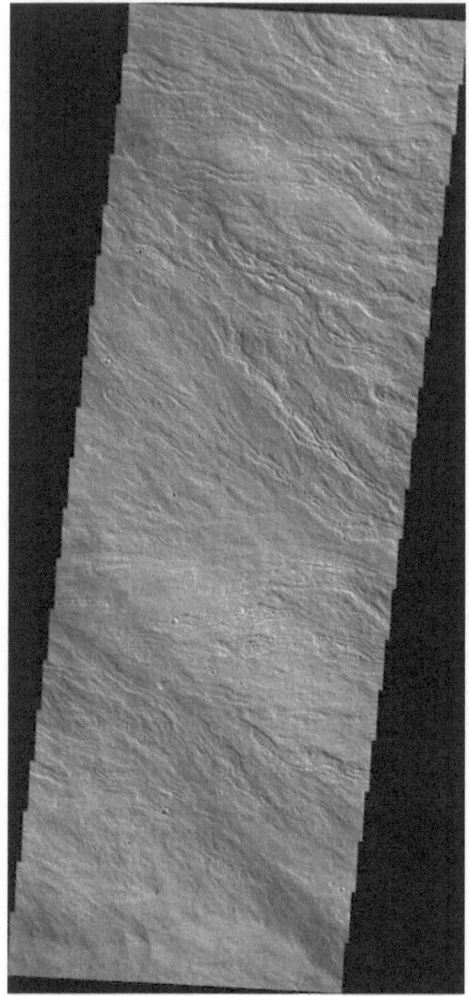

Figure 2.15 Olympus Mons lava flow: "The lava flows in this VIS image are located on Olympus Mons, the largest volcano in the solar system. Orbit Number: 49951 Latitude: 19.5453 Longitude: 223.009 Instrument: VIS Captured: 2013-03-19 01:22." Copyright: NASA/JPL-Caltech/ASU, Christensen, P.R., N.S. Gorelick, G.L. Mehall and K.C. Murray, THEMIS Public Data Releases, Planetary Data System node, Arizona State University, http://themis-data.asu.edu

Source: https://photojournal.jpl.nasa.gov/catalog/PIA17097

'dimensions of variation'" (ibid., 139). These dimensions, Lopes thinks, coincide with depictive styles that allow showing certain types of aspects as well as hiding others – and, consequently, with systems of pictorial representations. This is why, although a recipient might have mastered one pictorial system, she might still have difficulties understanding pictures of a different style, that is, of a different system (see ibid., 148).

Finally, Lopes mentions that another advantage of his theory of depiction is that it remains silent about the details of picture experience (see ibid., 175ff.). He considers this indeterminateness a virtue of the theory because, amongst other things, it is combinable in this way with all possible variations of twofoldness in picture perception (see ibid., 176). In his criticism of experience-based theories Lopes points out that the phenomenon of twofoldness should be regarded as a gradual matter and not as a necessary or even sufficient condition of picture perception. In this context, he highlights the fact that between the two extreme cases, namely *trompe-l'oeil*, which does not permit twofoldness, and painterly images (such as van Gogh's) that require it, there are all kinds of variants in-between with respect to twofoldness in picture experience. Thus, a theory that allows the accommodation of all these different kinds of pictorial experience seems to be clearly preferable to one that requires the exclusion of some of the possibilities of the spectrum of twofoldness.

Is the aspect-recognition theory, then, the theory of choice to explain depiction? Admittedly, it meets all four standards that Lopes mentions a theory of this domain should fulfil (see the list in section 2.2.6). Despite these virtues, however, there are still some critical points remaining. Which difficulties remain pressing for his approach becomes apparent on returning to Figure 2.14. Obviously, Lopes deals well with the aspect of *decoding information* presented by an image. His theory seems to be predominantly about understanding pictures, that is, about the cognitive tasks a potential viewer has to perform to grasp information visually presented. Furthermore, Lopes's remarks, namely his statements about the content of depiction which is structured by aspects (see ibid., 124ff.), also help to clarify what kind of information can typically be gained from visual representations alone. What seems to be neglected by, or at least underdeveloped,[82] in his theory, however, is the aspect of *encoding information*: the question concerning how information finds its way into the picture.

Lopes indicates two difficulties resulting from this neglect. Firstly, if it is the viewers' recognitive abilities that decide what a particular picture denotes, that is, to identify its referent, this might result in divergent identifications. The point is that each viewer has a stock of different background knowledge at her disposal which might or might not allow her to identify the object depicted. Thus, as depiction is based on these subjective resources, it can be asked, according to Lopes, whether visual representations ever allow for an *objective way* of information transmission – in the sense of objectively denoting a particular object by a certain image.

As a rejoinder, Lopes introduces what he calls a "suitable perceiver" (ibid., 151ff.). But what makes a perceiver suitable for the above task? In this respect Lopes states that "a suitable perceiver for a given picture is one who possesses a *suitably dynamic* recognition ability" (ibid., 152). Additionally, he claims that this ability is required to extend to both tasks of recognition relevant in picture perception, namely content-recognition and subject-recognition. This means that, on the one hand, a suitable perceiver is a person "adept at interpreting pictures" (ibid.). This presupposes that the viewer is familiar with picture interpretation in general, that she knows that pictures present content in their design. Additionally, she also has to be familiar with systems of depiction, that is, with depictive styles to interpret pictures of a given system correctly. On the other hand, the viewer also has to have background knowledge that allows her to recognise a particular subject in a given pictorial content. Only if she fulfils these three requirements can she count as a suitable perceiver in Lopes's sense. Clearly, he thereby defines an ideal case as a benchmark to assess which interpretations of a given image are correct and which are not. Two points have to be mentioned. Firstly, it can be argued that Lopes is in error fighting a straw man here. Relying on the subjective capabilities of a particular viewer to identify the referent of a picture correctly does not necessarily imply that the pictorial relation is also subjective. A comparison between different interpretations might reveal an inter-subjective agreement on what the picture shows.

Secondly, even if we concede that Lopes points out a real problem here, his solution of introducing the concept of a suitable perceiver can be questioned. Lopes does not insist that each picture has a suitable perceiver during all its lifetime, but only that at one point in its history is this picture recognised by a corresponding viewer (see ibid., 153f.). Yet can one be sure that such a person exists (existed)? After all, the abilities mentioned by Lopes still remain completely subjective. Thus, a potential viewer might simply *pretend* that she is able to recognise the content and the subject of a given picture. Who can tell that she does not? Moreover, imagine the case that there is a suitable perceiver of picture x in Lopes's sense who never talks about her interpretation. Hence, her interpretation can never be used as the benchmark needed for evaluating later interpretations of x.

Thus, what is still missing in Lopes's account is an *objective criterion* that tells us whether a particular interpretation of a given image is correct or not; that is, a criterion that determines what kind of information is encoded in the image.[83] One of the skills required by Lopes for his suitable perceiver consists in understanding depictive styles and thus systems of pictorial representations. However, as art history tells us, a variety of new styles of depiction has been invented by artists and it seems plausible to assume that they will continue this creative business in the future. Thus, it seems reasonable to seek advice from the artists involved about the kind of background knowledge required to interpret artistic pictures correctly. They might teach a perceiver what she needs to know about styles of depiction and, moreover, tell her what subject is supposed to be recognisable in the content of the picture.

This guides us to the next critical point of Lopes's account, namely his handling of the artists' intentions and their role in the meaning-constitution of a given picture. It is apparent that referring to *artists' intentions* might be a plausible explanation of how to objectively determine what an artistic picture is about, that is, what subject it is supposed to refer to, but Lopes does not make use of this proposal. On the contrary, he thinks that, although such intentions highlight what the artist has meant by the picture, they do not determine the meaning of the picture itself (see ibid., 159). He states that "[i]ntending to depict an object is not necessary or sufficient for depicting it; nor is knowing what the artist intended to depict necessary or sufficient for understanding her picture" (ibid.). The point he draws on is that, although the artist might have had the intention to depict x, she might have failed – for various reasons – to do so. Consequently, her image will not depict x, despite the fact that the artist had the intention to create a work of art showing x. Thus, artists' intentions can coincide with what the meaning of a depiction is about, but do not have to. However, contrary to Lopes, it has to be noted that his argument only lends plausibility to the artist's intentions not sufficing to constitute the meaning of a picture. It does not show that these intentions are also unnecessary.

Unaware of this difficulty in his line of reasoning, Lopes proceeds by making a distinction between two different kinds of meaning. "My view is that artists' intentions are relevant to what *artists* mean and to understanding what they mean, not what their *works* mean or to understanding their works" (ibid., his italics). Consequently, referring to artists' intentions as a standard of correctness, which is found lacking in Lopes's theory, is blocked by his considerations on this distinction between two kinds of meaning. The question about objective criteria, therefore, remains unanswered. Moreover, it can be asked whether these criteria are epistemically accessible or whether additional conditions are needed to indicate whether a certain criterion is fulfilled with respect to a particular interpretation. Assuming a coincidence between a criterion of meaning-constitution and conditions of correct interpretation seems to be informed by the (ideal) case of photography, although not all images are the result of a causally, let alone automatically working recording device.

Another criticism also directed at the role and precise nature of viewers' recognitive abilities in Lopes's theory is put forward by Katerina Bantinaki. She points out that "the relation between ordinary and pictorial perception might be a source of worry" (Bantinaki 2009, 239). The problem that arises due to Lopes's thesis that "identifying what a picture represents exploits perceptual recognition skills" (Lopes 2006, 144) is that it remains vague in Lopes's theory what exactly he thinks the "*pictorial* act of recognition" (Bantinaki 2009, 239, my italics) is about.

Robert Hopkins expands this objection (see Hopkins 2003, 666ff.). In accordance with Bantinaki's claim, he questions Lopes's thesis of an overlap between recognising x in reality and recognising x in a depiction of x. Both critics claim a difference here that, they think, Lopes does not satisfactorily explain. However, Lopes has to be defended here, as he explicitly points out that the *particularity* of pictorial recognition consists in its comprising two steps, namely content- and subject-recognition (see Lopes 2006, 144f.).

Moreover, Hopkins ties this critique to another concern, namely the relation of Lopes's theory of depiction to empirical theories and data, predominantly of "vision research" (ibid., 668). On the one hand, he calls into question in what sense an *empirical confirmation* can be expected to justify Lopes's definition of depiction. Here, Hopkins refers to Popper's famous insight that no hypothesis can ever be confirmed by empirical data, although it can be falsified by a single contesting observation (see ibid.).[84] On the other hand, Hopkins is also sceptical about the possibility that empirical sciences will reveal insights into human cognitive and perceptual processes that will allow a formulation of Lopes's thesis on depiction, especially, and more precisely, concerning the nature of pictorial recognition abilities (see ibid.). From Hopkins's point of view, such an argumentative strategy merely leads to "empirical speculations" (ibid.) and does not add the clarification needed.

I agree with Hopkins and Bantinaki that the virtue of Lopes's theory consists in drawing our attention to the particular cognitive abilities of viewers in understanding images. Obviously, we engage with depiction in a special way when decoding the information presented visually; this seems to be proven by the distinctive way in which we learn to understand images. In this sense, another positive aspect of his account is his list of requirements that a theory of depiction should meet (see the list in section 2.2.6). On balance, however, Lopes's theory faces no less serious difficulties than its rivals discussed above. A more thoroughly implemented combination of these different approaches could perhaps prove successful.

2.2.5 Mixed theories and image science

Regarding the results of our analysis, it seems reasonable to suggest new proposals for a theory of depiction as a combination of elements taken from the different accounts discussed above. Successful combinations would have the obvious advantage of pooling the strengths and eliminating the weaknesses of each particular theoretical framework.

Michael Newall (2011) and Klaus Sachs-Hombach (2006) are, for example, amongst the proponents of such mixed theories. Taking a closer look at both of them will show the potential fruitfulness of mixed theories, as they choose not only different theoretical frameworks as resources for their own approaches, but also differ with respect to the range of theories from which they pick. Their differences are therefore both quantitative (the number of theories involved) and qualitative (the kinds of theories involved) in kind.

Newall's proposal is the more ambitious one. He tries to bring together theses from more or less all of the above-mentioned theoretical frameworks (see Newall 2011, 19).[85] Yet he does not grant all of them the status of necessary conditions in the attempt to clarify what is meant by 'depiction'. Of course, such a procedure makes the project much easier to assemble.

Although his approach is a mixed one, there seems to be a certain preference with respect to one of the components chosen. That is, the central element of Newall's theory of depiction seems to be *experience-based*[86] and is reminiscent of Wollheim's thesis of "seeing-in". Within this context, he speaks about "veridical" and "non-veridical seeing" which he defines in the following way: "[v]eridical seeing is seeing as we ordinarily know it, whereas non-veridical seeing is that which occurs in visual hallucinations, dreams, visual imaginings, illusions, and visually mistaking one object for another" (ibid., 20). In the following, he argues that the understanding of a picture involves a kind of non-veridical seeing, namely one where:

> some item, not X, [is] present before the subject's eyes on which the non-veridical seeing of X counterfactually depends. Seeing X is thus dependent on the presence of [...] Y, before the subject's eyes, such that if Y were not so present [...] then seeing X would not occur.
>
> (Newall 2011, 21)

In this sense, picture perception is regarded as an instance of normal visual perception – of 'seeing' – by Newall.[87] This last fact also motivates Katerina Bantinaki to argue that Newall does not put forward a *particular phenomenology* in his account, as he models picture perception on visual everyday perception (see Bantinaki 2014, 946).

This, then, is the parallel to Wollheim's account. Similar to the "seeing-in thesis", Newall defends the claim that the viewer has the experience of non-veridically seeing the subject matter shown in the image at hand. The particular picture thereby functions as the causal stimulus of this experience (see Newall 2011, 41); it serves to fill in the Y-part of the above definition.

Having pointed out this apparent similarity to Wollheim's central thesis, the question arises how Newall conceives of the aspect of *twofoldness*, that is, the simultaneous awareness of design and content, that features so prominently in Wollheim's account. Does Newall also make use of this? Obviously, he thinks that the phenomenon of a twofold experience is relevant in picture perception. He states that "[p]ictorial seeing-in [...] involves the veridical experience of seeing the picture surface, and the non-veridical experience of seeing the depicted subject" (ibid., 30). Yet does this also mean that he grants twofoldness the status of a necessary condition of depiction? The answer to this question is in the negative. Although Newall admits that a twofold experience is often the case with regard to picture perception, he also thinks that there are crucial exceptions (see ibid., 42) – such as the case of *trompe-l'oeil*. Thus, what is of importance from his point of view is the aspect of non-veridically seeing the subject matter in the picture, but not a simultaneous awareness of content and design. Here his theory clearly diverges from Wollheim's account of depiction.

Moreover, as non-veridical seeing is not an exclusive feature of picture perception, Newall points out that, in addition to this, he needs a "standard of correctness" which demarcates picture perception from other kinds of non-veridical seeing such as, for example, hallucination (see ibid., 42). Furthermore, there are also instances "where pictures occasion the non-veridical seeing of things that are not the picture's subject matter" (ibid., 43). That is, a particular picture can cause different experiences of its assumed content in the viewer. Figure 1.1, for instance, might cause some viewers to see a cow, whereas it shows an aurochs. Thus, we are in need of a benchmark, so to speak, that tells the viewer whether her understanding of a particular picture is correct or not. Thus, Newall introduces a second necessary condition into his account of depiction, namely *standards of correctness*.

Right at the beginning of his discussion of possible candidates for such standards, Newall points out that, because of the diversity of pictorial representations, no single criterion will suffice. He therefore suggests two different criteria, one with respect to *manugraphically* and another with regard to *photographically produced pictures* (see ibid., 56). Consequently, he modifies his definition of depiction in the following way:

[a] manugraphically produced surface, X, depicts Y if and only if (i) X can occasion non-veridical seeing of Y, and (ii) the picture-maker intends X to occasion a non-veridical seeing of Y. A photographically produced surface, X, depicts Y if and only if (i) X can occasion non-veridical seeing of Y, and (ii) this non-veridical seeing of Y is causally and counterfactually dependent on Y's presence before the camera.

(Newall 2011, 61)

It is, on the one hand, the artist's intentions that determine whether the understanding of a manugraphically produced image is correct or not. On the other hand, it is the causal connection between camera and object that informs the interpreter whether her understanding of a particular photograph is correct.

This dual standard of correctness might be a bone of contention for those who strive to develop a unified approach to depiction.[88] This aspect, however, does not pose a problem for those more sympathetic to a Wittgensteinian approach that makes use of the concept of family-resemblance in this context instead of arguing for necessary and jointly sufficient conditions of 'depiction'.[89]

However, further problems can arise as a consequence of the standards suggested. One might object, for example, that Newall's rationale for his insistence on intentions as a standard of correctness for the interpretation of paintings confounds the different levels of understanding pointed out by Scholz. Newall claims that this standard is needed because the artist's intentions tell us what she wants to *communicate* to her audience with her picture (see ibid., 59). This way of formulating his thesis is, at the very least, ambiguous when different levels of understanding are taken into consideration.

As was pointed out above, the communicative purpose of a picture does not have to coincide with what the depictive content is about. Figure 1.1 is an example: the depiction of certain animals, for instance aurochs, can be seen in this image. However, it is not known what the artist's intention was when painting these images on the wall of the cave. It might have been the depiction of some mythical narration, the recording of a current success in hunting, or it might have had an educational purpose to teach his younger fellows to discern potential prey from dangerous carnivores. Does our lack of knowing the precise intentions imply that no correct interpretation of the image is possible today? Newall's account suggests an answer in the affirmative. I would admit that we might lose an informative dimension if we do not know what the artist intended to communicate with her image. However, this is only one level of understanding that evades our attention.

Defending Newall against this objection, one might argue that the viewer nonetheless grasps one level of meaningfulness correctly, namely that the artist had the intention to produce an image of an aurochs – or something that can be seen as a depiction of this kind. And as this seems to be the basic requirement to understand the painting correctly, Newall's standard of correctness seems to be adequate. Such a defence strategy, however, overlooks images appearing in much more complex activities. That is, one does not only have to interpret manugraphically produced images as works of an artist as one does in art galleries. They can be reused by other people with different intentions. Figure 1.1, for example, shows a reproduction of the original cave, hence it has to be asked what the intentions of the reproducer were when painting this fake cave. Scholz has made us aware of the point involved here, namely that the very same image can be used by different people for different communicative purposes. In this sense, Newall's standard of correctness leaves out the possibility of multiple usages of the same image, as not all of these uses have to be bound to the artist's original intention. Interpreting hand-made images in all the different contexts only in accordance with this one genuine intention might thus be misleading in the end.

Unfortunately, Newall's second standard of correctness does not fare much better. One problem can be regarded in close relation to Roger Scruton's account of photography here. Similarly to Newall's suggestion, Scruton regards the meaning of ideal photographs as solely determined by the causal connection involved. However, staged photographs seem to be clear instances that undermine such a *mechanical* conception of photography. Clearly, the correct interpretation of such pictures not only depends on the causal connection between object and camera – as Newall suggests – but partly also on background knowledge concerning the artist's intentions.[90] This is why Scruton argues that staged photographs no longer belong to the class of ideal photography. In accordance with this, it can be claimed that to understand photographs correctly, at least sometimes it is a combination of the artist's intentions and the causal connection that sets the standard of correctness here. Again this seems to be where Maynard's distinction between the *detective* and the *depictive* function of photography can be used to clarify this problem. Using photography as a detective device indeed puts the emphasis on the causal connection as suggested by Newall.[91] Using it to produce pictures, by contrast, might demand paying close attention to the artist's intentions in order to understand the resulting picture correctly.

Newall's approach has provoked further critique from Bantinaki. She points out that his thesis of a close connection between picture perception and everyday perception is not convincing (see Bantinaki 2014, 946). Two points of concern are mentioned. The first aspect concerns Newall's description of our awareness of design and content in the case of there being a twofold experience. Here, she points out that, in both instances, the viewer's experience is different regarding instances of face-to-face perception that Newall suggests as the guiding model. The second point concerns the phenomenology of picture perception. Bantinaki thinks, contrary to Newall, that pictorial experience does entail a particular phenomenology in contrast to everyday perception. She points out that what is missing in the former and thus making it different to the latter, is a "feeling of presence" (ibid.) caused by the object seen in a picture. Despite these critical comments, Newall's approach suggests some promising links to the analysis of visual representations in science which will be discussed in due course. However, before going into the details of how to make use of his theoretical approach with respect to our purposes, one final account from picture theory needs to be discussed. Klaus Sachs-Hombach pursues a less ambitious project than Newall's theory. He tries to defend a combination of perceptual and conventional accounts (see Sachs-Hombach 2006, 87). What is particularly interesting in his proposal is the more wide-reaching purpose intertwined with his account of depiction, namely the development of a complete *image science*.

What exactly is his definitional approach? Sachs-Hombach defends the claim that images are "perceptoid signs" (see ibid., 94).[92] He suggests two necessary conditions for something being an image. Firstly, images are signs (see ibid., 86). Secondly, their reception is perceptoid (see ibid., 92). What makes things slightly complicated is Sachs-Hombach's suggestion that, with respect to different types of visual representations, these two conditions can gradually vary in their relevance. For example, diagrams might put an emphasis on the semiotic part, whereas photographs rely more on the perceptual condition (see ibid., 94). Yet it remains questionable how this last claim can be related to the suggested definition. If the two conditions are necessary ones, then they cannot come in degrees.

Anyway, to start with the semiotic condition first, Sachs-Hombach claims that images as signs comprise a content, a physical basis (a canvas, for instance) and (sometimes) also a referent (see Sachs-Hombach 2006, 80). The content thereby concerns what we see in a picture and thus refers to phenomenological aspects of picture perception. Similar to Kjørup and Scholz, Sachs-Hombach argues that images can be parts of communicative acts and, in this sense, transmit even complex information (see ibid., 81). From his point of view, *understanding an image* implies grasping two different meanings. On the one hand, there is the content of the image, and on the other, there are the user's intentions when using this image for communicative purposes. Both meanings can, but need not, coincide (see ibid., 82).

To distinguish pictorial from other signs, especially from linguistic ones, Sachs-Hombach brings in the second theoretical approach, namely perceptual theories of depiction. He points out that the use of images depends on perceptual processes in particular ways (see ibid., 86),[93] but what exactly does he have in mind when referring to 'perceptual' theories of depiction?

Here he mentions two different aspects. On the one hand, he claims that the characteristic of 'being perceptoid' has much in common with Peirce's concept of iconic signs (see ibid., 88) – a claim that apparently indicates the relevance of resemblance relations. On the other, Sachs-Hombach regards this perceptual feature as being similar to Wollheim's theory of seeing-in (see ibid., 92f.). In this context, he discusses the twofoldness-condition, that is, the simultaneous awareness of design and content. Sachs-Hombach, however, does not make it explicit whether it is this feature that he wants to draw on or simply the fact that viewers are able to see x in y, i.e. a particular entity in a given picture.

Having said that, it still remains unclear how the two aspects borrowed from a perceptual theory of depiction, namely seeing-in and iconicity, belong together. The latter feature, however, is explained further by Sachs-Hombach when he discusses the level of semantics with respect to image science (see ibid., ch. 5). Perhaps the relevant hint can be found there?

In this context, Sachs-Hombach picks up the topic of resemblance again. He claims that resemblance is a necessary semantic condition of depictive representations and even a sufficient condition to discern pictorial from linguistic signs (see ibid., 124). More precisely, he states that the feature of resemblance can only be used to demarcate the latter two kinds of signs, but never signs as such (see ibid., 132). Being aware of the numerous counter-examples to resemblance-based theories of depiction, this method of integrating resemblance into his account is obviously meant to evade those objections by restricting as far as possible the scope of explanations related to resemblance.

Despite these precautions, Sachs-Hombach's account has nevertheless to face some difficulties with regard to the usage of resemblance relations within his own theoretical framework. What does he state about this concept? He defends an *internalistic concept* of resemblance, namely if a viewer perceives x as similar to y, then there is a correlated resemblance relation between x and y (see ibid., 124, 141). Moreover, the question arises which properties are relevant for comparison with respect to depiction. Sachs-Hombach suggests that x resembles y if x shows typical features of y in relevant property dimensions (see ibid., 144). Context and intentions influence what exactly the properties are that have to be considered (see ibid., 140). A final clarification is added when Sachs-Hombach suggests that, with respect to pictures, visual properties, especially concerning shape and colour, belong to the property dimensions relevant to be compared (see ibid.).

Vague as this definition might be – again for the sake of protecting his definition against counter-examples – there are nonetheless inherent problems provoking critique. Jakob Steinbrenner objects that the property dimensions of shape and colour pointed out above might be irrelevant to pictures that do not exhibit such similarities with this regard. Consequently, 'resemblance' defined in the above sense cannot be regarded as a necessary condition of depiction (see Steinbrenner 2009, 303f.).

Be this as it may, the question how resemblance and seeing-in are supposed to be related in Sachs-Hombach's theory has still to be answered, and it is not until he discusses the *pragmatics of image science* that he finally relates his concept of resemblance and the aspect of seeing-in (see Sachs-Hombach 2006, 173f.). Here he states that it is because of a perceived similarity between an object x and its depiction y that we are able to see x in y. Since people are, however, able to see different things in the same picture, he admits that resemblance relations of this kind are not sufficient to clearly determine the content of the picture and therefore its meaning. Nonetheless, he thinks that resemblance relations can at least set constraints on what people could rationally claim to see in a given picture (see ibid., 174).

Relating seeing-in and resemblance in this way, however, causes another problem for Sachs-Hombach's theory. Neither is his definition of the concept of resemblance safe from counter-examples, nor is resemblance a necessary condition for all kinds of pictorial representation. This latter objection undermines his attempt to make use of Wollheim's thesis of seeing-in, if he connects the two theoretical approaches in the above way. Eroding the basis for taking advantage of this concept then also poses the question in what sense Sachs-Hombach's approach can indeed be called a "mixed theory" of depiction, since its perceptual component becomes completely detached at this point.

Additionally, it can be asked whether his theory of depiction as a mixed approach is consistent. On the one hand, he proposes a definition of images that makes use of conventional as well as of perceptual theories of depiction. On the other, he states that there are *no intrinsic properties* that determine when an entity can be called an image (see ibid., 81). In this respect, he maintains the same claim defended by Goodman and Scholz, namely that the characteristics of the system to which it belongs allow one to decide whether an entity should be regarded as an image. However, contrary to the latter philosophers, Sachs-Hombach does not define characteristics of different representational *systems* – a strategy that allows Goodman and Scholz to claim that the same entity can be regarded as an instance of different systems relative to the concrete context of usage. Quite the reverse, Sachs-Hombach suggests a traditionally essentialist definition of images, and it is at this juncture that his two theses – the definitional and the pragmatic – clash. Either 'image' has an essence or it does not, but if it has, its tokens cannot be used in indefinitely many ways.

Setting these difficulties aside for a moment, it has to be noted that, contrary to the definitional projects discussed in the previous sections of this chapter, the approach at clarifying what the supposed meaning of the term 'picture' is supposed to mean is only a stepping stone to a much broader attempt suggested by Sachs-Hombach. He thinks that, just as semiotics is relevant to the analysis of languages, there is need of a comprehensive science – not merely a single theory – to deal with pictures in the same way. Thus, he suggests developing a *universal image science* (see Sachs-Hombach 2006).

Sachs-Hombach is not the only one to propose such an account. Jason Gaiger (2014) presents a sophisticated overview of competing conceptions of image science, for example, developed by Hans Belting or Gottfried Boehm, and adjacent disciplines such as art history and visual culture studies.[94] From the point of view of the philosophy of science, Sachs-Hombach's approach seems to be the most interesting one, as he strives for an *interdisciplinary and integrative perspective* on the topic. Thus, not only are meta-disciplines such as philosophy, semiotics, or art history met within his theoretical framework, but also sciences making use of images in a variety of ways. It is this latter aspect that makes Sachs-Hombach's work a valuable resource for investigations from the perspective of the philosophy of science. It is therefore of profit to examine the details of his suggestion. Firstly, which disciplines are considered in his account and why? In accordance with his interdisciplinary aim, Sachs-Hombach lists the disciplines involved in the following way:

- *Foundational disciplines*: these are disciplines that are especially important at the systematic level of image science (see Sachs-Hombach 2005c, 14). Within this category, he makes further distinctions. Firstly, he mentions disciplines concerned with perceptual and semiotic phenomena and thus supposed to shed more light on the core concept 'picture' defined as "perceptoid signs" (see Sachs-Hombach 2006, 69). Cognitive science, neuroscience, psychology, media studies, communication studies and semiotics are then subsumed under the label of 'foundational disciplines'. Secondly, Sachs-Hombach lists certain meta-sciences here such as mathematics, logic and philosophy. And, finally, art history attains its place among the foundational disciplines.
- *Historically orientated image sciences*: proponents of these disciplines investigate images of particular epochs or societies (see Sachs-Hombach 2005c, 17). Archaeology, ethnology, history and museology belong to this category (see Sachs-Hombach 2006, 70).
- *Image sciences related to social sciences*: here, Sachs-Hombach subsumes all disciplines that are concerned with the *usage of images* understood as a social process encompassing people who want to communicate via images in order to, for example, influence their recipients (see Sachs-Hombach

2005c, 17). Next to sociology, pedagogy, political sciences, cultural studies, jurisprudence and rhetoric belong to this category (see Sachs-Hombach 2006, 70).

- *Applied image sciences*: these are disciplines concerned with the more technical dimension of image production and usage, although there is no sharp distinction to the former category, as Sachs-Hombach admits (see Sachs-Hombach 2005c, 18). Proponents of these disciplines do not contribute to the theoretical clarification of the term 'image'. Cartography, computer science and public relations are exemplary elements of this category (see Sachs-Hombach 2006, 70).
- *Practical image sciences*: finally, there are some further approaches aiming at realising image-related projects which could, however, not be called academic disciplines, although they are related to them and are often featured at art colleges (see Sachs-Hombach 2005c, 19). Design studies or typography are examples in this context (see Sachs-Hombach 2006, 70).

This impressive list only excludes academic disciplines whose proponents make use of images as methodological tools, such as fMR images or PET scans used for diagnostic purposes in medicine (see Sachs-Hombach 2006, 68f.). Furthermore, it makes clear how wide-ranging Sachs-Hombach's approach to an interdisciplinary image science is, and yet how difficult it must be to find common results in this diversity. This dichotomy becomes even more marked when the aims pursued within this context are considered, as explained by Jason Gaiger:

> [t]his would not be a new discipline, distinct from the others, but a 'common theoretical framework that could provide an integrative research programme for the various disciplines'. Its 'minimal criterion' would be the provision of a model 'that allows all the other sciences of the image to be connected in a systematic manner without impairing their independence'.
>
> (Gaiger 2014, 210)[95]

Regarding the multitude of disciplines involved, this seems then to be a rather ambitious project. It will be seen in due course that this compilation of participating disciplines and the aspiration of universality has raised a number of critical concerns with respect to image science. However, before discussing the critical points involved in this, the project itself needs first to be better understood. Its structural features have hitherto only been mentioned, that is, a list of relevant academic disciplines and the requirement that the latter should not be replaced by image science, but rather that their expertise on image usage, production and theory should be compiled. Yet, beyond this pooling of knowledge, what is (a) the purpose of this new science, and (b) what is its methodology?

Sachs-Hombach's suggestions of how to understand image science remain relatively vague. He proposes that what is developed is simply a "theoretical framework" (see Sachs-Hombach 2006, 71). What exactly could such a framework be, though, if it has to integrate such a host of varying disciplines, and how is it supposed to achieve its intended goals? Here, Gaiger offers some more explanatory remarks:

[o]n this conception [Sachs-Hombach's hypothesis, N.M.], a universal image science should (1) integrate the perspectives, methodologies, and results of the different disciplines into a systematic body of knowledge; (2) analyse and define the basic concepts employed so that research can proceed on a common basis; and (3) develop strategies for effective inter-disciplinary cooperation.

(Gaiger 2014, 210f.)

Task (3) seems to be of obvious necessity if indeed all of the above-mentioned disciplines and approaches are to be integrated into image science.[96] Task (1) is the result of Gaiger's own concept of science as a "body of knowledge" (ibid., 209). Nonetheless, the question remains how such a pooling could be pursued in a way that an "integrative" perspective would be accomplished and not merely a more or less incoherent compilation of (in the worst case, mutually exclusive) concepts, methods, etc. Sachs-Hombach (2005b), for example, offers a juxtaposition of image practices and understandings of exemplary disciplines in the five categories above.

Obviously, task (2)'s importance is that of launching image science. This is the point where Sachs-Hombach's analysis of the term 'picture' is supposed to fit in. The above discussion of his approach, however, has already revealed many difficulties that, unfortunately, render any suggestion to regard images as "perceptoid signs" questionable. Moreover, a weakness in the definition also casts doubts on the project of image science as a whole, since the definition is supposed to play such a central role. Additionally, the initial definition is solely focused on depictive representations, as Sachs-Hombach admits (see Sachs-Hombach 2006, 74). Although this means a constraint on the disciplinary approaches that can be accommodated within this theoretical framework – engineering sciences, for example, would be excluded because images are often used here as models and not in the mimetic way suggested by Sachs-Hombach – he is also optimistic that this initial restriction can be overcome in the course of establishing image science (see ibid., 75, 77). Unfortunately, however, he remains silent about how this could be achieved. He simply does not point out in what way his definition might be open to modification or addition. Thus, next to the problems related to his proposal of pictures as perceptoid signs, this restrictive starting point seems to be another obstacle to accomplishing his interdisciplinary project. Nonetheless, Sachs-Hombach is right in demanding a clarification of the common understanding of the term 'image' used in the different disciplines. Only in this way can talking at cross-purposes be avoided and the expertise of different approaches synthesised.

Beyond this definitional project, he suggests modelling image science in analogy to semiotics. In this sense, areas of investigation are categorised as *syntax, semantics and pragmatics* (see Sachs-Hombach 2006, ch. 4, 5, 6). As a rationale for this analytical method, he refers to his definition of pictures as perceptoid *signs*. If, however, images can be defined in this manner as semiotic entities, then it can be said that they exhibit the same characteristics as other signs. In Sachs-Hombach's sense, this means that they are internally structured (syntax), that they are referential (semantics), and that they are part of broader contexts of semiotic usage (pragmatics) (see Sachs-Hombach 2006, 73). In addition to this, he claims that the three analytical dimensions just mentioned can also be used to categorise the different theoretical approaches on depiction (see ibid., 85). Unfortunately, it is not completely clear whether he refers to the list of disciplines above or to the different theoretical approaches on depiction discussed in the previous sections. Whatever the case, the suggested analytical dimension might indeed offer the opportunity to integrate the different academic disciplines of image science in order to pursue a common investigation on pictorial representations. They could contribute their understanding of images (syntax, semantics) and their practices of using them (pragmatics) so it might be that commonalities between the different approaches finally can be discerned or established. However, optimism concerning this matter aside, what Sachs-Hombach has in mind when suggesting syntax, semantics and pragmatics as analytical dimensions of image science requires closer scrutiny at this point. The topic of semantics has already been discussed above, but a brief reminder of the facts is in order: he thinks that the referential aspect of images can be explained by resemblance relations (see ibid., 144). Since the difficulties related to this approach have also been pointed out above, the syntactical level now needs examining.

The syntax of images concerns the formal level, already encountered in Goodman's theory of depiction. What is analysed here are relations between particular pictorial signs (see ibid., 105). Beyond the questions about signs and their correlations, the possibility of a pictorial grammar is also discussed here – a topic somewhat in dispute (see ibid., 103). In this context, Sachs-Hombach points out that there are apparently no criteria to assess whether a particular image is well-formed, yet even about badly produced images the claim would not be advanced that they are *grammatically incorrect* (see ibid., 104).

Within this syntactical context, he also discusses approaches of establishing pictorial languages. Here he mentions Neurath's ISOTYPE. However, although such accounts seem to suggest that a pictorial grammar can be realised, Sachs-Hombach relativises this positive result by claiming that, unfortunately, those approaches are not applicable to images in general (see ibid., 111). Summarising his results, he claims that, on the one hand, pictorial representations can usually be subdivided into smaller units on a formal level and, on the other, and contrary to linguistic units, those elements acquire grammatically functional roles only in particular contexts (see ibid., 119).

This takes us directly to the third level of investigation, namely *pragmatics*. The issue of using images in communicative contexts and for communicative purposes is discussed at the pragmatic level of image science (see ibid., 163). Scholz's theses on *image games* can be localised here as well as Kjørup's proposal of *pictorial speech acts*. Both approaches refer back to the problem of figuring out what functional roles are played by which parts of a particular image. The point is that, if the communicative acts that involve images are analysed, a distinction has to be made between illocutionary acts and (propositional) content (see ibid., 164ff.). Moreover, questions concerning the denotative function of images are discussed at this level, such as whether general denotation is possible or whether only specific entities can be addressed by images. It might be added that, although questions about the cognitive content of images and whether this can be regarded as propositionally structured also play a substantial role at this level, Sachs-Hombach obviously does not seem to see a problem here. He simply presupposes that pictorial contents are structured in a propositional way (see ibid., 183).

In sum, all three levels, namely syntax, semantics and pragmatics, constitute the theoretical framework that image science is supposed to provide to integrate all those different academic disciplines and projects mentioned above. A successful approach along such interdisciplinary lines would obviously be advantageous. Not only would it foster understanding of the role and status of visual representations in the sciences and in society in general, but it also seems to be the relevant foundation on which to establish and propagate a comprehensive pictorial literacy needed to deal (epistemically) appropriately with all of these different kinds of visualisations.

However, the very same features, namely interdisciplinarity and integrativity, reveal themselves not only as the virtues but also as the vices of Sachs-Hombach's approach. As Gaiger points out:

> [t]he sheer number of disciplines involved invites the question whether *Bildwissenschaft* should be understood as a single, unified enterprise, or whether it is simply an umbrella term for a multiplicity of overlapping and potentially divergent approaches that cannot be brought into alignment.
>
> (Gaiger 2014, 210)

Thus, it might be questioned whether it is at all possible to establish the theoretical framework that Sachs-Hombach is aiming at. The problems related to his definition of the core concept 'image', a concept needed in order to activate all the other theoretical components, are obviously a consequence both of the diversity of pictorial phenomena being subsumed under this heading and of the different theoretical approaches being combined – even in the weak sense intended by Sachs-Hombach. These problems then raise serious doubts about the chances of success of an interdisciplinary image science. Problems of this kind are also expressed by the art historian Horst Bredekamp, who questions the *additive* nature of image science as Gaiger indicates (see ibid., 211).

What seems to be especially difficult, however, is not the mere compilation of different disciplines, but the lack of a "realistic set of guidelines for how this [the realising of a synthetic view rather than a mere collection of individual practices, N.M.] is to be achieved in practice" (ibid., 211). This critique by Gaiger comprises two points.

- On the one hand, he highlights the fact that, among the various disciplines, no authoritative one is so distinguished that it could guide the others. On the contrary, in Sachs-Hombach's account, all disciplines are based on an equal level; only the foundational disciplines are somehow highlighted. However, there are also several of them and no guidance to co-ordinate their collaboration has thus far been put forward. Lambert Wiesing, for example, defends the claim that such a guiding position should be ascribed to philosophy. From his point of view, no empirical scientist can tell us anything about the concept of image that is needed to decide which phenomena and what disciplines are therefore to be considered in image science (see Wiesing 2006, 15f.). It is only philosophy that can pursue this conceptual task satisfyingly.[97]
- On the other hand, Gaiger also draws our attention to the possibility that a pluralistic approach might be more methodologically advantageous when considering the multitude of pictorial phenomena rather than the synthetic account favoured by Sachs-Hombach (see Gaiger 2014, 211). Here, however, we have to defend image science, because it is explicitly stated by Sachs-Hombach that it is not meant to replace those other disciplines. Pluralism will thus be maintained in his theoretical framework.

Beyond this critique, however, it seems worthwhile considering Gustav Frank and Barbara Lange's suggestion that image science might be able to connect the different theoretical branches which, as a consequence of scientific specialisation, have developed increasingly separately (see Frank and Lange 2010, 17).

But it is not only the feature of interdisciplinarity and integrativeness that is attacked by critics. They also contest the claimed *universality* of this approach. Furthermore, this seems to be the major bone of contention, puzzling especially to art historians. In particular, Bredekamp challenges the status of Sachs-Hombach's image science by calling into question the claim that it is the *unique* universal approach to the topic. Contrary to this, Bredekamp defends the thesis that art history has always pursued such a universal investigation of pictorial representations[98] and that it is only due to "a conscious amnesia" that this has been forgotten in recent discussions aimed at establishing a comprehensive image science (see Bredekamp 2003, 419).

I want to mention two last critical aspects. My first critique is conceptual in kind, but concerns also the precise status of image science. Sachs-Hombach refers to this approach by calling it a "theory", a "model" and a "theoretical framework". Moreover, not only does he make use of these terms interchangeably, he even relates them directly – for example when stating that *image science is a comprehensive model called a theoretical framework and that his theoretical approach on depiction* constitutes the latter (see Sachs-Hombach 2006, 95). This, at the very least, imprecise use of the concepts, namely of 'theory', 'model', 'theoretical framework', makes his account vague and open to criticism. It can be assumed that much of the puzzlement expressed above concerning the integrative and universal nature of image science is also a result of this wording, which has obviously been chosen to avoid adopting clear positions with respect to the status of image science in comparison with its disciplines.

Finally, it can be asked whether the term 'science' is really appropriate in this context. As far as I can see, the framework developed by Sachs-Hombach seems to be a *merely descriptive endeavour*, trying to systematise different practices and ideas. However, when it comes to explanations and predictions, the question arises how to realise this in *image science*, particularly when we take into account its rather vague conception. Admittedly, the well-known problem of demarcation in the philosophy of science highlights the notoriously difficult endeavour of stating what the particular characteristics of science are.[99] One should therefore not be too restrictive in this respect, although it remains problematic. Gaiger's suggestion of how to understand 'science' in this context could perhaps be followed namely as "a body of knowledge rather than a commitment to the methods of natural sciences [...]" (Gaiger 2014, 209). Another possibility might be to regard image science as a kind of auxiliary science akin to heraldry in history.

Let us now take stock of what our above analysis has shown so far.

2.2.6 Interim results: what can be learnt from picture theory?

Summarising the results, the first point to mention is the requirements that have to be fulfilled by proper theories of depiction. In the discussion about the different approaches, the recurrent difficulties have been encountered that such theories have to deal with. These points come down to a list of correlated requirements which Lopes has compiled:

1   Diversity constraint (see Lopes 2006, 32),
2   Phenomenological constraint (see ibid., 36),
3   Twofoldness constraint (see ibid., 42, 51),
4   Competence constraint (see ibid., 73, 75).

In detail, this means that a picture theory should be able to deal with all different kinds of pictorial representation (1). It has to offer an explanation with respect to the particular phenomenology of picture perception (2). Part of this will be (3) to take into account a gradual twofoldness condition, i.e. that we are simultaneously (more or less – depending on the depictive style) aware of the image's design and content. And finally (4), such a theory should also explain the phenomenon of natural generativity and transference within this context. That is, a valuable picture theory should explain why it is that inductive learning proceeds almost effortlessly with regard to systems of depiction and why we can recognise entities in the wild that we have learnt about only via depictions.

Furthermore, it can be assumed that, to some degree, these constraints are also relevant with respect to scientific images. Also, in this more specialised context, it is to be expected at least that a corresponding theory should be able to deal with the diversity of visual phenomena and shed some light on competences required for and particularities connected with learning from visual representations in science. As the emphasis is on information transmission in this context, for example of measurement data or of the contents of predictions, explaining our *cognitive access* to such information transmitted via visualisations seems to be of greater interest at first glance than accounting for a particular phenomenology in our perceptual processes. However, this latter aspect should not be too hastily dismissed, as it might be the case that it is precisely because of certain phenomenological features that a viewer can glean information from the image.

In addition, Sachs-Hombach's attempt at establishing an interdisciplinary image science is of benefit in that it informs about all the different practices, technologies and concepts of pictorial representations that predominate in the academic disciplines involved (see Sachs-Hombach 2005a). This diversity of phenomena and correlated epistemic practices seems to motivate the wealth of different theories of depiction indicated in the above analysis. This correlation has also been noticed by Keith Lehrer, who claims that the diversity of approaches to depiction suggests that philosophers ask and try to answer different questions concerning the topic (see Lehrer 2004, 1).

Even though it seems right to point to the diversity of practices, technologies and concepts of visual representations in different academic disciplines as a rationale for the multitude of approaches in picture theory, I also want to suggest another reason for this pluralism of theories. I think that a main cause contributing to the production of so many approaches consists in many of these theories being too narrowly conceived, for they focus either on the aspect of information encoding and neglect the aspect of information decoding and vice versa. Why do I think so?

The images discussed in the current context are predominantly regarded on the one hand as being caused or produced by something or someone and, on the other, as being perceived and interpreted by a recipient – even if this will be solely its producer evaluating her results. In this sense, images fulfil

Peirce's definition of *signs* as triadic entities (see Peirce 1983, 64). He thinks of signs as mediating parts between "objects" and "interpretants". Whereas the first *relatum* seems intuitively clear, the latter is more difficult to understand. Albert Atkin offers some helpful remarks here by stating that:

> [...] although we have characterized the interpretant as the understanding we reach of some sign/object relation, it is perhaps more properly thought of as the translation or development of the original sign. The idea is that the interpretant provides a translation of the sign, allowing us a more complex understanding of the sign's object.
>
> (Atkin 2013, sect. 1.3)

Thus, the interpretant refers to a kind of mental representation when grasping the meaning of the sign. It is this dual relation that I will borrow from Peirce's semiotics to make clear in what sense I think that some of the above approaches to depiction are, in some respects, too narrowly focused.

Table 2.1 shows a simplified summary of this critique.[100] It permits the identification of which theoretical frameworks emphasise which component of the dual relation mentioned above. The dimension of "information encoding" coincides with Peirce's object-sign-relation, whereas "information decoding" is related to the translation created by his interpretant.[101]

Eva Schürmann claims that there is no problem in regarding images as signs, since this would be compatible with many different approaches to depiction. However, just stating that images are signs does not solve the problem of how to discern pictorial from other, for example, linguistic, signs (see Schürmann 2006, 158). Although Schürmann seems to have a point with this, Table 2.1 shows that many theoretical approaches do not sufficiently deal with all the relational features of signs – whether pictorial or not. Proponents

Table 2.1 Summary: theories of depiction

	Information encoding	*Information decoding*
Resemblance-based theories	x (not applicable to all ways of image production)	X
Conventionalism	X	x (missing explanation of particular phenomenology)
Experience-based theories	-	X
Recognition-based theories	x (artist's intention)	X
Mixed theories	X	X
Causal theories	X	–

of conventionalist accounts obviously have difficulties in accounting for the particular phenomenology of picture perception. Affording pictorial and linguistic signs parity – simply regarding them as belonging to different symbol systems – seems to make it difficult to include perceptual aspects that exhibit a distinctive feature on one side or the other. Experience-based theories in particular lack an appropriate explanation of information encoding. Resemblance-based theories also show deficits because not all images can be said to be produced by mimicking an object of depiction, and recognition-based theories also exhibit a shortcoming in this respect. Within this context, the question has been discussed concerning what ensures that the content and subject that the viewer recognises in the image are the correct ones. Lopes mentions the intentions of the artist in this regard, although he also makes a distinction between two kinds of meaning – between meaning intended and the meaning of the work. In this sense, his theory also lacks an adequate account of how information is encoded in the image.

Obviously, proponents of mixed theories seem to fare much better with the dual requirements pointed out in relation to Peirce's semiotics. That is, they offer explanations for both information encoding as well as decoding. This better stance, however, clearly depends on what kinds of theories are combined in the particular account. Newall's approach, for instance, does well in this respect. Firstly, he suggests the criterion of non-veridical seeing as the typical way of decoding images. Secondly, his proposal of a dual *standard of correctness*, formerly indicated as a vice in his theory, now might emerge as a virtue instead. The duality does justice to the fact that the encoding of visual representations takes place in such a variety of ways that more than two such standards might be the end result.

The proposal that I will suggest with respect to the term 'scientific visualisation' pays close attention to these dual relations of signs. In the next section, it will be seen what motivates the different standards of correctness in this context by connecting the results from the discussion of the nature of depiction with the previously analysed categories of scientific images. It is this template of a combination of different standards of correctness with a particular way to glean information from the image that I take from Newall's account in order to elaborate on this for the case of scientific visualisations.

Before discussing details, however, I have to clarify one final aspect regarding Table 2.1. The last theoretical framework mentioned there is *causal theories of depiction*, which Newall does not consider within his five categories discussed previously. The reason for this neglect might be that, as Sachs-Hombach rightly points out, causal relations are of relevance to a multitude of phenomena. Thus, they do not account for the particular phenomenon of depiction, but might simply be used to select the object of reference depicted (see Sachs-Hombach 2006, 128).[102] Undoubtedly Sachs-Hombach has a point: causality plays a role in many different kinds of relation, not only in pictorial ones. Nonetheless, it seems equally wrong to exclude them from the realm of picture theory altogether since a variety of images – especially in the sciences – is produced in a causal way. Photographs are the most obvious example here, but also images produced by measurement and detecting devices, such as the Higgs image (see Figure 2.4). Consequently, I am convinced that causal theories should have a place in picture theory – even if their contribution to depiction is restricted to that part of information encoding shown in Table 2.1.

Causal theories of depiction state that a particular picture, y, is a picture of x because x caused the pictorial representation y. In this sense, a counter-factual dependency of y on x can be formulated in the following way: if x had revealed different properties, y would also have shown different ones.[103] This conception is compatible with Woodward's interventionist account of causality, introduced above. Obviously, the presupposition of an existing x causing y constitutes difficulties in the context of causal theories as well as in the context of resemblance theories. How are images to be dealt with that do not have a referent? However, I do not claim that causal theories are to be regarded as the ultimate solution to the topic of depiction. On the contrary, my suggestion is rather limited, namely to regarding causality as an important, though not exclusive, standard of correctness – to use Newall's term here – when analysing scientific images.

In the following, I shall now express in more detail what a mixed theory with respect to scientific visualisations will look like.

2.3 Summary: correlations between categories and theories?

Now that both the extension (categories) and the intension (nature of depiction) of visual representations in science have been considered, the question arises whether there is a way to overcome difficulties that are correlated with each approach in particular. The above analysis has revealed that neither way of approximating an understanding of what 'scientific visualisations' are has succeeded in providing an indisputable definition, although it would be equally wrong to dismiss our investigation as pointless. On the contrary, in the face of this variety of conceptual problems, the suggestion has arisen that a combinative strategy might be fruitful in this context. Pooling insights gained from both approaches separately might permit elaborating a mixed theory of depiction that meets both Lopes's constraints on picture theories and takes into account the particularities of scientific visualisations.

Admittedly, I am not alone in suggesting such a connection between perspectives on the topic (that is, considerations about categories and the nature of depiction) to achieve a better understanding of 'scientific visualisations'. Perini proposes a similar strategy (see Perini 2010). She makes use of a conventionalist conception to theorise about visual representations in science and suggests a way of developing a taxonomy within this context (see ibid., 146, Table 7.1). She begins with a particular picture theory – namely, conventionalism – and uses this as a framework to work out possible categories of scientific visualisations. Her criterion is to sort the diverse visual phenomena according to the different features of the symbol systems to which they belong. This is the level where categorisations play their part by guiding the refinement of a conventionalist approach to depiction. This strategy leads to three categorical labels which can be roughly characterised in the following way:

- *Compositional system*: images of this kind are similar to texts (see ibid., 141). This similarity is due to their internal set-up from "combinations of discrete atomic characters" (ibid., 140f.). Thus, contrary to Goodman, Perini thinks that there are visual symbol systems that do have a kind of alphabet. As examples, she mentions certain kinds of diagrams used in biology. Moreover, she points out that, within such compositional systems, the particular characters do not need to refer to the intrinsic properties of the entity depicted. An example might be Neurath's ISOTYPE (see Figure 2.11).
- *Pictorial system*: this category is characterised as syntactically and semantically dense in Goodman's sense (see ibid., 143). Photographs are amongst the examples that Perini discusses in this context. Moreover, she argues that images of this kind are regarded as paradigmatic instances of visual representations (see ibid., 144), namely those that show a resemblance relation to their object of depiction. She thinks, however, that the distinctiveness of images of this kind is not due to such a relation, but to the system's syntactic and semantic density (see ibid.)
- *Schematic system*: images in this last category that Perini mentions are not as precise as instances of the former two kinds of systems. Schematic drawings are used when properties of a more general kind are supposed to be visualised. "In these systems, relatively abstract visual features like continuity, inside/outside, etc. are used to represent generic, rather than specific, properties" (ibid., 145). Maps might be an example here, where line drawings refer, for example, to streets, such as in Figure 3.1.

Interesting though this suggestion might be, Perini only presents incipient ideas about the possibility of such a taxonomy. There is no claim to completeness involved. Nonetheless, her account shows that Lopes's thesis that theories of depiction have to deal adequately with the wide-ranging diversity of visual representations also implies a clue as to how to develop or modify such accounts to fulfil this task.

In a similar fashion, I want to suggest a refinement of Newall's mixed theory of depiction to account for scientific visualisations. As indicated above, I agree with him that such a theory should consist of two parts: one explaining how information is encoded in the image, thus offering a standard of correctness for further interpretation, and another explaining how information is to be decoded later by a recipient.

The broader framework of my proposal consists in the assumption that a *theory of information transmission* constitutes the suitable background for an account of scientific visualisations. This is not a surprising hypothesis if it is taken into consideration that all sciences are (more or less) about gathering, synthesising, evaluating and distributing information. In the next chapter, the functional roles of images within such informational processes will be analysed in more detail.[104] Anticipating some of the results of my analysis, I refer here to Fred I. Dretske's theory of information (see Dretske 1999).

Within this context, images can be regarded as signs in the Peircian sense. Thinking about images along these lines clarifies that they mediate between objects – conceived in the broad sense – and interpretants, that is, the understanding of their meaning. This corresponds to the relational features of images indicated in Figure 2.14. In order to undertake the left-hand side of the mediating task accomplished by visual representations, the results of the discussion of paradigmatic instances of scientific visualisations can now be utilised.

Firstly, it has been noted that there is a variety of methods to encode information. In science, there are hand-made images as well as technically produced ones. Additionally, the latter category can be subdivided further by pointing out that some of these technical procedures of information encoding are causally related to an existing object, whereas others are not. That is, some images are the results of measurement or detecting processes, whereas others are produced by, for example, computer simulations. Moreover, if images belong to the former category, a distinction has to be made between direct and indirect measurement. That is, some images show properties that their object of depiction also directly possesses, such as colour properties or outline shapes. Other images, however, show only indicators: features that are caused, but not possessed, by the related object. The Higgs image (see Figure 2.4) illustrates this case. It does not show the Higgs boson but the detection of the traces of other subatomic particles that the unstable Higgs decayed to. Finally, we have to take into account that images produced in one of these ways can be modified by other procedures later on. Hand-made images might, for instance, be scanned and integrated into computer graphics or vice versa: computer graphics might be printed and further details added by hand, etc.

What should be clear is that these categories provide different standards of correctness. The producer's intentions will be relevant in the case of hand-made images. A causal connection is essential to information encoding in mechanically produced visualisations, and algorithms have to be taken into account when computer graphics gained by simulations are considered. Similar to Perini, I do not claim completeness here, as technological developments might result in further methods of image production. Thus, allowing for different standards of correctness does justice to advancements in technology which are often deeply intertwined with progress in science.

Consequently, a mixed theory of scientific visualisations can meet Lopes's diversity constraint on picture theories by offering suitable explanations of information encoding. In the next chapter, it will be seen in what way social processes in science play a role and can thus be used to account for diversity in an even broader sense, namely not only with respect to different designs, but also to types in a more general sense, and with respect to the multitude of functional roles that visual representations accomplish in scientific processes. Adding the results of this analysis will allow us to easily accommodate the fact that standards of correctness might overlap when different methods of picture production are employed.

Let us now turn to the second part of a mixed theory of scientific visualisations, namely information decoding. It seems to me an open question how far my proposal diverges from Newall's initial suggestion to take the criterion of non-veridical seeing to account for the particular method of information decoding. Non-veridical seeing obviously seems to be meant as a way of gaining information about the object of depiction. Thus, Newall introduces the epistemic source of *perception* here to fulfil the decoding part.

What I have in mind is a *combination of perception and reason* performing the same function. However, the exact nature of the intertwining of perceptual and (other) cognitive processes is still a controversial topic even in the empirical sciences. When does vision stop and interpretation start? Can a distinction of this kind even be made? Gombrich's claim that there is no innocent eye, for example, speaks in favour of interdependence. Similar suggestions have been put forward in the philosophy of science in which it is claimed that the process of scientific observation cannot be described as a neutral and passive recording practice. On the contrary, observation has to be regarded as an activity. The observer makes extensive use of formerly acquired background knowledge – particularly of concepts – when engaging in this activity (see e.g. Chalmers 1999, ch. 1, 2; Kosso 1993, ch. 6). Thus, depending on the stance chosen in this debate, perception is regarded as being more or less detached from (or intertwined with) reason, that is, with our logical capacities of drawing inferences relevant in interpretation. Referring to both epistemic sources, then, means being secure in this debate and acknowledging Gombrich et al.'s insights.

Nonetheless, it seems advisable to be more precise about what I have in mind when considering the part played by information decoding in a mixed theory of scientific visualisations. The first point I wish to make is that I regard images as signs in this context and assume that they do what signs are supposed to do, namely transmit information. In this sense, I agree with Hopkins's thesis that *images are artefacts* and that one of their main purposes is to serve in communicative acts (see Hopkins 2003, 654). Visualisations in science are products of intentional processes such as the measuring, detecting and recording of information. And as their genesis is intended, they are artefacts and not natural signs, although they might be causally produced. Moreover, within those processes, they are the bearers of information, a fact that constitutes their suitability as components of communicative acts – a fact that is proven by their ubiquity in scientific journal articles, books and presentations. Regarding scientific visualisations in this way then gives rise to another thesis, namely that we should also allow for the possibility of developmental processes in this context. What does this mean?

It can be assumed that images were once genuinely based on resemblance relations, with early artists trying to record visually what they saw. This genuine form of picture production relied on how *human vision* works – the dominant sense of human beings to cognitively access the world. And I think it is this connection between images and vision which explains why we are so adept at deciphering visual representations – this is Lopes's competence constraint.[105] Peirce's iconic signs can constitute the starting point of a developmental process – the process of learning – in the following sense: artefacts are subject to modification, elaboration and replacement, and I think this is what has happened to images in the arts as well as in the sciences. Art history tells us that there is a continuity of innovations in picture production, of developments of new depictive styles. Why should the sciences differ in this respect? Why should their repertoire of expression have remained static? There is a crucial similarity here, namely in the same way that we learn about new depictive styles in the arts, we learn and teach our fellows about new ways of picture production in the sciences. A part of any scientific training consists of learning how to read, for instance, diagrams about the statistical significance of measurement data (see Figure 2.5) or how to understand data visualisations showing indirect measurement results such as the Higgs image (see Figure 2.4). Scientists learn how to interpret them correctly, that is, to understand in what sense they are informative about the object of research, and are able to comprehend almost immediately the meaning of similar images encountered later – although the same image might be more or less unintelligible to an untrained observer. For this reason, background knowledge and drawing inferences (that is, the epistemic source of reasoning) come into play when deciphering scientific visualisations.

Consequently, I agree with Newall's appraisal in regard to visual representations that conventions, understood in the above sense as solutions to co-ordination problems, can play a role, but that the conventionalist thesis that "depiction is entirely a matter of convention" is false (Newall 2011, 10). Moreover, as seen above, conventions are especially relevant when it comes to understanding the intentions that image games are based on, namely the illocutionary acts in which images play a role. But even though conventions can become crucial in this sense to a correct understanding of images, I think that resemblance nevertheless is genuinely basic to all of these decoding processes, not least because scientists normally want to learn something, not about the image, but about its object of reference, namely their object of research. Only if they are convinced that the image can tell them something about this object, that the image is *about* the latter, does it make sense to regard this as data, as information in the relevant sense, and it is this broad sense of resemblance that plays a role in scientific visualisations.

Let me add one final remark about the nature of this relation between image and object when it comes to decoding the former: Goodman's term *exemplification*[106] seems unsuitable here (see Goodman 1976, 52ff.). Goodman defines 'exemplification' as "possession plus reference" (ibid., 53). That is, the respective symbol not only refers to a particular property of an entity, but possesses this feature itself. Colour swatches, for example, function in this way. I think, however, that this does not work for the majority of scientific visualisations. Quite a few of them are the results of indirect measurements, such as the Higgs image (see Figure 2.4). In these cases, the resulting images are evidence or indicators of the relevant phenomenon, not an equivalent on a smaller scale. On the contrary, scientists have to make inferences to reason from the evidence to the phenomenon in which they are interested. This is why I think that Goodman's concept of exemplification is not the appropriate explanation of how to conceive of the typical relation between a scientific image and an object of research.

In the next chapter, I will develop in detail a mixed theory for what seems to me the most important case of contemporary scientific visualisations, namely those that result from measurement or detecting processes. The standard of correctness in this case will be causality based on the sensitivity of the instrument involved. The decoding process, however, relies on a mixture of resemblance and convention. Usually, mathematical functions map the data onto object properties when instruments are used in scientific investigative processes and guide the scientists in their interpretations. Moreover, these functions are usually a part of theoretical networks in science. That is, they are somehow anchored in an empirically tested framework and are thus not arbitrary by nature in the way that linguistic conventions are. It is in this way that I think that resemblance and causal connection can both play a role in science (see also Mößner 2013c).

However, we have to keep in mind that two questions with regard to scientific visualisations should be firmly kept apart. Firstly, what is relevant to consider when *producing* such an image? This can be a causal process as indicated above, but other methods of image production are possible as well. Secondly, what is needed in order to *understand* such a visual representation?[107] It is only with regard to this second question that I suggest that resemblance relations play an important part due to the cognitive set-up of the human mind.

Notes

1 For more examples see Kemp (2003).
2 A brief note to the reader: people who are more interested in the epistemological questions of the topic might skip the following analysis and directly move on to Chapters 3 and 4 of this book.

3 On the notion of 'artefact' see Reicher (2013).

4 The camera used for those purposes was part of the Hubble Space Telescope, which orbits our planet. This location allowed scientists to take photographs not affected by distortions due to atmospheric effects.

5 A comprehensive collection of them is, for example, presented in Walden (2008).

6 Wilfried Wiegand presents a collection of essays written during those initial days of photography whose authors (e.g. William Henry Fox Talbot) discuss the particular characteristics of this technology, the difficulties involved in producing their first photographs, and how they were affected by those images as recipients (see Wiegand 1981).

7 Roger Scruton tries to create such an ideal understanding of the photographic picture in aesthetics when defining it as a *relatum* of a causal and intention-independent relation between object and picture (see Scruton 2008, 149). He thinks that the ideal photographic picture does not rely on artistic intention for it to be a picture of the object depicted, but only on the causal process at hand.

8 Of course, both ways to manipulate the photographic process *can* be performed by one and the same person.

9 Elsewhere, I argued in more detail how to deal with the question of whether and, if so, in what sense photographs can be regarded as scientific evidence (see Mößner 2013c).

10 Obviously, this is different on the theoretical level, for instance, of the philosophy of science. Here, scholars indeed discuss what the inventory character of scales and measurement devices might imply for the data obtained (see e.g. Chang 2007; Mößner and Nordmann 2017; Schlaudt 2009).

11 He speaks about "imagining" – and not "imaging" – as he adopts Kendall Walton's conception of depiction. Walton's approach will be discussed in section 2.2.3 of this book.

12 In her book about photography in science, she offers more examples when stating that "[o]ften images made to be measured, like earthquake tremors or spectrograms, aren't considered photography at all. They use photography, but produce images that don't appear to depict anything recognizable. Even when they are recognizable as pictures, the images are often strangely distorted" (Wilder 2009, 35).

13 Walton's theory will be discussed in detail in section 2.2.3 of this book.

14 See Klaus Hentschel's (2014) discussion of the topic for more details (see ibid., ch. 5.4).

15 On the usage of false colours in astrophotography see e.g. Müller (2007). A discussion of the decisions of NASA scientists about the colours used in photographs taken by the Hubble Space Telescope is presented by Adelmann (2009a, 166ff.).

16 See e.g. the Sloan Digital Sky Survey project whose participating scientists produced three-dimensional maps containing more than 930,000 galaxies, www.sdss.org/, accessed August 20, 2015.

17 See for instance the case study about the *Mars Exploration Rover* presented by Janet Vertesi (2015).

18 See for instance the case study about the *Mars Express* orbiter presented by Ralf Adelmann (2009b).

19 It has to be added that MRI technology is not only used in medical science, but also in biology, biophysics and the like.

20 A similar point is made by Peter Kosso who discusses the distinction between observables and unobservables in the context of the debate between scientific

realists and anti-realists. He argues for a more precise concept of observability and, as a consequence, makes a distinction between four dimensions of observability. One of these categories he calls the "amount of interpretation" which seems to be coextensive to Roskies's concept of inferential distance (see Kosso 1988, 456). I will come back to Kosso's point in more detail in section 3.1.3 of this book.

21 This process is explained in detail by Roskies (2008, 23ff.).

22 I owe this point to Maria E. Reicher.

23 More information about this is provided by Roskies (2008, 27ff.).

24 Of course, there are important exceptions where this assumption does not hold, for example, in the case of chronophotography (see Figure 2.3), or in respect to false colour composite photographs depicting kinds of radiation invisible to the human eye (see Figure 2.2).

25 CERN is the abbreviation for Conseil Européen pour la Recherche Nucléaire, i.e. for the European Council for Nuclear Research.

26 For more details about the Standard Model see, for example, Glenn Elert's "The Physics Hypertextbook", http://physics.info/standard/, accessed April 4, 2016.

27 ATLAS is the abbreviation for A Toroidal LHC ApparatuS.

28 CMS stands for Compact Muon Solenoid.

29 For more information about the detectors see http://home.cern/about/experiments/atlas and http://home.cern/about/experiments/cms, accessed April 4, 2016.

30 For a definition and critical discussion of the term 'theoretical entity' see Andreas (2013).

31 For more information about this topic see e.g. Mayer-Schönberger and Cukier 2013).

32 He claims that "[t]o function as evidence in science, the act of observation must offer reason to think that we believe truly what we see" (Kosso 1993, 119).

33 Vertesi points out that these visual techniques turn photographs into maps to navigate the rover on Mars. "But the power of maps is not merely in their making: it is in their deployment. [...] They work [...] also by presenting possibilities for our interactions with the mapped territory and suggesting potential for wayfinding within a space. [...] Rover maps are annotated rover images: marked up, circulated among team members, and used daily to structure conversations about how, where, and why to next interact with Mars" (Vertesi 2015, 106).

34 As the graph is specially tailored for this example, a labelling of abscissa and ordinate is deliberately left open.

35 An introduction to Goodman's approach is offered in Scholz (2009b, 128ff.).

36 I owe this point to Maria E. Reicher.

37 Sun-Joo Shin, Oliver Lemon and John Mumma (2014, sect. 2) present an insightful overview about those kinds of diagrams and their developmental history.

38 Sun-Joo Shin (2015) offers an interesting discussion of the question why diagrams are cognitively useful in logic.

39 For more details see also Christian Suhm's discussion of the different theses defended in scientific realism (Suhm 2005, ch. 2.2).

40 Of course, it is not an impossible project, since it depends on how the term 'observable' is defined (see van Fraassen 1980, 13–19).

41 A critique of van Fraassen's theses is offered, for example, by Suhm (2005, ch. 4.2.).

42 Elsewhere, I also argued for the intertwining of different representational modes (see Mößner 2013b).

43 See Wittgenstein's explanations of the topic in his *Philosophical Investigations* (1995, §§ 7, 23 on p. 65ff.).

44 This historical example, however, also shows that science is a dynamic process, that is, no facts are taken for granted. Today, we only speak about the eight planets of our solar system, as meanwhile – more precisely in August 2006 – Pluto has lost its status as a planet and been categorised as a 'dwarf planet' instead. This was due to the discovery of the so-called Kuiper Belt with its many asteroids, some of them even bigger than Pluto.

45 The limits of this assumption are nicely illustrated by the example of the *Pioneer plaque*. Obviously, there is still some background knowledge relevant to understand the drawings and graphics on the plaque correctly. For instance, some of its critics objected that, instead of a gesture of greeting, the raising of a hand might be misunderstood as a gesture of aggression. In a similar vein, Ernst Gombrich argued that the arrow in the lower graphic is an artificial sign that was developed in human culture and thus could not be understood by non-human beings as an indicator of origin (see Gombrich 1972).

46 Examples of original ISOTYPE symbols are provided by the web archive of Gerd Arntz who was the responsible designer in Neurath's project, see http://gerdarntz. org/isotype, accessed July 2, 2016.

47 As a semiotician, Eco regards images as signs.

48 Eco here suggests nearly the same problem solution as put forward by Kjørup and Barthes, discussed above.

49 I will discuss the topic of conventions in more detail in section 2.2.2 of this book.

50 Klaus Sachs-Hombach and Eva Schürmann (2005) present a helpful overview concerning the topic of pictures in philosophy.

51 Of course, there are also pictures which do not depict in this way or even not at all. Most of all, abstract paintings seem to be of this non-depictive kind. Nonetheless, some philosophers hold the view that even these pictures represent and, therefore, depict something (see e.g. Newall 2011, ch. 8).

52 I do not want to make a claim of completeness here as a variety of mixed theories has been developed in recent years.

53 A variety of different objections is discussed by Scholz (2009b, ch. 2.4, 2.5).

54 The relevant passage is derived from Plato's dialogue "Protagoras" (see Platon 1994a, 331d-e).

55 This critique is presented in Plato's dialogue "Cratylus" (see Platon 1994b, 432a–c).

56 Amos Tversky (1977) analyses the concept of similarity in psychology. Within this context, he suggests rejecting this commonly held thesis of symmetry with respect to similarity. He discusses several empirical studies that highlight the fact that test persons performing comparative experimental tasks tend to asymmetric similarity judgements. Usually they describe only the variant (that is, the less salient stimulus or entity) as similar to the referent (that is, the more salient stimulus), but not vice versa. This finding motivates Tversky to state that "[...] it appears that proximity data from both comparative and production tasks reveal significant and systematic asymmetries whose direction is determined by the relative salience of the stimuli. Nevertheless, the symmetry assumption should not be rejected altogether. It seems to hold in many contexts, and it serves as a useful approximation in many others. It cannot be accepted, however, as a universal principle of psychological similarity" (ibid.,

338). Thus, at least in the context of psychology, a different understanding of similarity seems to be called for. I thank Raoul Bussmann for making me aware of this study. Moreover, considerations such as Tversky's might also be the reason why Richard Wollheim states that "experienced resemblance, unlike resemblance itself, is nonsymmetrical" (Wollheim 1998, 219).

57 On the concept of convention, also in comparison to mere speaker intentionality, see von Savigny (1983). I thank Christoph Diehl for suggesting this source to me.

58 Goodman cheerfully admits that his approach undermines Peirce's categorisation. "The often stressed distinction between iconic and other signs becomes transient and trivial [...]" (Goodman 1976, 231). In this sense, Sachs-Hombach points out that Goodman's approach is restricted to the difference between *natural* and *non-natural signs*, connecting the former to indices and the latter to symbols. Thus, icons are no longer available as a third independent category here (see Sachs-Hombach 2006, 48). One might object that Goodman's suggestion erodes Peirce's theory even further. Eventually, he suggests that pictures are symbols, though he seems to be reluctant to agree that their denotational relation is conventionally established as in the case of linguistic symbols.

59 In Chapter 3, I will suggest that, in many instances, *the comprehension of* such an image requires more than mere causal connections.

60 Moreover, it seems reasonable to assume that some information on the object depicted is also of relevance in order to decipher the picture correctly.

61 Newall suggests this example especially as a critique of John Hyman's refined resemblance-based theory (see Hyman 2006).

62 I would like to thank Joachim Bromand for making me aware of this difficulty in Newall's examples.

63 With regard to Newall's classificatory scheme, Peacocke's theory might better be subsumed under the label of 'experience-based approaches'. Nonetheless I will briefly discuss his theory here, as he also presents a refined conception of how to understand resemblance within the context of depiction.

64 A critical analysis of Peacocke's approach is offered by Lopes (2006, 20ff.) and Gavin McIntosh (2003).

65 Scholz's analysis makes us aware of the numerous questions involved in clarifying pictorial representation (see Scholz 1993) and with respect to their understanding (see Scholz 2009b, ch. 5.4).

66 Lopes himself notices this tension in his discussion of Goodman's theses (see Lopes 2006, 73f.).

67 I use the term 'representation' as referring to all three kinds mentioned above: numbers, words and images. Goodman makes a distinction here (he never speaks about representation especially with respect to linguistic symbols) which I do not follow.

68 A critical introduction to Goodman's theory is offered by Lopes (2006, ch. 3). Further helpful comments on his theory can be found in Scholz (2009b, ch. 4) and in Giovannelli (2010).

69 It might be objected that common measuring devices do not show exact values but only approximations. However, this seems to be a consequence of pragmatic reasons and therefore does not constitute a problem in principle. A digital scale can show the exact measurement result also with respect to its decimal places, although it might sometimes be more convenient to round them up. Yet two

aspects have to be admitted. On the one hand, the accuracy of a digital device depends on its sensitivity. That is, some digital scales for example are only able to show their results in kilograms and grams whereas others are exact to the level of milli- or even micrograms. On the other hand, errors in measurement might occur and thus detract from the corresponding result. In this sense, an exact match between measurement result and actual weight can be called into question even when using a digital device. I thank Maria E. Reicher for making me aware of this difficulty regarding Goodman's terminology.

70 Lopes states accordingly that "[...] repleteness is a matter of degree: pictures are replete relative to other forms of representation, such as graphs, because some properties have representational significance in pictures but not in graphs" (Lopes 2006, 68).

71 It has to be added that in hermeneutics there are also theoretical accounts of interpretation whose proponents would deny the privileged position of the artist.

72 A helpful introduction to colour perception is offered by Karl Schawelka (2007). Dominik Groß et al. (2006) discuss the role of colours of visualisations used in the medical sciences.

73 Furthermore, Gombrich is convinced that the very same mechanism is also in place in normal perception. That is, we make assumptions about what we perceive and test these assumptions by comparing them with our experiences (see Gombrich 2004, 231).

74 Gombrich points out that, at different times, pictures were, for instance, predominantly regarded as cult objects and not as replicas of real world entities.

75 A similar point is put forward by Newall: "[t]he trouble with this theory [Gombrich's illusion theory, N.M.] is its claim that pictures can deceive us in this way. Many pictures cannot under any conditions prompt an illusory experience of their subject matter" (Newall 2011, 24).

76 From Lopes's point of view, twofoldness is a gradual matter. He writes that "[p]ictures may be thought of as arranged along a spectrum, at one end of which trompe-l'oeil pictures lie. Visual experiences of them are experiences of their subjects, but which generally preclude, or at least suppress, experiences of decorated, marked, and painted surfaces. At some intermediate point on the spectrum lie pictures which afford one kind of experience or the other, but not simultaneously. [...] At the other extreme lie pictures typical experiences of which are simultaneously experiences of their subjects and experiences of flat, pigmented surfaces" (Lopes 2006, 50).

77 Newall puts forward a similar critique when complaining that Wollheim's theory does not explain *how depiction functions* (see Newall 2011, 53f.).

78 From Lopes's point of view, the phenomenon of seeing-in is, in principal, also compatible with a conventionalist theory of depiction. Within such a theoretical framework, this feature of picture perception would then be cognitively explained, but not as being based on the phenomenology of perception (see Lopes 2006, 44).

79 A wealth of fascinating medical case studies is presented by Oliver Sacks (2010).

80 As a definition of 'picture', Lopes states that it "is a representation whose content presents a 'spatially unified' aspect of its subject" (Lopes 2006, 126).

81 Of course, this ability is not unlimited. Lopes explains accordingly two different constraints restricting recognisability (see Lopes 2006, 138f.).

82 It can be assumed that this might be where alternative approaches such as resemblance- or convention-based theories might fit in, as Lopes claims that his approach is combinable with such accounts of depiction (see Lopes 2006, 111).

83 Robert Hopkins also highlights this difficulty, not only with regard to Lopes's theory, but also with respect to experience-based theories such as his own. He states that "[...] what is needed is appeal to a standard of correctness, something making it right to respond to a given set of marks, whether that response is experiential or recognitional, in one way rather than another" (Hopkins 2003, 658).

84 It has to be added that this somewhat mundane critique with respect to a philosophical project of concept analysis is motivated by Lopes's own striving for a close connection of his theory of depiction to empirical sciences analysing human sense perception.

85 It is only conventionalism that he thinks can be dismissed as a source of insights concerning an adequate theory of depiction (see Newall 2011, ch. 1). I discuss his reasons for rejecting accounts that follow Goodman's ideas in section 2.2.2 of this book.

86 It has to be added, however, that, although Newall thinks that experience-based theories of depiction are in principle preferable to other accounts of this topic, he claims nonetheless that such theories still lack an explanation of *"how depiction works"* (Newall 2011, 52). Moreover, he is of the opinion that his theory of veridical and non-veridical seeing can fill this explanatory gap and thus constitutes a better theory in comparison to merely experience-based ones (see ibid., 54f.). He thinks that what he states about non-veridically seeing the picture's subject matter explains the information decoding step highlighted in Figure 2.14. What he has to say about the aspect of information encoding will be discussed in due course in correlation with the "standards of correctness".

87 By using the concept of seeing here, Newall combines *experience-based theories* with *recognition-based* ones. This becomes clear when taking a closer look at his definition of 'seeing'. From his point of view, seeing is a process comprising three different steps: "(i) the stimulation of the visual system, (ii) engagement of the subject's ability to recognize X, and (iii) a visual experience of X, which is the experience of seeing X" (Newall 2011, 44).

88 I discussed this point in Mößner (2013d).

89 On Wittgenstein's concept of family resemblance see Wittgenstein (1995, § 67 on p. 278).

90 This is also pointed out by Bantinaki as a possible objection to Newall's theory (see Bantinaki 2014, 947).

91 He states that "[p]hotographs are made by relatively simple optical, mechanical and chemical processes. As a consequence of this, a photograph's depictive content is counterfactually dependent on what is present before the camera. A photograph of X indicates that X was present before the camera, and if X had not been so present, then the photograph would not depict it. This is why photographs are reliable conduits of information about what is present before the camera" (Newall 2011, 60).

92 "Perceptoid signs" is the translation that Schirra and Sachs-Hombach (2013, 144) suggest. The German term is *wahrnehmungsnahe Zeichen* (see Sachs-Hombach 2006, 94).

93 Moreover, he suggests that on all three analytical levels, namely syntax, semantics and pragmatics, a particular intertwining of semiotic and perceptual approaches can be discerned in image science (see Sachs-Hombach 2006, 87).

94 Gustav Frank and Barbara Lange point out that competing conceptions are also a consequence of mutual neglect due to close connections between the theoreticians' mother tongues and their theoretical approaches (see Frank and Lange 2010, 12f.). Gaiger, for example, starts his analysis by stating that the German expression 'Bild' does not coincide with the English term 'image' (see Gaiger 2014, 209). Bredekamp defends the even more drastic claim that the German expression 'Bildwissenschaft' (here and elsewhere translated as 'image science') "has no equivalence in the English language" (Bredekamp 2003, 418).

95 The text written in single quotation marks is Gaiger's translation of Sachs-Hombach's text in Sachs-Hombach (2005c, 11).

96 Eva Schürmann gets to the heart of this necessity when stating all the divergent theories, methodologies, concepts, presuppositions, etc. of the disciplines involved (see Schürmann 2006, 155).

97 Wiesing's account is critically discussed by Gaiger (2014, 217ff.).

98 On the one hand, art historians have analysed mundane as well as works of high art and, on the other, also taken into account new pictorial media such as photography and film.

99 On the problem of demarcation see, for example, Gordin (2012) and Curd, Cover and Pincock (2013, sect. 1).

100 Of course, the table only offers a rough conception of this, as we have seen above that there is a variety of different approaches connected to each category. Particular theories – also ones perhaps not considered above – might therefore diverge from this general scheme.

101 I introduced these terms in Figure 2.14.

102 Further difficulties related to causal theories of depiction are discussed by Scholz (2009b, ch. 3).

103 Admittedly, we do have to presuppose that y is also sensitive to the properties shown by x. A photograph of a man may show his height and the size of his body, but not his weight, as cameras are not sensitive to measuring weight.

104 Robert Hopkins also points to the centrality of informational tasks of images even more generally when stating that "[p]ictures are artefacts, and essentially used in acts of communication, and these facts are reflected in what it is to grasp pictorial content" (Hopkins 2003, 654).

105 We will discuss the suggestion that visual representations exploit our evolutionary set-up as visually orientated beings in more detail in section 4.2 of this book.

106 I thank Alfred Nordmann for making me aware of this possibility as an alternative to claims about resemblance.

107 Scholz has made us aware of the misunderstandings evoked by confusing the diverse questions in picture theory (see Scholz 1993).

3 Functional roles, appearances and the problem of diversity

Obviously, scientists make use of visualisations in different contexts and for different purposes. They are found, for example, employed in educational processes as well as in genuine research activities. In the previous chapter, their functions in science have already received cursory attention. However, what has been discussed so far is primarily related to the method of their production, that is, the techniques and instruments scientists use to create an astonishing diversity of visual representations in science – a diversity that seems to prevent any coherent categorisation.

Moreover, this diversity of production processes also makes it difficult to subsume scientific images under the label of any one of the discussed theories of depiction. As a solution to this difficulty, I proposed a mixed theory paying attention both to the different ways of visually encoding information and its later decoding by the recipient. By allowing for different standards of correctness, referring to the encoding part, such a mixed theory pays attention to the main source of problems to be clarified in this context, namely the inventive character of science itself. A huge variety of different imaging technologies is at the scientists' disposal – a stock that is growing steadily in correlation with the development of new experiments and instruments in science.

Stephen M. Downes connects these two aspects of pictorial diversity and points out that no "unified representation-based account" can explain how scientific images play their role in epistemic and non-epistemic processes in science (see Downes 2012). For Downes, this is a consequence of his conviction that it would be wrong to claim that all scientific images *represent* entities we find in the world around us. He thinks that at least some images function as models and thus do not represent real entities, but are merely hypothetical (or fictional) in nature. I agree with Downes that scientific visualisations can play different epistemic roles and sometimes no epistemic role at all (see ibid., 117f.). Furthermore, our previous analysis also supports his thesis that a *unified* account of depiction, in particular theories referring to only one method of production, cannot explain the diversity of functions and appearances.

In the following, a new turn will be taken in the approach to categorise scientific images. My starting point will no longer be the different modes of visual representation, that is, explanations of depiction, but the *functional roles* that visualisations play in science. This way of approaching the topic takes into account the fact that scientific images are artefacts, that is, produced entities. Classifying visualisations as *artefacts* is meant as a neutral claim with respect to ontological and epistemological questions concerning their capacity to tell scientists something about related objects or not.

To systematise their functional roles, the first task will be to point out and describe the contexts in which scientific visualisations appear. I will follow a distinction put forward by Felice C. Frankel and Angela H. DePace (see Frankel and DePace 2012). From a graphic design point of view, they suggest that scientists should bear in mind the purpose of their visualisation and the audience intended to receive them, guiding them to propose a distinction between the "exploratory context" and the "explanatory context" (see ibid., 13). Whereas the former context is about genuine research activities, the latter comprises activities of communicating research results to scientific peers, laymen and within educational settings.

In addition to describing the role of scientific images in both of these contexts, I will argue in favour of two theses. Firstly, I will show that the confusion of exploratory and explanatory contexts of use explains to a significant degree problems in accounting for the epistemological and also the ontological role of visualisations in scientific processes. Moreover, an advantage of this new approach to systematising visual representations in science will be that analysing their functional roles in these two contexts will give us, secondly, a clue to why visualisations are so diverse in their appearances. It will turn out that the constitutive part of the relevant explanation lies in the connection to the social setting of science.

We will approach this latter topic from the broader perspective in which science is seen as a social endeavour to acquire knowledge. Here, Ludwik Fleck's theory of the relevance of social processes to epistemic achievements in science will be applied. Fleck's theory of "thought styles" and "thought collectives" (see Fleck 1979) seems to be well-known amongst people theorising the epistemic function of images and pictures both in the scientific and the everyday context (see e.g. Bredekamp, Schneider and Dünkel 2008; Pörksen 1997; Vögtli and Ernst 2007; Zimmermann 2009).[1] However, contrary to many authors using Fleck's theory in their attempts to analyse scientific images, I will not emphasise the relativistic tendencies of his ideas concerning our way of perceiving the world along the lines of the thought-style to which we adhere. In this chapter, I will remain silent about this undoubtedly challenging aspect of Fleck's theory, which many theorists use to support a constructivist account of scientific images (see e.g. Bredekamp, Schneider and Dünkel 2008; Zittel 2014).

My analysis focuses on Fleck's considerations concerning *scientific communication processes*. His studies elaborately *explain* what common sense and graphic designers' guidelines, such as DePace and Frankel's, already seem to suggest, namely that target audience, purposes and aims of communication all play a significant role with regard to the appearance and the concrete function of visual representations in scientific discourse. Fleck presents a cautious and practically informed[2] examination of communication processes, especially with respect to the publication of and subsequent discussions about scientific results (see Fleck 1979, ch. 4.4). Although he develops his thoughts most of all with regard to linguistic representations, there are hints in his analysis that he also wants his theses to be applied to visual representations. It will be examined whether this transfer is possible and, if so, what it can tell us about scientific visualisations. It will become clear that the *social origin* of visual representations can contribute an understanding of the diversity which representation-based accounts alone cannot explain. Fleck's theory will be used to unravel what exactly is meant by *social origin* in this context.

3.1 Context-orientated approach

Common sense tells us that not all visual representations are equally suitable to all the different contexts of possible usage, an example of which is afforded by the computer graphic from the LHC showing the (indirect) discovery of the Higgs boson (see Figure 2.4). Of course, there are numerous contexts in scientific discourse in which it seems appropriate to use this image, as in journal articles communicating the latest research results to scientific peers or as parts of presentations documenting scientific success in acquiring new external fundings. However, there are also contexts in which showing this image seems inappropriate, for example, teaching pre-school children about what is going on at CERN. It is clearly the target audience, respectively *the assumed background knowledge of the target audience*, that informs the assessment of suitability. It comes as no surprise that professionals from the realm of graphic design give similar advice when asked how to improve scientific representations. Consider as an example the following statement made by Frankel and DePace:

> [s]cientific graphics are used in many contexts [...]. Each of these contexts makes different demands on the graphic. For example, in an oral presentation, a graphic needs to make the point quickly and clearly because the audience does not have the time to contemplate a graphic in nearly as much depth as it might in a research article. Graphics may have to appeal to different types of audiences as well. A graphic that is intended for a colleague may be quite different from a graphic intended for a student in your course, and still different for a program officer or a congressperson without a background in your field.
>
> (Frankel and DePace 2012, 13)

With regard to the above example, we are inclined to think that scientific peers will obviously understand what the Higgs image shows, that they can grasp what is depicted. However, we do not expect children to have the same intellectual resource at their disposal. Consequently, presenting research results to different target audiences also means taking into account their background knowledge concerning the issue at hand. Assuming a lack of such knowledge would usually motivate altering the image presented, for example, by simplifying it, or by adding didactic elements such as explanations, or even by choosing a completely different image.

However, there are instances where, although it is not supposed that the target audience understands the graphic entirely, it is nevertheless deemed reasonable to use the same image as in the professional setting. What differentiates between denying the reasonableness of presenting the same visualisation to colleagues and to children, and agreeing to the same procedure concerning colleagues and programme officers deciding on external funding? The point is that it is not only considerations about the audience's background knowledge that play a role in deciding how to present research results. Another relevant aspect is the *communicative aim* which the scientist pursues with her presentation. Not all of these aims presuppose a complete understanding of the information offered. With regard to programme officers, the scientist's intention of showing the image might simply be to catch the attention of potential sponsors.

It seems useful to make a distinction between two notions of meaningfulness at this point. On the one hand, there is the meaning of the pictorial content. This is the kind of meaning that was mentioned in the previous chapter when discussing the encoding of information into the visual format. On the other hand, there is the meaning ascribed to the visual representation in the wider context of communication. Of course, these two kinds of meaningfulness can be connected, and often are, but not by necessity. Klaus Hentschel makes a similar point in his discussion of the role of scientific images in the history of astronomy. He points to the fact that the effect produced by an image on its viewer is not determined by its concrete scientific content alone. Consequently, the meaning extracted from such a visual representation may vary across different audiences.

> These various representations of the lunar surface should remind us that 'images' can change their appeal over time, not only by simply deteriorating physically, but also by becoming obsolete. Hence their appeal is different at various stages in their 'life cycle' and towards different audiences. Even where the *image* (understood to be the visual message or information) stays the same, the *picture* (i.e., the image concretized in specific print media and embedded in a wider culture of print technology and reception) may change.
>
> (Hentschel 2000, 22)

Hentschel's labelling of the two different notions of visual representations might be controversial and I will not defend his terminological choice. However, he highlights a crucial distinction that should be borne in mind while analysing the epistemic roles and the status of scientific images. Next to the initial meaning encoded in the visual representation, there is another connected to the context of usage that might also vary in correlation to different usages. This meaning can clearly be quite independent of the visual content intentionally produced by the image designer. It is this second context, the context of intentional usage, in which scepticism concerning a fixed representational content of visualisations seems to be appropriate, as expressed by Nelson Goodman and in the same vein put forward by Laura Perini for the realm of scientific images. Both refer to the fact that "almost anything can denote or even represent almost anything else" (Goodman 1976, 89; Perini 2010, 136). For example, a scientist might use the Higgs image (see Figure 2.4) to refer to, say, the elegance of geometrical forms or, even more far-fetched, to the processes of political decision-making, just by declaring that the image represents those things. Setting that aside, what is implied by Hentschel's distinction for the analysis of scientific visualisations will be discussed in more detail when the cognitive content of visual representations is scrutinised in Chapter 4 of this book. For the time being, his finding corroborates the relevance of my approach to clarify the contexts in which visual representations are used in science. If meaning, ascribed to a particular visual representation changes due to context, knowing the different contexts seems to be a crucial precondition for correct understanding. Thus, how many contexts of image usage are there in science?

My proposal will be to follow Frankel and DePace's suggestion that we can discern *two* broader contexts in science, namely the exploratory and the explanatory context. The next section is meant to explain what these contexts are about and why this distinction can help circumvent problems which appear in the epistemological assessment of visual representations in science.

3.1.1 *Exploratory vs. explanatory use*

It has already been mentioned at the beginning of this chapter that the explanatory context relates to communicating research results and hypotheses, whereas the exploratory is about genuine research activities such as experiments and observations. A feature that visual representations have in common in both contexts is their *mediating role*, namely that they play a connecting role. But what are the *relata* to be connected? And are those related parts the same in both settings? To answer these questions, we have to examine the precise functional roles of visualisations, starting with the exploratory context.[3]

In this first setting, visual representations can come into play in two different ways. Firstly, visualisations might be the *object of research*. Of course, art history is a notable example in this context. However, it would be a mistake to think that it is only in the humanities where images play such a prominent role in research. There are many other disciplines in the social and natural sciences that rely no less on visual representations than some disciplines in the humanities. To illustrate this point, think of the paintings in the *Altamira cave* (see Figure 1.1).[4] Archaeologists, such as Steven J. Mithen (see Mithen 1998; 2007) or David Lewis-Williams (see Lewis-Williams 2004), examine these images to reveal our predecessors' cultural life, their ways of thinking and the constitution of their minds; they analyse the pictorial content of these paintings and make use of their results as evidence to support their theses about the cultural meaning of these pictures.

Furthermore, as the discussion of Adina L. Roskies's studies on the status of fMR images in neuroscience in the previous chapter shows, visual representations are parts of genuine research activities. Here, they are an object of research even if the imaging technologies are still not completely understood and thus the evidential status of the respective images is still in dispute. For example, Roskies points out that scientists still argue about what exactly fMR images show and whether they can be used to support certain hypotheses about processes in the brain.[5]

Prominently, visual representations are also the object of research in pure sciences such as mathematics, where they are, for instance, analysed in the context of knot theory (see e.g. Sazdanovic forthcoming). A second setting comprises images as tools for thinking.[6] Here, they might play either the role of auxiliary means or present procedural steps of reasoning itself. Visual representations appear, for example, in developmental processes of the design and invention of models and theories. Sketches and blueprints belong to this group. Most prominently, they appear in different disciplines of *applied sciences*. Architecture and engineering sciences offer rich sources of examples here: design plans, for instance. They come both as hand-made images and as technically produced ones. Technical drawing, for example, is still taught as a freehand skill in these domains, but also as a technically supported activity such as in the case of computer-aided design (CAD).[7]

A common feature of those images is their *direction of fit*, which is different to that of pictures that resemble certain phenomena (see Scholz 2010, 45ff.). Contrary to the latter, they do not depict pre-existing entities. They serve rather as models to bring into existence what they depict: for example machines or buildings – although it is not guaranteed that the latter will be produced using those blueprints. Throughout the design process, it is then the image that the designer is working on, probing different features of the machine to be built, calculating relations and components and manipulating others to infer the effects on the whole system. However, it is not only in the applied sciences that sketches play a role. Quite a few scientists use visualisations as tools of thinking, that is, as means to elaborate on hypotheses created either as explanations or predictions of particular phenomena. A striking example of such *visual thinking* is presented by Edward Tufte (see Tufte 2010, 27ff.).

Tufte describes Dr John Snow's discovery of the transmission route of the cholera disease in 1854. He points out that this finding was to a significant degree enabled by Snow's working with graphical displays, namely maps (see Figure 3.1), of epidemic victims and their environment.

[...] Snow had a good idea – a causal theory about how the disease spread – that guided the gathering and assessment of evidence. This theory developed from medical analysis and empirical observation; by mapping earlier epidemics, Snow detected a link between different water supplies and varying rates of cholera [...]

(Tufte 2010, 29)

Figure 3.1 shows Snow's final map that allowed him the relevant inference to explain the spreading of the disease in London. As Tufte points out, this

Figure 3.1 Cholera map: a variant of the original cholera map drawn by Dr John Snow (1813–1858); original map published by C. F. Cheffins, Lith, Southhampton Buildings, London, England, 1854, in: Snow, John: *On the Mode of Communication of Cholera*, 2nd ed. John Churchill, New Burlington Street, London, England, 1855.

Source: https://commons.wikimedia.org/wiki/File:Snow-cholera-map.jpg

is but one of the visual displays of data that Snow produced to enhance his understanding of the assumed correlations. Finally, these visualisations let Snow literally see the relevant correlation between the apparently polluted water supply of the Broad Street well and the number of disease victims in the vicinity of the pump. More precisely, visually mapping the data facilitated not only the inference of the relevant correlation, but also the exclusion of rival explanations. In this sense, visual representations served to elaborate his theory on the disease's transmission route. Manipulating and further enhancing those maps revealed the correlation that he was looking for. Thus, working with the map – the visual representation – examining it and asking questions about it, supported solving the mystery of the source of the disease to a significant degree. Scribbling incipient ideas is often an important part of exploratory practices in science. Some of these visual working devices even attain the status of canonical depictions as in the case of the famous Feynman diagrams in quantum electrodynamics.

Henk W. de Regt explains this example of visual thinking and cites Richard Feynman in order to emphasise the importance of the visual mode of representation in elaborating his ideas of how elementary particles interact (see de Regt 2014, 392). In de Regt's analysis, it becomes clear that, for Feynman, drawing those diagrams was the only way to develop his ideas further, as neither words nor mathematics allowed him to express his theory adequately. It was these visualisations that were inspected and manipulated in order to learn more about elementary particles in physics. These diagrams, namely the initial form for theorising the correlated processes, later became standard in theoretical physics and, today, they are taught in university courses (see ibid., 391).

Whereas it is still disputed whether the Feynman diagrams represent real elementary processes,[8] Snow's map of the transmission routes of the cholera epidemic clearly represents particular observational results, namely death rates and locations of water pumps within the city. Although it seems right to categorise this visualisation as reasoning with images, by displaying observational results, it nonetheless also makes an appeal to a third way in which visualisations come into play in the exploratory context.

In addition to images that are scrutinised *de re* either as direct objects of research or as tools of thinking, visualisations are, thirdly, often the *output of measurement or observation processes*. What is of relevance to the scientific investigation is not the image *de re* but what it represents and, of course, whether the image represents correctly what it is supposed to show. Here, visualisations are regarded as a *surrogate* for the genuine research object.

The difference between using images as objects of research and surrogates for such objects can be illustrated nicely by means of the Altamira cave and the Higgs boson. Scientists analysing the Altamira cave paintings are not interested in the animals depicted, but in the cultural practice of cave painting. Thus, they do not want to find out something about aurochs (although this might be an unintended side-effect), but about the ways in which the paintings were produced by our predecessors and why. Scientists analysing the LHC images, on the other hand, are interested in detecting results of their particle accelerator, not in the resulting graphic. For them it is, for example, a completely arbitrary choice which colours and forms are used in these graphics as long as their usage is consistently related to certain particles or their properties in the tube.

To better understand what is meant by the thesis that visual representations can serve as surrogates for genuine research objects, another subdivision must be taken into account here. On the one hand, measurement data might be visualised. Consequently, there is a moment of choice inherent to this situation, in the sense of deciding whether to take a direct look at the (numerical) data, or to plot them as diagrams or graphs, or to convert them into graphics.

On the other hand, there are instances where no converting steps are involved. Thus direct visual representations are the output of the measurement device or instrument. Photofinishes in sports or X-ray images used for diagnostic purposes in medicine may illustrate this point. Patrick Maynard elaborately shows how, for example, cameras are used as detecting devices in this sense in science (see Maynard 2017). An example where photographs are used for the purpose of detection is the so-called *blink comparator* in astronomy. Here, two (or more) photographs of the same region in space taken at different time slots are compared in order to detect, for instance, asteroids or variable stars. By comparing them at short intervals (that is, by blinking at them), moving objects in front of the fixed stars or changes in brightness of particular celestial objects become apparent.[9] This example illustrates a case where the data being compared are the respective photographs and not visualisations of previously registered data.

What complicates matters is that many images that scientists deal with today are digital ones: computer-generated images.[10] The use of analogue images seems to be more an exception than the rule these days. It might be objected that, as digital images are *almost always* a visualisation of numerical values, the latter are of a more basic nature than their visualisations. Consequently, the increasing use of digitalisation seems to constantly blur the above distinction between *data visualisation* and *visualisation as data*.

As a rejoinder, I want to highlight two points. Firstly, despite the abundance of digital images, analogue images are still available as well. Moreover, it has to be borne in mind that digitalisations of analogue images are only copies of copies, so to speak. An important consequence of this in the epistemic domain is that, if information has already been lost in the first place – for instance by taking a picture (for example, in black and white, to take a very salient case) of the research object – digitalising such an image can only be a reproduction of the first image: an image of the image, so to speak. The digital image cannot provide information (for example colour) about the research object that the first image did not contain. In this sense, it is still the former visualisation that plays the informative role, not the newly digitalised image. The latter is just a new way of encoding and preserving the information engraved in the first visual representation. Hence, scanned analogue images do not tip the scales in support of a new majority of digital images. Thus, although more and more images might be digital ones today, our initial distinction between data visualisation and visual representations as an output of measurement processes still makes sense.

I will only hint here at the second and maybe more significant rejoinder in this context and will return to it in more detail in section 4.2 of this book. The above objection not only questions whether images can be data or whether all data is numerical in kind which can be visualised afterwards, but also seems to suggest that numerical representations are *in some sense more basic in kind* than visual ones. Thus, it has to be explained what 'more basic' means in this context. Apparently, the basicness expressed in the above objection can be either *ontological* or *epistemological* in kind. As digital images are produced by algorithms interpreting binary values,[11] it has to be admitted that, in these instances, numerical representations are ontologically more basic than visual ones. But what does this disclosure imply for my epistemological analysis of scientific visualisations? Does ontological basicness imply an epistemological one as well?

I will defend the thesis that this is not the case. As a brief anticipation of the discussion to come, let me recapitulate what the exploratory context in science is about. It deals with extracting information, with discoveries and learning about the world – about phenomena, their properties, their behaviour and their correlations, etc. Visualisations belong to the set of phenomena being investigated, serve as tools for thinking, or are used as surrogates for the object of research in this context. The whole setting is about gaining and justifying knowledge (claims). Thus, the interesting question is not so much about ontology, but about the epistemological role of the different kinds of representation. Can numerical data yield the information that scientists are looking for in their analyses?

The answer to this question is mixed. Obviously, there are many instances where numerical data – as measurement results – are essential to the scientists' work. However, as Alan F. Chalmers has pointed out (see Chalmers 1999, ch. 1), it is *relevant* data that scientists are in need of. Thus, they are challenged to discern between relevant and irrelevant numerical data that they obtain via their experiments. They have to sort out measurement errors and find the correlations they are interested in. All of these activities presuppose the possibility of *cognitive access* to the data given. That is, human observers have to analyse and understand the data. Yet in many instances, the raw data available from experiments and observations would serve neither of the above-mentioned purposes. This difficulty becomes especially apparent with regard to today's "big data sciences"; that is, scientific experiments and observations yielding huge amounts of data such as the measurements at the LHC in Geneva. In such contexts, information technology is an inevitable part of the process of data analysis. Furthermore, it comes as no surprise that the output of such programmes are visual representations of the data. It is only in this format that significant patterns or deviations produced by, for example, measurement errors become apparent to the human observer. This epistemic feature of scientific visualisations is also highlighted by Frankel and DePace: "[e]xploratory graphics [...] invite the viewer to discover information. Many scientific disciplines generate enormous [data sets]. New graphical approaches are required to make sense of the data and to organize and communicate the main points" (Frankel and DePace 2012, 13).

John V. Kulvicki picks up this point and elaborates on it (see Kulvicki 2010a). He shows how the divergent ways of data representation facilitate different epistemic functions in science. In this context, he explicitly discusses the differences between numerical and visual representations. Amongst others, he highlights the fact that lists, numerals, but also descriptions, have to be decoded first by the observer before further, more abstract questions – concerning, for example, correlations amongst the data – can be asked. The role of images, on the other hand, he sees in accordance with what Frankel and DePace point out. Kulvicki claims that:

> [b]y and large, however, images have the advantage of presenting much information across many levels of abstraction in an immediate manner. Images and graphs are tools for discovery and diagnosis, interestingly enough, because they present a wealth of information in such a way as to allow us to ignore what simply does not matter.
>
> (Kulvicki 2010a, 307)

That images provide cognitive access to the information presented in the various methods of data representation serves as an explanation of these distinct epistemic features in Kulvicki's analysis.

Considering instances of big data sciences supports my thesis that, although data visualisations might be ontologically inferior to numerical representations, they can be epistemically superior in comparison to them. Otherwise, immediate cognitive access for human observers would in most instances be denied. The fact that mankind's approach to the secrets of nature is heavily determined by the evolutionary development of our senses, and thus by privileging vision, is highlighted by cognitive scientists.[12] De Regt emphasises this point by stating that:

> [a]s seeing is for humans arguably the most important way of grasping the world around us, it is not surprising that when we want to extend our grasp beyond what we directly observe, we prefer to rely on our well-developed visual skills and employ visualization as a tool for understanding.
>
> (de Regt 2014, 394)

Bearing this in mind, let us now turn to the second context of usage of scientific images, namely the *explanatory one*. Here, the most relevant point to remember is that visual representations are always parts of communicative acts. In the last chapter, the point has already been mentioned that analysing visual representations in complete isolation from their embedding in communicative acts would deprive the recipient's understanding of some important dimensions of their meaning. Søren Kjørup's study on visual representations especially has shown the necessity to take the broader communicative context into account for a proper understanding of the visualisation in play.

Yet communication itself is a more diverse activity than a first glance might suggest. Thus, it makes sense to unravel the particular roles that visualisations play in this environment. Taking into account insights from speech act theory makes this apparent. Philosophers of language have pointed out that communication serves various purposes, i.e. linguistic expressions are used for different speech acts (see e.g. Austin 1998; Searle 1983). Communication taken in this way is not only about transmitting certain propositions, but also about fulfilling different *illocutionary acts* by the use of words. For instance, we can use linguistic expressions to warn or to ask someone about something, to explain or to declare something to someone, etc. Taking visualisations as proper components of communicative acts implies the assumption that they can yield at least similar contributions to perform the intended act. Kjørup highlights this aspect of visual communication. He mentions, for instance, the possibility of warning or advertising with the aid of visual representations in communicative acts (see Kjørup 1989, 312). In this sense, visualisations can make contributions to illocutionary acts within the broader context of communication.

This basic idea of speech act theory, namely that usually one act of communication comprises different parts – propositional, illocutionary, locutionary and perlocutionary acts (see Searle 1983, 40ff.) – also explains Scholz's suggestion of "image games" (see Scholz 2011). Obviously, visual representations can be used to express much more than is genuinely encoded in their cognitive content.

Frankel and DePace describe the parts played by visual representations in communicative acts in the following way: "[e]xplanatory graphics are used to communicate a point or call attention to patterns and concepts. Explanatory graphics can be used as evidence or proof in research, and can be teaching tools for colleagues and students" (Frankel and DePace 2012, 13). What becomes apparent in this quotation is the diversity of roles played by visual representations in the explanatory context. Enhancing our understanding of contributions made by visualisations requires us to be more precise about the communicative contexts in which they appear.

From my point of view, there are several aspects that are relevant to this analysis. One point already appears in Frankel and DePace's quotation. Although the main purpose is communication in all the different settings, the people addressed by the communicative acts vary. Thus, there are different *target audiences* that scientists, using visual representations in communicative settings, have to take into account when designing their information supplies. A rough distinction can be made at this point between laymen and scientific peers. However, as will become immediately clear, this differentiation has to be more fine-grained.

The point is that both categories are of a rather gradual matter. Laymen may be students as well as funding-programme officers. The difference between both is that the former are supposed to be future experts in the respective domain.[13] The latter, on the other hand, are not expected to develop their knowledge about the correlated epistemic domain in any sense. They will remain what they are – laymen.

Less obvious, though equally true, is the case for academic peers. Here, too, there are people whose professional domain is related much more closely to that of the communicating scientists than the domains of others might be. Nonetheless, all of them might share certain academic standards, for example, how to quote correctly or how to record their results in a lab log. Moreover, this more fine-grained description of expertise reveals that there are instances where people are both scientific peers and laymen. A physicist and a sociologist might share certain academic virtues and follow common standards in assessing scientific competence and behaviour or progress – in this sense, they can be regarded as peers. On the other hand, there is usually no overlap between their domains of research – neither with regard to their theoretical interests nor with respect to their scientific practices. They will remain laymen with respect to each other's scientific domain.

Moreover, the dimension of time has also to be taken into account. Scientific communication not only aims at the distribution of information but also at its preservation, that is, recording and storing in diverse media. All of the above-mentioned distinctions are located on a synchronic level. However, there is the diachronic level as well, which can turn today's experts into the laymen of (past pre-) scientific (or future) domains. Consequently, there might appear problems of understanding with regard to particular scientific concepts or theses as well as with respect to scientific illustrations.[14] Brian S. Baigrie, for example, discusses why succeeding scientists omitted the illustrations originally published in René Descartes's book, the Latin edition of *Principia Philosophiae* (see Baigrie 1996a).

Baigrie thinks that what explains this omission is that Descartes's successors in science and philosophy no longer shared his world view as the background to their research:

> they [the illustrations, N.M.] vanished because historians and philosophers who have been trained to regard science in the form that Newton gave to it could no longer see their point in science. As we have lost touch with Descartes's particular conception of mechanism and the take on imaging that goes hand in glove with his conviction that nature is to be grasped through artifice of craft, we can no longer see the relevance of Descartes's pictures to his natural philosophy.
>
> (Baigrie 1996a, 129)

Descartes's successors did not understand the reason for publishing the illustrations together with the text. They were no longer able to interpret the complete communicative act and, therefore, thought that no information would be lost if they removed the images from the text in later editions.

Moreover, Frankel and DePace point out in their design guide to scientific visualisations that addressing different target audiences means taking into account their different background knowledge of the entity depicted (see Frankel and DePace 2012, 18). That is, depending on the intended audience scientists might have to simplify their images or add didactic elements such as labels and arrows. Thus, they will literally *design* their graphics to enhance their audience's later interpretative processes. They present a wealth of case studies that are supposed to demonstrate the relevance of the question 'who is your audience?' as one of the main aspects to consider when choosing or producing visual representations (see ibid., ch. 2, 3).

However, the diversity of target audiences alone does not suffice to explain the different roles played by visualisations. What is also of relevance is the purpose or *aim of communication*. Here we come back to the initial ideas taken from speech act theory. More than one example can easily be imagined to illustrate the point that, although the target audience stays the same – laymen for instance – the aim (that is, the scientist's intention) of communication might vary enormously. The illocutionary acts performed will be different ones. Visualisations included in those contexts and contributing to these divergent acts will fulfil different roles. The Higgs image provides a concrete example (see Figure 2.4). The same image can be used either for teaching students some essential characteristics about events of particle decays yielding Higgs bosons or for merely illustrative purposes, for example, by making an article in a popular science magazine visually more attractive. As will be seen below, these two different aspects – namely, target audience and communicative aim – that account for why visual representations can play so many different roles in the explanatory context, are sophisticatedly explained by Ludwik Fleck's theory of scientific communication (see Fleck 1979; 1986c).

However, before going into the details, an answer to the question posed at the beginning of this section should first be supplied. Visualisations play a *mediating role* in the two indicated contexts of scientific practice. But do they also connect the same *relata*? The analysis above shows that this is not the case: the *relata* that are linked in the two contexts are different. In the exploratory context, where visual representations are means of thinking, the objects of research, or function as their surrogates, they make the latter epistemically accessible to the investigating scientist. The visual representation connects the object of research with the scientist's cognitive capacities to reason about it. In the explanatory context, what is related by visual representations are on the one side the target audience that can differ with respect to its level of background knowledge. However, it is only rarely the case that the object of research is the second *relatum*. Scientists are well advised to take into account who will later have to interpret their images. Thus, they (either themselves or with the aid of public relation officers) will *design* what they want to communicate. In this sense, the second part of the relation is the scientist's intended interpretation of the research object.

In the next section, I will defend the thesis that this difference of *relata* in the two indicated contexts plays a crucial role when analysing the epistemic roles of scientific images.

3.1.2 Context-related problems

There are at least three different kinds of problems to consider with respect to the assumed epistemic roles played by visual representations in science:

1 the visual mode of representation, that is, the apparent lack of a propositionally structured content, which does not fit to the demands of verbal arguments;
2 the problem of theory-ladenness, which is a consequence of visualising otherwise unobservable entities and making use of computer algorithms in doing so;
3 the possibility of image manipulation, which is a result of a steady increase in digital imaging in science.

In the following, these three problems will be discussed in more detail.

(1) We have already mentioned the worry highlighted by Perini that visualisations simply belong to a mode of representation which is considered inappropriate to making proper contributions to scientific arguments. Scientific knowledge is distributed and increased by the use of arguments, and the latter are constituted by linguistic expressions. Traditional ways of testing the validity and soundness of arguments depend on characteristics that are traditionally associated with verbal expressions only – such as bearing truth values. In this sense, visualisations simply belong to the wrong mode of representation, apparently incapable of revealing features essential to argumentative contexts. Understood in this way, visual representations can never make proper epistemic contributions to scientific discourse. This difficulty and suggestions concerning its solution will be discussed in Chapter 4 of this book.

The second concern (2) is related to the assumed *aboutness* of scientific visualisations. Anjan Chakravartty argues in an even wider sense that "[...] the least controversial feature of scientific representation is the idea of intentionality: a representation is something that is about something else, and it cannot be a representation unless there is something that is represents" (Chakravartty 2010, 206). Emphasising that scientists use visual representations as surrogates for their objects of research presupposes that the respective image is connected to the object in such a way that the former is about the latter. Visualisations should be informative about their source. They should be able to answer questions that scientists ask about the latter. This can easily be achieved if visualisations share certain properties with their object of depiction or allow scientists to infer them indirectly.

However, as pointed out in section 2.1.5 of this book, there are two major sources of concern which foster scepticism about such a relevant connection between object and image. On the one hand, in many instances the objects of research belong to the set of unobservable entities, for example, elementary particles such as the Higgs boson. Thus, we need instruments such as particle accelerators and detectors to make them visible in the first place. On the other hand, information-technology devices used in many of these settings seemingly remove data visualisations even further from their assumed source, as Martina Heßler points out (see Heßler 2006, sect. 10). The fact that quite a few visualisations are based on algorithms might add to the problem of understanding and interpreting correctly the causal connections between image and instrument as well as between instrument and object.

As a consequence of these technological prerequisites in many scientific observations and experiments, gaining epistemic access to the objects of research depends on a variety of theories. Theories of the domain of research play a role, as well as theories about the instruments in use. And, as Roskies pointed out, the more theories are involved, the more difficult the interpretation of the resulting image becomes.

The problem that is addressed here is known as the *theory-ladenness of observation*. Martin Carrier and Peter Kosso make us aware of the fact that this is a multifaceted problem (see Carrier 1994; Kosso 1993). Theoretical assumptions enter at the level of *observation*: theories tell us where to look for our object of research. At the level of *observation statements*: theories yield the vocabulary to describe our findings. And at the level of *measurement processes*: theories tell us how to evaluate and interpret our data. Consequently, observational data become questionable in their justificatory role and the same goes for data visualisations as the common output of scientific experiments. Kosso states that "[t]o function as evidence in science, the act of observation must offer reason to think that we believe truly what we see" (Kosso 1993, 119). But how can we assume that we can rely on observational processes if they are so highly dependent on previously held theoretical assumptions? How can the fatal conclusion be precluded that scientists simply create the phenomena they think they observe?

Put in a nutshell, then, the problems derived from these considerations are the following: as visualisations are highly theory-dependent, their status as evidence of observations can be questioned. At least, we cannot simply accept them as neutral, objective evidence of assumed facts of the world. Furthermore, if we are not able to show that, despite their theory-dependence, visualisations can serve as evidence, scientists are not allowed to use them as surrogates for research objects. This is because the relevant aboutness of visualisations can no longer be presupposed; it has to be shown first. And if we cannot make a convincing case for the relevant connection holding between a real object and its visual representation, we would have to admit that the latter is not suitable as a means of telling scientists something about a mind- (or theory-) independent world. Relying on them would thus amount to mere self-deception. Before discussing a solution to this problem, which will be presented in section 3.1.3, the related third problem should be examined.

The last problem (3) can be reinforced, as it might be suggested that such deception can also be intentionally pursued. That is, some might assume that scientists use the possibilities of data visualisation to intentionally manipulate their results. This last point is motivated by the same two particularities of visualisations as observation results, namely making the *unobservable* visible with the aid of *information-technology devices*. The latter especially fuels the debate, as digital images seem to allow any kind of easy manipulation. Did not Frankel and DePace suggest that scientists should design their visual results in accordance with their target audience's cognitive capabilities? And if so, who will guarantee that such interventions are revealed and explained in order to let others reason about their utility? Problems of this kind are broadly discussed with regard to digital photography in aesthetics (see e.g. Mitchell 1994; Ritchin 2010; Walden 2008). Here, many theorists point to the negative consequences that the invention and distribution of digital technology may have. Barbara Savedoff, for instance, emphasises that, in the long run, understanding the mechanisms of digital picture production – and its referential vagueness implied by the ease of its manipulation – will destroy our confidence in photography in general (see Savedoff 2008, 134ff.).

> Digital manipulation is relatively fast and easy for anyone with the appropriate software [...] As a result, not only are we finding ourselves surrounded more and more by images that have been altered in some way, but it is also becoming impossible to tell which images are straight and which have been altered. [...] In a world where digital manipulation – digital collage – has become the norm, we may simply come to assume that a photograph has been altered if it is at all challenging to read it as straight.
>
> (Savedoff 2008, 136f.)

Hence, Savedoff thinks that, in the long run, we will lose our trust in photographs as representations of real entities in general because digital technology will diminish the epistemic status of photography as evidence.

It is not only photographs whose evidential status is questioned in this way, but also any other kind of data visualisation that allows quick and easy intentional interventions, that is, altering their appearances with the aid of software devices such as *Photoshop*™. "When images are 'the' experimental result, photo editing opens up the distinct possibility of overmanipulation and misrepresentation" (Ottino 2003, 474), as Julio M. Ottino puts it.

A loss of reference will normally be unintended in cases of theory-ladenness. Scientists simply might not be aware of the missing reference as they are blinded by their own theory, so to speak. In the case of digital images, however, it might be assumed that at least some researchers consciously undermine the referential relation by manipulating their results intention-ally. Their results might be pure constructs – solid fakes. So, how could these findings be reconciled with the widespread use of visualisations in science? Is it still possible to ascribe a significant epistemic role to visualisations in science? It is this question that will be investigated further in this chapter, and I think that a positive answer will be found for the following reasons.

Worries about image manipulation seem to spread into the debate about scientific visualisation as a consequence of a similar discussion within media studies. In the latter field the question is also about the evidential status of photography and how it might be undermined by the possibilities of post-processing, especially of digital images.[15] As our above analysis has shown, there are indeed instances of the intentional manipulation of scientific visualisations. Design guides explicitly ask scientists to take care of the visual appearances of their images, to reflect on their audience's capabilities of understanding and on the purpose of using particular images. Hence, they ask for intentional inter-vention regarding the (visual) method of information transmission – which can be regarded as manipulation in the broader sense, namely as adaptations to the cognitive setting of presentation, although without an intention to deceive. That is, we *are* confronted with the fact of photoshopping on the sub-ject of scientific visualisations, too. Considering this fact is of relevance in the explanatory context, in which scientists deal with the question of how to pre-sent their data best to different target audiences. Yet what apparently happens in the debate about scientific images is that the same problems are transferred to the exploratory context. That is, no distinction is made between these two contexts of usage, although, as seen above, there are significant differences between both when pursuing an epistemological analysis.

In this sense, I think that problems involving the images' constructivist character or, even worse, involving intentional fraud, which are motivated by considerations concerning the possibilities of intentional alterations to images, are somewhat exaggerated in the debate at hand.

A more likely explanation for this development in the philosophical debate can be found in that philosophers predominantly analyse visualisations presented in the explanatory context and not those in the exploratory setting. They analyse what is presented on websites and in journal articles, what they see in science documentaries and in books. This is because philosophers rarely have access to laboratories and raw data. They are normally not engaged in empirical studies such as ethnological projects in the more sociologically orientated science studies. Thus, it is not surprising that philosophical results are more or less focused on only one part of the phenomenon.

Yet this at first glance innocuous close attention of interest results in the epistemically problematic confusing of those two contexts: the explanatory and the exploratory; it is a hotchpotch that draws its sceptical potential from an only partially appropriate analogy to discussions of image manipulations in media studies. From an epistemological point of view, the interest lies in clarifying why one can rely on visual representations in science, why they can play an evidential role. I think that there are good reasons to rely on visualisations and that sound criteria to assess the epistemic virtues and vices of particular visual representations in science do exist. These reasons, how-ever, are closely connected to the contexts of usage, as I will show below. Thus, confusing the two settings causes people to lose sight of the distinctive methods of justification connected with these two contexts.[16] The last two problems expressed above, namely concerning theory-ladenness and an assumed general constructivist character of visualisations in science, can be alleviated by keeping those two contexts apart. More details about how to deal with theory-ladenness will be discussed in the next section. Moreover, paying close attention to the contexts involved will aid an understanding of what really constitutes the reliance on visual information in science. It will transpire that the reasons why we rely on their evidential status are not only and exclusively based on representational criteria, but are often also derived from the social setting of science.

What reasons, then, allow scientists to rely on information presented visu-ally in these two contexts? A start can be made by answering this question with regard to the exploratory context. Why are scientists justified in regarding visual representations, that is, the output of their measurements, as surrogates for their research objects? What reasons might justify the relevant assumption that they can learn something about the latter by investigating the former?

3.1.3 Reasons (1) – causality and informativeness

Two questions seem to be of importance in the exploratory context: how does the instrument bring about the respective image? In what sense can this image be regarded as evidence for the hypothesis under consideration? These two aspects have already appeared in the discussion of Roskies's analysis of fMR images in neuroscience (see Roskies 2007; 2008). These images play a crucial role, on the one hand for diagnostic purposes and, on the other, as evidence for theoretical reasoning about the human mind. Roskies points out that professionals also seem to lack a robust understanding of their data in two different ways: they lack relevant background knowledge both with regard to the "causal stream", namely how the instrument works, and to the "functional stream", namely what the resulting image can tell them about the human mind.

What makes it so difficult for professional users is the *correct interpretation* of the image at hand. Perini highlights this difficulty when she states that "[t]he problem with MPIs [mechanically produced images, N.M.] stems from the fact that they could be interpreted in a variety of ways" (Perini 2012b, 166). The reason why there are apparently different ways of interpretation possibly lies in the fact that scientists have to have knowledge about the two domains indicated. That is, they need to know what their instrument is sensitive to, how it works causally. In addition to that, they need to know how to bridge the gap between image, that is, the measured datum and theory. The question is in what sense the image plays an evidential role with respect to the theory at hand.

These are the two domains that foster insecurities in epistemically dealing with visualisations in science. As a corollary to this, these are also the two domains where reasons are needed to justify the practice of inferring information about the research object from its visual representation. To begin with the "causal stream", as Roskies calls it, what might be the reasons in this domain to justify knowledge claims based on visualisations?

Obviously, regarding visualisations as surrogates for research objects only makes sense if scientists are somehow convinced that the former can yield knowledge *about* the latter, that is, that the image can tell them something about relevant aspects of the object depicted. Anjan Chakravartty puts this "aboutness" in the following way: "[...] successful representation contains information regarding the thing it represents" (Chakravartty 2007, 184). However, considering the example of the Higgs boson (see Figure 2.4) highlights the fact that, in many instances, measurements and thus visual data are only indirectly related to the object of research. The Higgs is not seen, but rather its decay into two electrons and two muons depicted on the corresponding image. In what sense can it thus be stated that this image is *about* the particle that scientists at CERN are searching for? If the concept of representation is to be adhered to in this context, the proposed *aboutness* of visualisations has to be qualified in such a way that instances referring to indirect measurement processes can be convincingly accommodated. Accordingly, Chakravartty refines his initial claim that a representation of x contains information about the represented object x in the following way: "sometimes this information is rather minimal: in the limit, it may be exhausted by the fact that the subject of representation exists [...]" (ibid.). With regard to the example above this means that – by using theoretical background assumptions – the visualisation in question can be interpreted as evidence of the apparent existence of the Higgs boson.

However, as Kosso points out, there is still a difference between *observation* and *evidence*, namely:

> the necessary role of inference in the latter [case N.M.] [...] In the best circumstances, this inference is supported by a good understanding of the link between what is seen and the interesting thing that is not seen, between effect and cause. [...] This understanding is theoretical in this sense of being not amenable to a direct observational check. That does not make it bad or suspect, but it does point out that evidence is *a bit riskier* than observation itself.
>
> (Kosso 2010, 217f., my italics)

This makes plain why the evidential status of scientific visualisations might be called into question. Obviously, there are numerous inferential steps involved in claiming the existence of the Higgs boson by analysing visual data apparently showing the traces of two electrons and two muons. Supporting a claim in this way seems to be "riskier" than supporting a claim by direct observational data. The likelihood of errors or defective reasoning might be higher in the former than in the latter case.

Thus, two aspects are of importance to enable the relevant inference. (1) The scientist has to understand how datum and theory are related – why the radiation patterns of two electrons and two muons can be an indicator of a Higgs boson – but also (2) how her instrument – the particle accelerator and the CMS detector – works; how it brings about the image that we are analysing, and that gives rise to the former question. This second aspect is what Roskies investigates as the "causal stream" with regard to fMR images. It is concerned with the causal connection between entity and measurement device and with the instrument's reliability in indicating this connection.

If such a causal connection holds, it will support the following counterfactual reasoning: *if the entity under investigation had shown different characteristics, the measurement results would also have been different.* Thus, in order to regard a visualisation as a surrogate for the measured entity, it has to be presupposed that it has been produced by a causal connection between the object and a reliable instrument. However, both aspects – the causal connection and the instrument's reliability – might be called into question. In this context, Roskies claims that not only laymen in general but also professional neuroscientists often lack a relevant understanding of the causal processes with regard to fMR images (see Roskies 2007, 871).

Yet Roskies not only indicates a particular difficulty in interpreting and understanding scientific visualisations. She also, albeit inadvertently, indicates a solution to the problem. Comparing fMR images with photographs, Roskies states that people have no difficulty in comprehending how photographs are produced due to "our familiarity with photography and with the use of photographic images in science" (Roskies 2008, 21).[17] However, as Kelley E. Wilder rightly shows in her elaborate analysis of photography in science, this familiarity is the result of a long-term process (see Wilder 2009, 78). When photography was first introduced into the scientific context, it was confronted with the same sort of difficulty that Roskies highlights in the case of brain scans. Its evidential status was far from clear. Nonetheless, as Roskies's thesis suggests, scientists were able to overcome this difficulty so that nowadays the technology of photography can be called a *familiar* one.

Wilder explains this development by pointing out that experiments making use of photography in science were constantly accompanied by experiments to find out more about the photographic process (see Wilder 2009, ch. 2), most of all about the characteristics and particularities of emulsions used on photographic plates (see ibid., 52). She speaks about a "symbiotic relationship of photography and science in the twentieth century" (ibid., 78). This interrelationship she spells out in the following way:

[e]ach small step in building an experiment involving photography required an equally small step in forming and moulding the photographic emulsion. The more controlled and predictable the emulsions became, the more reliable their evidence. As photographic evidence gained reliability in one photographic method, it was more likely to be used by another scientist in another method.

(Wilder 2009)

Consequently, Wilder shows that it took quite a long time before the photographic method became a reliable detective device to measure and record certain processes and entities, and it took no less time for scientists to understand this method and to use it correctly and successfully.[18] Thus, it is only today that we are used to this technology – or "family of technologies" as Maynard sophisticatedly describes the nature of photography (Maynard 2000, 3) – and regard its mechanism of picture production as familiar.

Yet Wilder's analysis not only makes plain that today's familiar technologies were epistemically challenging in the past, but also makes clear that there are methods to establish the reliability of an instrument or technology to show and record the results of a causal connection between instrument and object. Moreover, now that the initial process of understanding the instrument, the camera, has been completed, findings of these experiments can be used to *calibrate* the instrument – that is, to ensure its reliability. Calibration means repeating experiments or measurements with known results. This allows the testing of the instrument's working mechanisms. If the instrument, e.g. the camera, fails to reproduce the known result, we then have a good indicator of its unreliability as a measuring device.[19]

By pointing to the development of the technology of photography, Roskies also suggests a solution to the problem of interpretation in this realm. Once the relevant causal connection is sufficiently understood, the calibration of the instrument will help to guarantee the relevant relation and, consequently, the accuracy of the resulting image. Furthermore, this shows that the interpretation of images produced by technological devices can become common practice – not only among scientific experts but also among laypeople.

Finally, if a reliable causal connection is available, as Wilder shows with respect to photography, the transmission of information that was required as a presupposition to using an image as the surrogate for the object is possible. In this sense, Kosso makes plain why the detection of unobservables is possible in such a case, when he states that:

[...] a causal account [,] has the existential claim built in in the sense that if you accept the fact that there is information about x which comes from x, then you also get the fact of the existence of the source of information.

(Kosso 1988, 464)

Only one final ingredient is missing, then, to support the claim that visual representations can serve as evidence in this sense.

Both Kosso and Roskies highlight the fact that what makes things so difficult is the amount of interpretation correlated with the diversity of theories involved in the act of reasoning from the image to its effect in nature. But even though the theoretical reasoning cannot be eliminated – theory-ladenness is unavoidable – it can, however, be ensured that it does not result in a vicious circle. Kosso's insistence on "independent evidence" is the key element needed here.

This means that the evidence has to be independent in the sense that the theory of constructing and understanding the instrument does not converge with the theory to be tested by the data yielded by this device (see Kosso 1988, 456f., 463ff.). This is Kosso's fourth dimension of observability, which he calls "independence of interpretation". By fulfilling this requirement, problems concerning the epistemic significance of the evidence are avoided. Escape from one of the three dimensions of theory-ladenness can at least be achieved, namely the one related to the instrument – "measuremental theory-ladenness" as Carrier calls it (Carrier 1994, 9). If the theory about how the instrument causally works is identical to the theory that is supposed to be tested by observations made with this instrument, one might assume that scientists are merely producing artefacts, that is, instances of wishful thinking, to confirm their hypotheses.[20] If the two (or more) theories do not converge, such problems are undermined because it can then be assumed that there is a real entity out there in the world that has caused its visualisation.

Although this is good news for the scientific realist defending the *ontological thesis* that there is a mind-independent world, there is still the question of whether there can be *epistemic access* to this world – that is, the second part usually connected to the topic of scientific realism.[21] Is it possible to know more about the unobservable part of the world than merely claiming its existence? This brings us to the second interpretive project mentioned above – Roskies's "functional stream".

A closer look at what has been discussed so far, namely the scientist's background knowledge about the causal process and the reliability of her measuring device, makes plain that she is in need of further criteria that support her interpretation of the visual data to justify knowledge claims about unobservables such as the Higgs. The scientist does not only want to know *that* there was something that has been detected by her instrument, she also wants to know *what*. Hence, she has to know how to bridge the gap between theory and datum, that is, how her measurement results are correlated with the theory in question.

Perini draws our attention to this very question (see Perini 2012b, 154) and suggests that what is needed are certain "interpretative 'rules' [...] that are at least partly conventional" (see ibid., 163). This emphasis on conventions in her characterisation of those rules is due to her agreement with other philosophers – especially Goodman[22] – that visualisations can often be used to represent a variety of things and, therefore, can also be interpreted differently (see ibid., 166). Thus, to avoid complete arbitrariness and, correlated with this, misunderstandings of visual signs, conventions are used to fix the meaning of pictorial symbols.[23]

Although following Goodman thus far, Perini thinks that his account is not totally suitable to explain the interpretation of scientific visualisations. She adds that those conventions are not completely a matter of choice in this context. Here, she refers to the same distinction between information encoding and decoding that was introduced in discussing Figure 2.14. From her point of view, the fact that MPIs (mechanically produced images) could be interpreted in arbitrary ways, that is, the decoding of the image, is restricted by the way those images are encoded. The *usage* of such images might be guided by conventions and thus reveal the arbitrariness that seems to spark off from Goodman's account, but taking into account the way the image was produced in the first place will set certain limits to such interpretive endeavours. Scientists are interested in the specimen the visual representation is about and it can thus be assumed that they will pay attention to the connection established during the encoding of the image (whereas others might not). Perini writes:

> [r]epresentation with MPIs is conventional in the sense that images are objects that could be used in alternative ways, including using them to refer to something besides the specimen that contributed to its production, or to convey different content about the specimen. But when representational interpretation is based on the imaging technique, it is not conventional in the sense associated with Goodman's extreme early view of pictures; it is not a completely arbitrary matter that an image refers to a particular specimen and represents the specimen as having the features it does.
>
> (Perini 2012b, 167)

The causal connection, which has been pointed out as an important feature of measurement devices above, will therefore set constraints on possible interpretations of its visual data output. Perini states that "[t]he selectivity of the imaging technique provides a constraint on image interpretation that preserves a link with the specimen" (ibid.). Thus, the causal connection restricts what we can correctly claim to see in a respective image. It functions as a standard of correctness here. If there is no entity, no corresponding image can be produced. However, as mentioned above, this still seems to be a vague limitation to what can be properly stated as an interpretation of a given image. Moreover, Perini's discussion of the conventionality of some visual representations in science neglects to mention directly that it is not only usage that is ruled by conventions but that there are also methods of depiction itself – for instance the use of false colours in the images of Olympus Mons and M51 (see Figures 2.12 and 2.2) – that are based on conventions. Consequently, even though I fully agree with Perini on the constraint mentioned by her concerning interpretative rules, I think that one can and should be more precise about the latter – which would also mean reducing the alleged conventional element even further.

The natural way to find further relevant criteria to assess the correctness of interpretations is by observing scientific practices, how scientists proceed when interpreting visualisations as surrogates for research objects.

Photometric methods to measure the brightness of stars might illustrate this point. For example, the light from a red star will appear bigger in an image produced with a red filter than a hot blue star. Scientists are told this fact by a corresponding *mapping function*. Such a function helps the scientist to read the data. It tells the scientist what kind of information the image can provide *about* the object, in Chakravartty's sense of 'aboutness'. Aboutness is not tied to resemblance as, for example, linguistic expression is, although arbitrary signs can also be about an object, etc. Nonetheless, I think that functions mapping properties of visualisation x to properties of entity y also tell the scientist in what sense x *resembles* y. One might say that resemblance can be regarded as a kind of aboutness, but this relation does not hold vice versa. The important part of the mapping in the example is that the brightness of any star measured depends on the bandpass[24] of the filter through which it is observed.

A similar case can be made for the Higgs image (see Figure 2.4). In this instance the relevant function tells the scientist, for example, how the amount of energy in the spreading particles and the length of the green towers depicted are correlated. It thereby enables a correct interpretation of the corresponding image.

Moreover, the example of the Higgs reveals that there are parts of the image which are, so to speak, more conventional than others. These are domains in which more local definitions are at work. Colours are a salient example in this context. Neither the green nor the yellow traces in the Higgs image correspond to any colour properties of particles. The choice of colour is totally arbitrary. Scientists could have chosen pink and blue instead. However, what matters is that, in this case, different colours indicate different phenomena. Yellow lines represent a retrodiction, whereas green lines mean measured energy. Another example of such more local mappings are false colours in photography. Consider the picture of Olympus Mons as an illustration (see Figure 2.12). In this case, different colours indicate different altitudes. Red colours mean low altitudes, blue ones represent heights.

However, decisions such as how to use colours, etc. have to be stated clearly by the image producer.[25] They are meant to ensure that other people working with the same image will understand which colours have been chosen and for what reasons. Captions or legends are common tools employed to point this out. Those explanations make sure that others will interpret the image in the same way and thereby make the interpretation *repeatable*. Correspondingly, these interpretive rules are fixed for the image in question. They cannot be varied at will; thus, conventions are restricted further in the scientific context. Moreover, such local decisions can even become long-lasting conventions that are shared by the whole community. Colours indicating altitudes are a common example here.

Moreover, mappings are not only part of scientific discourse and therefore, in many instances, inter-subjectively fixed. Many of them are also part of the theoretical network of science; they are *law-like*. Thus, questioning the connections between visual datum and research object also means questioning the validity of these laws which are normally embedded within the wider context of a theoretical network. Kosso discusses the *coherence* of a theoretical network as an indicator of its approximation to the truth (see Kosso 1993, 8). Thus, considerations about the coherence of our theoretical network restrict possible image interpretations further.

In the above example of photometric measurement, such relevant components of the theoretical network are, for instance, considerations about the correlations between age, composition and the altering distances of stars and their colours. Yet there are other theoretical considerations belonging to this network that perhaps do not immediately strike the eye, such as correlations between wavelengths of radiation and colours, the temperatures of bodies and colours, and considerations concerning spectroscopy.

Summing up my considerations so far, background knowledge about the instrument's causal connection to the object of research, about the former's calibration and a relevant mapping function that often explains in what sense the image resembles the depicted entity, constitute reasons to rely on visualisations in the exploratory context. Knowledge of these aspects reduces the arbitrariness of necessary interpretations and thus makes the scientific practice of using visual representations as surrogates for research objects rational.

Of course, this does not mean that no errors can occur. Especially in the context of basic research, as Perini shows (see Perini 2012a, 155), there often remain problems in discerning artefacts from data. I take this, however, to be no special problem of images but a difficulty concomitant with scientific experiments and observations in general. This is probably why scientists and philosophers of science take the current scientific theories not as ultimate rationales, but as *approximately true hypotheses* which still leave room for improvement.

It might be objected that the above considerations are the result of too narrow a focus on the exploratory context in science. Only the case of visualisations as measurement results, that is, as surrogates for research objects. As a rejoinder, I want to highlight the point that this seems to be the *most common* case of visual representations in the exploratory context, namely where technological devices are their causal source. Moreover, the detailed discussion of this phenomenon also seems to be justified because it is instances like these that fuel sceptical scenarios.

Admittedly, however, visualisations also play another role in the exploratory context. Images might also contribute to the development of scientific hypotheses because they can serve as tools for thinking. In such instances, the respective image does not necessarily represent a pre-existent entity; they *do not refer* to something in the usual sense, as Scholz points out (see Scholz 2009b, 37). Some might nonetheless show referential relations to reality, as does Snow's map of cholera transmission routes (see Figure 3.1). However, if no reference can be presupposed, causality and informativeness are the wrong dimensions to assess the epistemic significance of such images. As they are part of the developmental process of thinking, what seems to be more important is the theoretical network they are embedded in. Do they fit background assumptions? Do they fit adjacent theoretical hypotheses? Consequently, considerations of *coherence* seem to play the crucial role before embarking on the empirical testing of hypotheses. This was exactly the procedure demonstrated in the example of Snow's cholera map. In the end, it was Snow's suggestion to remove the handle of the Broad Street pump that confirmed his thesis that this public water supply was the source of the spreading of the disease. Consequently, as tools for thinking and representations of ideas, visualisations are subject to the same criteria of reliability assessment as are verbal or numerical expressions. They are not an exception to the normal method of theory development and testing, but a proper part of this process.

The second context – the one of explanation, that is, communication – will now be examined.

3.1.4 Reasons (II) – trust and reputation

Following Kjørup's suggestion of taking the whole communicative act into account and not regarding visual representations in isolation, it becomes clear that the *epistemic source* addressed in the explanatory context is *testimony* rather than *perception*. Traditionally, epistemologists discuss five different epistemic sources, namely perception, memory, reason, introspection and testimony. *Testimony*[26] thereby refers to all instances of communication, taking face-to-face communication as its paradigmatic instance. As I have pointed out elsewhere (see Mößner 2010b, ch. 2.4), analyses of testimony as an epistemic source should pay close attention to the two perspectives involved, namely speaker and hearer: one person who wants to transmit information and another who wants to learn it.

Approaching the topic this way seems to be plausible for the usage of visualisations in scientific discourse as well. On the one hand, there is the scientist using visual representations to transmit information, that is, she exhibits the intention to *transfer knowledge* encoded in those representations. She wants her assertion, either written or spoken, that contains the image to be taken as an epistemic source. On the other hand, her audience (readers of her text or listeners to her talk) normally wants to *learn* something from her assertion. They want to be informed about her research results, about her measurement methods, her interpretation of the data, etc. Learning, however, aims at acquiring knowledge via an epistemic source, namely via testimony. The audience is convinced that they can receive the relevant information from the scientist's communicative act – which, amongst others, means from the visualisation at hand.

Visual representations in this context are usually presented as *supporting evidence* by the speaker and are accepted (or rejected) as such by her audience. In this respect visual representations can be independently assessed by the recipient in the sense of questioning whether they adequately depict what they are supposed to represent by the author. Some philosophers question this widespread epistemic practice in science. They point out that only representations conveying propositional content can fulfil such an epistemic function, whereby they normally mean verbal sentences, as it is arguments that play the relevant role here.[27] I will discuss this objection in detail in Chapter 4, when we will investigate the possibility of *visual arguments*.

But even though the problem of propositionality is not solved at this point, the evidential role of visualisations in the explanatory context can nonetheless be evaluated. This is due to the fact that their *status as evidence* is an effect of their being part of the testimonial act as a whole. It is the speaker who presents them as evidence, and it is the hearer who assumes that they play this part. Thus, in the search for reasons to rely on visualisations in this context, what is actually sought are reasons that justify speaker and hearer in the epistemic practice of testimony. It is *social epistemology* that furnishes at this point an appropriate account of the evidential role played by visual representations. Embedded in an epistemology of testimony, the corresponding questions are the following: *under what conditions is the recipient justified in accepting the scientist's testimony (including visual representations)? When is she allowed to form a corresponding belief? And what reasons might justify her knowledge claim based on this source?*

As a concrete example, consider again the announcement of the discovery of the Higgs boson by scientists at CERN in 2012. A corresponding article, also containing the image discussed (see Figure 2.4), was published on the CERN website by Lucas Taylor.[28] So why should it be believed that the announced detection had taken place? Why should it be believed that a new particle had been observed showing the relevant, namely the theoretically predicted, mass to qualify as a candidate for a Higgs boson? What are the reasons to accept this message simply on the strength of being told this and shown images of apparently decaying Higgs bosons detected at the LHC? The debate about *knowledge by testimony* revolves around this very problem.

Taking into account that visual representations in the explanatory context are parts of communicative acts serving as testimony also means that social epistemologists' proposals can be utilised that deal with the question of justifying knowledge claims based on testimony. Normally, the *speaker's sincerity and competence* with regard to the topic of her assertion are held to be crucial in this context. These two conditions are the default criteria, albeit their theoretical framing varies within the philosophical discussion.

There is a lively debate about whether it should be required that the recipient has good and independent[29] reasons to rely on the word of others or whether there is an epistemic principle – a presumptive right – involved that allows the hearer to trust the speaker. This principle works in the absence of positive reasons speaking in favour of the speaker's credibility, only presupposing that there are no negative reasons, that is, reasons that would defeat her presumptive right to trust.[30] These are the two traditional main accounts in the epistemology of testimony.[31] The former, requiring positive reasons, is known as *reductionism*. The latter is labelled *anti- or non-reductionism*.[32]

Proponents of both approaches try to explain why it can be rational to *trust* a certain speaker and thus to use her assertion as an epistemic source.[33] However, both sides agree that mere trust is not enough. Most philosophers stress the point that blind trust will not do, as this would mean *gullibility*.[34] The question, then, is what might justify our epistemic practice of trusting a speaker? Do we have reasons to do so? Or are there other forms of justification available? And are these forms the same in any possible context, or do they vary, are there situations in which we have better reasons to trust the speaker than in others? And, if so, what makes the difference?

To find an answer to at least some of these questions, I will follow an account developed by Paul Faulkner (see Faulkner 2011). He proposes to take the *attitude of trust* as the rationale for our reliance on the word of others (see ibid., 151). Moreover, he points out that not all testimonial situations are epistemically equal. That is, some contexts require justifying reasons in addition to the attitude of trust which might suffice for others. In this sense, he suggests two different strategies based on the notion of trust to justify testimonial beliefs.

Firstly, he presents a *strong concept* of trust: "[a] trusts S (in the affective sense) if and only if (1) A depends on S ɸ-ing; and (2) A expects (1) to motivate S to ɸ (where A expects this in the sense that A expects it *of* S that S be moved by the reason to ɸ given by (1))" (ibid., 146, his italics). Obviously, Faulkner's definition is meant to deal not only with instances of testimony, but also with situations of trust more generally. That is, situations where we have to trust another person in regard to a particular form of behaviour or action, such as trusting someone to keep her promise, are also addressed by Faulkner's definition.[35]

In the case of testimony, the first condition of Faulkner's definition states that the recipient is dependent on the speaker's assertion for acquiring knowledge about a particular event, etc. Thus, the audience's dependency is epistemic in kind (see ibid., 144). This trust-based account implies that there is no other way for the recipient to get the relevant information (immediately) and to form the corresponding belief. Obviously, this dependency seems to be a gradual matter. Sometimes it is only due to practical reasons, such as limitations of time or other resources, to acquire a belief without depending on testimony. On other occasions, however, such *checking of the facts* is not possible in principle. Many people underestimate how wide-reaching our dependency on testimony is. How do you know, for example, about your own name? You can ask your parents or godparent – this is testimony. Looking it up in your passport or in the church register, however, is also an instance of testimony. On such occasions, our dependency on others to gain the relevant information cannot be avoided.

The second condition of Faulkner's definition is recursive. It states that the recipient expects the speaker to notice her dependency on his assertion. Moreover, she assumes that it is this very fact that motivates the speaker to tell her the truth. Now, in what sense does this suggestion differ from the concept of blind trust that was rejected because of its apparent licensing of gullibility, and hence of irrationality in epistemic endeavours?

The crucial difference consists in the assumed individual relation between confider and confidant that Faulkner highlights in his approach. Affective trust means to trust *somebody* not something – in the case of testimony, this means to trust the speaker, not the asserted content. Obviously, Faulkner's concept of affective trust is based on the assumption that both parties involved (speaker and hearer) are epistemically close. As Faulkner points out, affective trust is a mutual relationship (see ibid., 149): it is not only the hearer trusting the speaker to tell the truth; the definition of the concept also requires that the speaker can realise that the hearer depends on his sincere and competent assertion and, hence, acts correspondingly. This, however, seems to presuppose that the speaker knows his audience or, at least, that he has an interest in a positive outcome for her regarding his behaviour (that is, knowledge about x in the case of his assertion that x). In this sense, it is due to the speaker's intention that the hearer does not end up being gullible.

Faulkner then *combines* his notion of trust with the "assurance view" of testimony (see ibid., 169). Proponents of the latter thesis state that what justifies the hearer in adopting a particular testimonial belief *that x* is the (implicit) assurance of the speaker that, by asserting *that x*, he takes up the responsibility of the hearer's believing *that x* is true. Faulkner thinks that trusting the speaker in question is due to corresponding internalised social norms (see ibid., ch. 7.3).

However, it might be objected that scientists are rarely in a position to trust their peers in such an affective sense. On the contrary, taking the globalised nature of today's science into account, it can be assumed that, except for a few cases, they will not know their peers in person. Moreover, it is not only the spatial dimension that speaks against the relevant close relationship between individual scientists, it is also a question of time. Scientists cite their colleagues' works. They connect their own work to a stock of established theories, either in a cumulative sense of adding knowledge or in a critical sense of trying to falsify particular hypotheses and of suggesting alternative views. If they could not rely on their peers' words, they would have to begin anew *ad infinitum*. Scientific progress as we know it today would simply not be possible.

Relying on the words of others because of these two contextual requirements – with respect to a spatial and a temporal level – means admitting that the *division of cognitive labour* plays a major role in science.[36] Both levels, however, which necessitate the collaborative structure of the scientific endeavour, preclude the individual relationship – the emotional bond – that Faulkner's concept of affective trust presupposes.

Obviously, this concept is not suitable for the context at hand. Faulkner's epistemology of testimony is fortunately not restricted to just one concept of trust. He acknowledges the diversity of contexts of epistemic processes in which testimony plays a significant role and the different requirements that therefore influence the respective process of justification (see ibid., 198). I therefore propose taking into account Faulkner's second, weaker notion of trust which he calls *predictive trust*. "A trusts S to φ (in the predictive sense) if and only if (1) A depends on S φ-ing and (2) A expects S to φ (where A expects this in the sense that A predicts that S will φ)" (ibid., 145). In this case, to trust the testifier in question means that the recipient knows about her dependency on the latter's sincerity and predicts that the speaker will act accordingly, namely that he will tell her the truth. Yet why should she predict such a positive outcome? Are there not plenty of reasons which speak in favour of more cautious behaviour? Quite a few people seem to lie regularly. Many pretend to be competent testifiers, although they are not. Why should we trust?

It becomes clear that predictive trust alone is not sufficient to justify a testimonial belief as being true. It has to be supported by further reasons. In accordance with this, Faulkner adds that "[...] predictive trust is reasonable just when there are grounds for judging a cooperative outcome [...]" (ibid.). Consequently, predictive trust that a hearer might refer to in order to justify her belief has to be based on reasons that rationalise this attitude. But does this not imply that the notion of trust is redundant in this context? Does it not suffice to have good epistemic reasons – as the reductionist requires?

Indeed, Faulkner admits that this approach coincides with classical reductionism as stated above *and* he seems to be reluctant to endorse this as a correct epistemology of testimony. Nonetheless, there are paragraphs in his book suggesting that he thinks that in some instances such a way of justifying testimonial beliefs might also be an option.[37] In this sense, it seems not unreasonable to adhere also to the second concept of trust and a correlated reductive way of justifying testimonial beliefs – also in the framework of Faulkner's approach.

Such a context-sensitive approach to an epistemology of testimony meets our common sense experience that there are many situations in which recipients have good epistemic reasons to justify testimonial beliefs. Moreover, as I have pointed out elsewhere (see Mößner 2010a, ch. 5.5.3), there are also contexts where recipients have good epistemic reasons to expect this to be the case as they belong to the same social group as the speaker. Notably, these are contexts where the members of the group are subjected to particular forms of training in how to assess information and sources, such as journalists for example. In due course, I will show that this is also true in the scientific realm, namely that scientists often do have good epistemic reasons to rely on the words of their peers. Prior to this, however, I will rule out the objection mentioned above. It might be questioned why we need the notion of trust at all if we can rely on reasons instead, but I think that the former plays a role and should not be dismissed because of assumptions of redundancy. As I have pointed out elsewhere (see Mößner 2010b), it is the presumptive right to trust that inaugurates the whole epistemic practice of testimony in the first place. It constitutes a *prima facie* justification to trust others until there are reasons defeating this right. Moreover, that a kind of trust still plays a role when positive reasons are available to rationally support such a trust-based exchange becomes apparent when instances of betrayed trust are taken into account. As Faulkner and others have pointed out, it is characteristic of such instances that recipients show particular emotional resentments (see e.g. Faulkner 2011, 24). No such emotions play a role if the recipient discovers that her reliance on others was not appropriate. Whereas reliance is based on reasons alone, trust moves beyond them.

Regarding particular social groups it seems correct to say that the hearer *trusts* the respective speaker will tell her the truth. What is of importance are certain social norms, for example *professional ethics*, shared by the members of a group, for instance by scientists or journalists. Though no particular emotional bonds are in place – as there may be in a relationship between a child and her parents or between close relatives in general – peers in these social groups trust one another in the sense of expecting that they will follow the social norms of their groups. In the scientific context, this seems to imply that a testifier also cares about his recipient in the sense that the latter gets the facts right by his assertion. That is, he has an interest in a positive outcome of the latter's epistemic endeavour, at least because he wants his discipline to flourish. Although no personal bond is in place, sharing common professional ethics might still establish a link between people that could supply the grounds for a trust-based relationship.

Moreover, combining trust and reasons in this case does not pose an epistemic problem. People will mention reasons if challenged why they accept the testimony of a particular speaker. Nonetheless, their initial motivation to believe what they are told is their attitude of trust towards the speaker. As Peter J. Graham has shown, such instances of *epistemic over-determination* do not preclude the possibility that each particular part (trust and reasons) plays a relevant role (see Graham 2006b). He argues that testimonial beliefs can *prima facie* be justified by a respective presumptive right and, nonetheless, *on balance* be justified by positive epistemic reasons (see ibid., 95).[38] I think this insight can also be applied to Faulkner's second (weaker) concept of trust. This means that a hearer believes *that p* because she trusts the speaker who has asserted *that p* and is *prima facie* justified in doing so. On balance she is justified in her testimonial belief *that p* because she has additional reasons that speak in favour of the speaker's reliability as a testifier.

Two possible sources of such supporting reasons in the scientific domain will be considered in the following: the first one is correlated with the concept of reputation and, therefore, occupies the individual level. The second is connected to the social setting of science and, as a result, is situated at the social level. The latter was also hinted at by the mentioned social norms above. Both aspects belong to the scope of experiences that a particular scientist has concerning her working environment.[39]

Taking the individual level first, it can be noted that, with regard to peer-to-peer communication in the scientific context, the position offers some advantage, as there are socially operating mechanisms providing empirical evidence of the speaker's credibility. An interesting contribution to clarifying these processes and thus making them accessible to an epistemological analysis is put forward by Gloria Origgi (see Origgi 2012). She draws our attention to the concept of reputation, which seems to be a promising candidate in the context of scientific communication to help us evaluate the testimony in question. As Origgi states: "[r]eputation serves the cognitive purpose of making us navigate among things and people whose value is opaque for us because we do not know enough about them" (ibid., 411). Origgi thereby addresses the difficulty of applying a thick notion of trust, so to speak, in the scientific domain. In a globalised community and with regard to international contributions to the scientific endeavour via journals and web-based communication, however, it is seldom the case that there is sufficient background information about individual researchers to afford an assessment of their credibility as testifiers. Plainly, most members of the scientific community do not know each other. How, then, can the concept of reputation be useful in this context?

The idea is that reputation is a social phenomenon that spreads through a given community. That is, although I may not know the respective testifier in person, others might do so and share their experience with their peers accordingly. "Reputation is commonly seen as the informational trace of our past actions: it is the credibility that an agent or an item earns through repeated interactions. Reputation is a relational property: it is the informational value of our interactions" (ibid., 401). Furthermore, a scientist's reputation is also an *evaluative concept*, namely a description that is a result of her former research successes, but which also takes into account her conformity to certain academic virtues like honesty in citing sources correctly (see ibid., 409). Having spelled out the notion of reputation in this way, it appears natural that scientists seek to achieve and to maintain a *good reputation* in their community. Eventually, it is due to her reputation that she will be invited to a conference or obtain employment. Conversely, this also means that scientists will try to avoid behaviour, such as intentionally publishing false results, that would jeopardise their reputation.

The concept of reputation therefore allows scientific peers to learn from others about the credibility of the testifier in question.

> Judgements of reputation involve always a "third party" – a community of peers, experts or acknowledged authorities that we defer to for our evaluations. Reputation is in the eyes of the others: we look at how others look at the target and defer, with complex cognitive strategies, to this social look.
>
> (Origgi 2012, 403)

Obviously, a scientist's esteem is a factor that she will preserve so that her peers have good reason to believe that this will motivate her to behave in accordance with academic virtues. Being a competent and honest testifier with regard to her research results belongs clearly to these academic virtues.

Moreover, as we are concerned with scientific publications, another aspect of reputation comes into play. Origgi states that "reputation leaks through a social network of deference relations. A good person who judges you good makes you earn status. A bad person who judges you good makes you lose status" (ibid., 410). It is, however, not only such a direct ranking that is of relevance here. Indirect assessments of your reputation are possible as well. Having published results in a *high-ranking journal* might, for instance, enhance the testifier's reputation further. In a similar vein, Martin Kemp suggests that the "authority of the source" (Kemp 2012, 44) can play an essential role when it comes to evaluating the credibility of the testimony in question. Obviously, information about the author and the publishing media concerned belong to this category. The ranking of scientific journals is commonly regarded as an expression of their esteem in the respective scientific community. This means articles published in those journals are assumed to be of high scientific quality. Some of them are often cited in the corresponding field of research[40] and in this sense may crucially contribute to its further development.

It is, however, not only the number of citations that constitutes the prestige of a particular journal, it is also its participation in the *peer review system* (see e.g. Goldman 1999, 176ff.). An established peer review process normally implies that articles submitted to the journal will be critically checked by experts with regard to their content. Usually, this is an anonymous procedure, i.e. the reviewer does not know the author and the author will not be told the name of her reviewer. It is assumed that such a critical check of the article's content will ensure (or also enhance) its quality. Furthermore, knowing about such mechanisms of peer criticism will also enhance the credibility of the paper and the reliability of its authors in the eyes of her recipients.

Yet it can be questioned whether the system of peer review really is the ultimate form of unbiased quality checking. Every now and then there are scandals, such as the following, which reveal apparent shortcomings in the system. In 2015, the press reported that one of the major science publishers, namely *BioMed Central*, had to withdraw 43 surveys published in its different journals because of manipulation in the peer review process.[41] Elizabeth Moylan, senior editor of *BioMed Central*, discusses this incident critically on the publisher's blog (see Moylan 2015). She admits that a manipulation of peer review reports had taken place on a grand scale. Obviously, third parties sold positive reviews to potential authors of the publisher. As a means to prevent future shortcomings of this sort, Moylan has announced that the editors have cancelled the facility for potential authors to suggest the names of reviewers for their articles directly in the publisher's online submission tool.

Obviously, peer review is not the panacea that it is made out to be. By lowering our expectations, however, through the peer review process not being expected to carry the whole epistemic burden, it might nonetheless contribute to the stock of positive reasons that recipients in science can cite to justify their testimonial beliefs. Probably no particular epistemic procedure is infallible, as all of them are performed by fallible human beings.

There are, however, situations where no information about the specific reputation of the scientist in question is available and, clearly, problems concerning reliable criteria to assess the reputation of publication sources were expressed above. Fortunately there are alternatives: a second evidential source which can support the justification of testimonial beliefs in science. Elizabeth Fricker shows that there are further empirical reasons available to the recipient in the scientific community which are correlated with the social setting of science (see Fricker 2002). Fricker's suggestion is similar to Origgi's reputation-based account. Nonetheless, she approaches the topic from a different angle, and thereby obviates the necessity of *knowledge about the individual*, namely that the recipient has to have knowledge about the reputation of a particular testifier.

Fricker suggests that scientific peers can use background knowledge about the *social role* of the testifier in question to assess her contribution (see ibid., 382). She writes: "[...] I suggest, scientists' basis for trusting each other lies in their knowledge of each other's commitment to, and embedding within, the norms and institutions of their profession" (ibid., 383). Fricker here refers to the same social mechanisms as Origgi does, namely those that create a scientist's reputation. Contrary to Origgi, however, she emphasises the social level. Fricker thinks that, in the context of peer-to-peer communication, the recipient adheres to the background knowledge about shared norms and commitments in the scientific context that are relevant to achieve and sustain such a reputation. To put it bluntly: A trusts B not because A knows that B has a good reputation. A trusts B because A knows that B fulfils a certain social role, i.e. that B belongs to a particular community C which is governed by social norms. Moreover, A prima facie assumes that B behaves – to the best of her knowledge – in conformity with these norms to fulfil her social role correctly.

In addition, Fricker highlights the fact that, within the context of science, our reliance on background knowledge of a speaker's social role is supported by further considerations about the consequences of academic misconduct. Jeopardising one's reputation might be but one consequence of inadequate behaviour here. Cuts in research funding or even losing one's job might be even more serious outcomes. Fricker states that "[u]nreliability is likely to be subsequently discovered and highly penalised in such a setting, and this gives one strong empirical reason, amongst others, to expect informants to be trustworthy" (ibid., 383). As pointed out above, she draws our attention to the fact that, within the scientific community, there are certain mechanisms at work to sanction the misbehaviour of scientific peers. In a nutshell, then, Origgi as well as Fricker shows potential sources of empirical reasons that might allow a recipient to trust a scientist's testimony comprising visual representations. This is not to deny that scientists may be mistaken. Errors can occur. However, the social mechanisms described and the threat of losing one's reputation may, at least, reduce the risk of the recipient being deliberately misled.

Information about scientists' reputations and about social roles often belong to the recipient's background knowledge. There is, however, another way in which the background knowledge may join in the process of testimonial evaluation, which is briefly mentioned by Kemp. He highlights the broadly acknowledged fact that new items learnt by the word of others must be *consistent* with our background assumptions (see Kemp 2012, 44). Consistency – or better, coherence[42] – can then be regarded as a further criterion to evaluate the credibility of the testimony in question. Of course, what belongs to a scientist's background knowledge are not only considerations concerning reputations and social roles but also information with regard to theories, instruments, calibrations; that is, amongst others, all the different reasons discussed in the exploratory context will also influence coherence assumptions here.

In sum, it can be stated that, with regard to visual representations, there are often good reasons to rely on in the explanatory context. Most of them are a result of their being embedded in the epistemic practice of testimony, but some are also specific with respect to visual representations. This is a result of taking our background assumptions, gained in the exploratory context, into account to check for coherence among our beliefs. Of course, none of these reasons guarantees knowledge with certainty. Testimony is as fallible as any epistemic source. However, it seems that we are clearly better off in justifying our epistemic practices in this context than sceptics initially suggest.

Yet these positive results should not obscure the fact that there is another level of communication involved in the explanatory context which makes matters more complicated. Disseminating scientific results not only affects scientific peers, but also laypeople. Admittedly, only seldom will they read genuine research articles published in professional journals and the like, as they lack the relevant professional training (knowledge and skills) to understand them.[43] Nonetheless, they are addressed by a diversity of science journalists and numerous public relations officers employed at universities and research centres.

Obviously, a novice's initial epistemic situation is different to the one considered above. Moreover, this change is not only due to the recipient's lack of expert knowledge, but comprises also a difference in communicative aims and, last but not least, different communicative agents as hinted at above. Usually it will not be the scientist who transmits her results to the broader public, but a public relations officer or a science journalist. In this sense, the whole communicative situation can be seen to have changed when compared with the case analysed above.

Thus, we have to pay proper attention to this altered setting and, especially, to the changed communicative aims in these new surroundings. Communication is a multifaceted activity, as speech act theory teaches us. Transmitting information is but one aspect of communication; beyond that, a variety of other purposes are possible. With respect to scientific discourse, Ludwik Fleck has elaborately shown which different aims might be involved (see Fleck 1979, ch. 4.4; 1986c). In the following, I will adhere to his approach.

Fleck emphasises the same distinction between communicating *with scientific peers* and *with laypeople* that became apparent in our above analysis. He suggests that communicating with scientific peers might be intended as *information* or *legitimisation* (see Fleck 1986c, 86f.). The former just means neutrally transmitting research results, whereas the latter implies that the particular scientist asks her peers to acknowledge her results as being in accordance with the prevalent "thought style" of her community.[44] Communication with laymen, however, follows different aims, namely either *popularisation* or *propaganda*, depending on the audience's degree of ignorance concerning the thought style of the communicating author (see ibid., 85f.). We will discuss Fleck's theory in more detail in section 3.2 and thereby shed some light on his technical terms already mentioned. Yet what should by now have become clear is that visualisations as tools of communicating with laypeople serve other purposes than those shared among scientific peers. Whereas, in the latter context, they predominantly play an evidential role, in the former context, they might not serve any epistemic purpose at all. Public relations officers might have added them as mere eye-catchers, for example.

Thus, it comes as no surprise that – returning to the question posed at the beginning of this section – what is related by such visualisations is *not* the genuine research object offered to an inexperienced audience, because what recipients of such visualisations are mostly confronted with are the scientist's – and more often also the journalist's or the public relations officer's – deliberate decisions about how to present her data and, even more importantly, her interpretation of the data in a certain way. Modifications of data presentation may be performed for different purposes. They may be motivated by the intention to enhance the comprehensibility of the visual representation, for instance, by guiding the audience's perception or by adjusting the image to certain visual habits. They may also be motivated by an intention to raise public interest, for example, by attracting the recipients' attention. Often such modifications may also be intended to increase the visualisations' persuasive function.[45] Anyway, what should have become clear is that the scientist's (or whoever is responsible for the particular presentation) intentions have to be taken into account as an essential factor when evaluating the data. The visualisation at hand is purposefully used by the scientist to convey her *opinion* about the object of research.

From an epistemological point of view, it is a problem that, in such contexts, visualisations are often presented in a manner that raises misleading expectations by appealing to their known role as evidence in other settings. Julio M. Ottino expresses respective misgivings about the pseudo-realism of visual representations in scientific journals (Ottino 2003).[46] His example is nano-robots. Ottino states that pseudo-realistic representations might mislead laypeople in the sense that they think that entities such as those depicted have actually been constructed by scientists – although what is shown in the image violates the laws of nature and is, therefore, impossible to construct (see ibid., 476). The audience's confusion of the fictional for a real scientific development is thereby a consequence of the evidential role that they expect the image to play in this context. They apply this interpretative guideline also to images that do not play this role. Consequently, Ottino urges scientists to be aware of their responsibility not to mislead their inexperienced audience.

Sceptical misgivings concerning the epistemic functions of visual representations in science are easily derived by generalising cases of popularisation and propaganda. Constructivist interpretations of images seem to gain a lot of support here. But even though Ottino is right in demanding a responsible handling of design software by scientists and artists and, although popular science illustrations call for a more cautious epistemic handling than they are normally subjected to, there is still no reason for taking these instances as licensing general distrust in scientific visualisations.

What reasons may support a layperson's trust in a potential testifier in this context? From my point of view, there are at least two different aspects that a novice could refer to. On the one hand, she might point to (1) the *professional training of science journalists* that should enable the latter to critically check the reliability of their source, namely the scientist. On the other, it can be assumed that (2) they are also bound by a kind of *professional ethics* that contains rules about truthful reporting. Based on considerations like these, which are frequently used as a shortcut to evaluate the credibility of such testimonies, recipients might refer to the *reputation of the source*. Both individual experience as well as socially distributed knowledge will tell a particular recipient that some journals, blogs, etc., are more trustworthy than others and why. A more careful interpretation of images seems to be advisable here. Helpful clues to fulfil this task might be derived from comparisons with background knowledge, that is, from coherence assumptions. Instances where public relations and other kinds of promotion play a role seem to be epistemically more difficult. It can be assumed that here a particular 'wow-effect' might be more important to the information transmitter than a careful consideration of the facts. Furthermore, recipients cannot expect that testifiers of this kind will have a training which would enable them to pay attention to the criteria of epistemic significance hitherto discussed. Usually, other standards are more relevant to such professions. Amongst these are design criteria which take a particular target audience into account, usability criteria and, last but not least, economic considerations play significant roles. None of them has to have an intrinsic connection to truth or competence with respect to the information transmitted. Despite this sounding so epistemologically discouraging, it does not imply that recipients are never allowed to trust testifiers of this sort.

What does play a role, however, and might offer a recipient some reason to trust nonetheless, is that public relations officers often professionally belong to the same institution, for example the same research centre, as the scientist whose results they promote. In this sense, they are bound by the requirements of *institutional credibility*. That is, maintaining the reputation of their common employer will also affect their job. It can therefore be assumed that not transmitting unreliable information in the company's name is also motivated by their own interest. Yet again, although this will give the recipient some reason to trust testifiers of this sort, it does not mean that visualisations presented in this context can be regarded as supporting evidence. On the contrary, recipients should be aware that propaganda of this sort will normally exploit the ability of the images to persuade rather than transmit relevant information.

3.1.5 Interim results: what can be learnt about reasons?

Summing up my results so far, the starting presupposition was that contexts matter when it comes to justifying our epistemic practice of using visual representations in science. As visualisations fulfil different functions, keeping apart the exploratory and the explanatory context has permitted the insight that there are various different reasons that support beliefs based on visual representations in those surroundings. Table 3.1 summarises the results of the above discussion.

This fine-grained analysis of functional roles played by visualisations, taking into account the relevant target audience and purposes, reveals a variety of reasons that allow a reliance on visualisations used in the epistemic practices of science. To forestall misunderstandings, my thesis is *not* that, in order to justify a belief based on visualisations in one of the two contexts, the recipient has to have reasons for all the kinds mentioned. Some beliefs might be perfectly justified by mentioning reasons of only one kind. If I do have background knowledge about the testifier's reputation, I do not have to refer to her social role. Some reasons are meant as alternatives. Yet in most cases, the more reasons a recipient can cite to explain her reliance on a particular belief, the better her belief is epistemically supported.

Moreover, the obvious disparity of the different settings supports my thesis that some of the major problems discussed by philosophers with respect to the epistemic status and roles of scientific images are derived from confusing these contexts. Especially, misgivings about image manipulation seem to be a result

Table 3.1 Summary: reasons to rely on

Context	Function	Aim	Target audience	Reasons for reliance
Exploratory context	1. Surrogate for object of research a) Data visualisation b) Visualisation as data 2. Tool of thinking 3. Object of research	Genuine research	Scientific peers	1. Background knowledge about a) Reliability of the instrument (causal connection) b) Informativeness (e.g. resemblance) defined via mapping 2. Coherence
Explanatory context	1. Communication of research results (evidence)	Information/ legitimisation	1. Scientific peers	1. Trust based on background knowledge about a) Scientist's (testifier's) reputation b) Social norms (testifier's social role) 2. Coherence
	2. Education (evidence)		2. Laypeople to become peers	
	3. Communication to acquire grants (evidence/ persuasion)		3. Laypeople (scientific background)	1. Trust based on background knowledge about
	4. Communication as infotainment (eye-catcher, persuasion, simplified presentation)	Popularisation/ propaganda	1. Laypeople (addressed by science journalists)	a) Professional training b) Professional ethics 2. Coherence with background knowledge about a) Reliability of the instrument (causal connection) b) Informativeness (e.g. resemblance) defined via mapping 1. Trust based on background knowledge about
			2. Laypeople (addressed by public relations officers)	a) Institutional credibility 2. Coherence with background knowledge about a) Reliability of the instrument (causal connection) b) Informativeness (e.g. resemblance) defined via mapping

of generalised concerns arising predominantly within the explanatory context in which laypeople are addressed by public relations officers. Additionally, the problem of theory-ladenness can be relieved by highlighting the divergent evidential sources of the exploratory and the explanatory contexts. It turns out that, although a tailoring of representations can occur, they normally do not belong to the exploratory context, but are more a characteristic of communication with laypeople.

Thus, although errors and cases of fraud might occur, epistemic practices that presuppose our reliance on visualisations are far from being irrational.

As a concluding remark, I want to relate my results to Chakravartty's proposal on theorising about *scientific representations* in the broader sense – comprising not only visual, but also numerical and linguistic representations (that is, theories, models, measurements, etc.) in science (see Chakravartty 2010). From this more general point of view, he defends the claim that requirements of *information-based* and of *function-based accounts* of scientific representation can be combined to clarify the notion of the latter term. These two approaches – usually presented as mutually exclusive (see ibid., 198) – are characterised as follows by Chakravartty. Proponents of information-based accounts claim that scientific representation presupposes an *objective relation* to the entity for which it stands. It is only in this way that the representation can *be informative about* the latter. Moreover, such a way of transmitting information is established best by some kind of *similarity* relation (see ibid., 198).

Contrary to this, proponents of function-based accounts lay their stress on the recipients' *cognitive activities* (interpretation, etc.) performed with the aid of scientific representations (see ibid., 199). It is via the representations' functional roles (for example exemplification) within these activities that permit an understanding of what *scientific representations* are about.

Chakravartty thinks that the above dichotomy is false, as it is motivated by confusing the question *what are scientific representations?* with *what do scientific representations do?* (see ibid., 208). He defends the claim that *both* approaches contribute equally to our understanding of scientific representations (see ibid., 199). In this sense, he summarises the conditions of 'scientific representation' as "intentionality,[47] relations of similarity (or more specific versions),[48] and capacities to facilitate interpretations and inferences regarding target systems" (ibid., 209). He defends this combination of requirements by pointing to the fact that it is only through the representation's informativeness that it can serve the purposes which are relevant to the recipient's cognitive processes. If there were no objective relations between entity and representation, the latter could not be interpreted as transmitting information about the former.

The above analysis of visual representations supports Chakravartty's thesis. Scientific images are epistemically potent if they contain information about the object depicted. And in most instances in science, they become informative via a causal connection encoding information in the image and a resemblance relation that allows the scientist to decode the image.

Especially with regard to visualisations gained by instruments and information-technology devices, a correlated function will tell the scientist which parts of the image are meant to resemble which parts of the real entity. This coincides with Chakravartty's thesis that often the relevant similarity relation has to be learnt by the scientists to allow its decoding (see ibid., 203).

Furthermore, the above analysis also made plain why there is a need for both knowledge about the similarity relation plus the image's aboutness as well as knowledge about the functional roles to understand scientific visualisations correctly. Taking into account the contexts pointed out above, the different aims and participating agents, it can be inferred that informativeness of visualisations is a rather gradual matter. In the exploratory context, visual representations are valued most because of their objective relation to the entity depicted (at least in the natural and the life sciences), whereas in the explanatory context they can also fulfil their functional roles if they are less informative, also because not all functions are epistemic ones. These latter insights, however, can only be gained by broadening the focus of analysis from semantic aspects of representation (the content) to its pragmatic aspects (the usage).

Last but not least, it is also this shift of focus that has brought to light the kinds of reasons that can justify epistemic practices involving scientific images. Both representational (semantic) aspects – knowledge about the causal connection and informativeness – and functional (pragmatic) aspects – knowledge about the context of usage, for instance the testimonial setting – are of importance.

Thus, with respect to the explanatory context, my argument has been that the relevant units to be evaluated are whole communicative acts, not visualisations in isolation. Utilising the results from the debate about knowledge by testimony, the relevant question to answer has turned out to be about what reasons there are that allow us to rely on the words of others in this context.

The above analysis has shown that, concerning scientific peers and laypeople accustomed to the rules and habits of scientific discourse, the recipients' trust in the words of others is based on reasons derived from an individual and a social level of science. Background knowledge about the reputation of the source and about the social norms of the community are the relevant criteria in this context.

Moreover, it has turned out that the more complicated case concerns the testimonial beliefs of laypeople who do not belong to the scientific domain. Not only do they lack the background knowledge mentioned above, but they also often interact with other informants. Public relations officers, especially, play a crucial role. Evidential status being but one functional role that visual representations can pick up in scientific processes was highlighted here. This is in accordance with Downes's finding that visualisations can serve multiple epistemic functions and, of course, sometimes no epistemic function at all (see Downes 2012, 117f.). Downes thinks that this functional diversity as well as the diversity of appearance of scientific visualisations cannot be explained by a unified representation-based account. Taking Chakravartty's insight into account that we need conditions, representation and functional roles in order to understand scientific representation completely, Downes's claim comes as no surprise. However, what is still missing is an adequate explanation of the diversity stated by him. This takes us directly to the next point of analysis. In the following, I will suggest an account of the diversity mentioned that pays close attention to the contexts indicated above and thereby takes into consideration the social setting of science.

3.2 A social explanation of diversity

A contextual approach to the epistemic practices of science, as elaborated above, offers initial clues which highlight the relevance of social interactions in the scientific quest for knowledge. It was pointed out, for example, that there are different agents interacting in those contexts. Moreover, it was indicated that, with respect to justificatory questions, mutual expectations and dependencies play a role, as well as background knowledge and the intentions of the different people involved.

Within the last decade, proponents of the debate about knowledge by testimony have made it clear that the ideal of an individual epistemology, that is, an epistemology that focuses on an autonomous epistemic subject, is misleading (see e.g. Coady 1992; Faulkner 2011; Gelfert 2014; Goldberg 2010; Lackey and Sosa 2006; Lackey 2008; Matilal and Chakrabarti 1994; Mößner 2010b; Scholz 2001). They have convincingly shown that testimony is a social source of knowledge and, thus, that social interactions are relevant to consider when theorising on the genesis, distribution and acquisition of knowledge.

The relevance of the social, however, is not restricted to the epistemic phenomenon of testimony. Social epistemologists, most prominently Alvin I. Goldman (see Goldman 1999; Goldman and Whitcomb 2011), have shown that social aspects and dynamics – such as esteem, reputation, deference, but also oppression, suspicion, bias – affect epistemic processes more broadly. In this sense, they have extended the focus of the epistemological research. Additionally, they also discuss whether the epistemic subject has to be an individual or whether groups could also be accredited this status.[49] The collaborative character of scientific cognitive processes has often been highlighted within this context. Famously, John Hardwig posed the question, who exactly the epistemic subject is if scientific papers show dozens or sometimes even several hundreds of authors (see Hardwig 1985). Who knows what has been stated in papers of such a kind? Obviously, the role of the social by far exceeds mere teamwork in science – especially in the natural sciences.

The Higgs image (see Figure 2.4) is, for example, a result of the so-called *CMS*[50] *experiment*. Yet, although the CMS detector is located at CERN in Geneva, the project is international and globalised in its nature.[51] Distributing visualisations such as Figure 2.4 in correlated publications thus implies that more than 2,000 people are involved in just the one facet of this communicative act.

Yet what seems to be even more impressive is the other facet, the audience. As the above analysis had made clear, it is not only scientific peers that are the recipients in this context, but also laypeople of different degrees who will regard this image in different communicative contexts. One example might be the CERN website, which explains the CMS experiment to the interested public. Another instance is this very book, theorising philosophically about the epistemic role of scientific images. And not only does the audience vary, but also the meaning ascribed to the image. Scientific peers will understand the details and thus understand the image as a measurement result. Laypeople (namely, the interested public) will regard the image as evidence of the successful detection, as this is the way that the corresponding text interprets the image for them – otherwise those readers would not understand how to decipher the image. And, finally, in the context of this book, the image simply exemplifies a set of visual phenomena in scientific discourse and will be regarded as such.

Social mechanisms of communication shape the notion of the image and thus its function in a specific context. Moreover, just as laypeople rarely read papers written for the scientific community, they seldom face visualisations produced for peers. The social context therefore not only influences the function and meaning of scientific visualisations, but often also their appearance.

This is then a rough sketch of my answer to Downes's question of how to explain the diversity of functions and appearances of scientific visualisations, if representation-based accounts are clearly predestined to fail in this respect. Moreover, I am convinced that there can be more precision in the respective answer than just stating the responsibility of *social mechanisms*. It can be made plain how exactly to conceive of this apparent social origin of scientific images and thus how to grasp the social impact that creates functional roles and design in the explanatory context. What is needed to fulfil this requirement is a better understanding of communicative processes in science. Here I will make use of Ludwik Fleck's theory of thought styles and thought collectives. Fleck emphasised the role of the social in any cognitive activity and, explicitly, demonstrated this with regard to scientifically epistemic practices (see Fleck 1979; 1986a).

Despite my making use only of Fleck's analysis of scientific communication, it seems advisable to start with some introductory words concerning his theory on the influence of the social on cognition in general.[52]

3.2.1 Preliminaries on Ludwik Fleck

In order to understand Fleck's conception of scientific communication correctly it is necessary to introduce some terminology.[53] "Thought collective" is the first relevant concept to examine (see Fleck 1979, 102ff.). Such a collective comprises people being involved in a joint conversation (see ibid., 39, 44, 102).

Because of the breadth of this definition, such collectives are facets of scientific as well as of everyday contexts (see Mößner 2011, 367). As we are interested in scientific images here, I will focus my considerations on the former and take it that Fleck's term "thought collective" can, amongst others, be understood as denoting a *scientific* community, referring to a group of people (see Fleck 1979, 39, 102f.). Such a community comprises different parts. There are the experts who belong to the "esoteric circle" – forming a group around the specialists of a certain topic. Additionally, there are the more or less informed and interested laymen as members of the "exoteric circle" (see ibid., 111). These realms are not strictly separated; expertise comes in degrees.

In science, there is more than one such collective, especially as they are not congruent with formal disciplinary boundaries, for example, between biology and physics. Actually, Fleck thinks that they are innumerable due to their origin in joint conversations (see ibid., 102). To give an example, there may be a thought collective of researchers exploring the formation of galaxies, one that is interested in comets, one that is concerned with the developments of stars, etc. – and all of them belong to the realm of astrophysics. Furthermore, a single individual can be a member of diverse thought collectives at different levels of expertise at the same time (see ibid., 105).

This idea of thought collectives as groups of people clustered around a common topic of interest – or a set of beliefs – can be regarded as a first approximation to what Fleck holds to be the collective's unifying element, namely its "thought style" (ibid., 99). Asking what distinguishes between different thought collectives, Fleck draws our attention to the divergent thought styles of those groups.

This central concept of Fleck's theory is not easily understandable as it consists of different components varying with different contexts.[54] Within the scientific context, some of these elements are epistemological in nature – such as shared background knowledge, theories, etc. – and some are of a more psychological nature – such as a certain mood.[55] Beyond this, the usage of certain instruments is also part of the thought style, as is the particular way to represent and to publish research results (see ibid., 99). To be concise, Fleck states that:

> [w]e can therefore define thought style as [the readiness for] directed perception, with corresponding mental and objective assimilation of what has been so perceived. It is characterized by common features in the problems of interest to a thought collective, by the judgements which the thought collective considers evident, and by the methods which it applies as means of cognition. The thought style may also be accompanied by a technical and literary style characteristic of the given system of knowledge.
>
> (Fleck 1979, 99)

Finally, a collective's thought style gives rise to certain consequences with respect to the work of the particular researcher. Fleck thinks that the individual's perceptions and actions are influenced and directed by her collective. I do not agree with him about all the details of this claim. Most of all, I think that Fleck's theses concerning those consequences are far too wide-ranging. Thus, in the following, I will try to cite his theses on this topic neutrally and will afterwards add some critical remarks when applying his ideas to scientific images. Admittedly, this might dissatisfy those interested in Fleck's philosophy as a whole, but I think it is legitimate to proceed in this way as long as it can be shown that the ideas that I will borrow from his work are not indebted to this broader – some would say *relativistic* – framework.[56] To meet this requirement, I will discuss potential difficulties in Fleck's theory in this respect in the following analysis of scientific visualisations.

I shall begin with the descriptive part. A serious consequence of the social determination of cognitive processes is its impact on scientific observations. Fleck thinks that the individual shows a specific liability to observe only what is of interest to her collective. Consequently, he claims that "[w]e look with our own eyes, but we see with the eyes of the collective body [...]" (Fleck 1986d, 137). This way of perceiving things is a result of the background knowledge provided by the collective to which the particular scientist belongs. The particular researcher learns what is relevant to observe and how to perceive correctly, namely in accordance with the prevalent thought style, during a period of practical training. Learning by example plays a crucial role in this context. Wojciech Sady explains this relation between training and observational practices by stating that "[a]n expert sees differently than a layperson because she went through a special training during which she was familiarized with many examples [...]" (Sady 2012, ch. 5). Fleck thinks that professional training preconfigures the student's mind in a certain way. That is, by becoming familiar with what is of importance to the thought collective, the individual's mind becomes equipped with certain perceptual patterns, so to speak, which will guide her in finding a starting point for new observations later on.

Fleck argues that the particular researcher will look for specific starting points in the chaotic beginning of a scientific observation which are provided by those very patterns (see Fleck 1979, 92; 1986d, 139ff.). Ilana Löwy explains this phenomenon in the following way:

> [i]solated researchers may at first be bewildered by the chaos and disorder of their observations and may have great difficulty in separating fact and artefact. The shared conceptual framework of their scientific communities and the techniques and methods elaborated by these communities provide them with cognitive and material tools that enable them to distinguish signal from noise.
>
> (Löwy 2008, 375f.)

Such a concept of scientific observation implies that the scientist will be somehow blind to deviations from the prevalent style-determined content of her perception. She will be prone to omit some details and to add others (see

Fleck 1986d, 137). Fleck thinks that observations which are in accordance with the prevalent thought style can be termed 'objective'. Yet traditional concepts of objectivity are hardly in line with this definition. The same difficulty also appears with respect to 'truth'. From Fleck's point of view, problem solutions that are in accordance with the prevalent thought style are true (see Fleck 1979, 100), but whereas traditional philosophers of science would infer that Fleck's concepts of truth and objectivity speak in favour of a relativistic inter- pretation – that is, being relative to a particular thought style – Fleck denies this relativistic tendency. He claims that within a thought collective an idea is either true or false. Moreover, as a comparison between different thought collectives in this respect is not possible from his point of view, relativism is therefore absent (see ibid.). Yet it should be clear that this is an epistemolog- ically problematic statement. Consequently, it is here that controversies about Fleck's theoretical approach arise in philosophy.

As my own approach is indebted to scientific realism and the relativistic tendencies of Fleck's theory are, charitably speaking, at least unclear, one might wonder why I refer to Fleck's theory at all. The reason is that I think that Fleck has put forward an excellent description of scientific communica- tion processes which can be used without making concessions to the more problematic parts of his theory. Thus, in the next section, I will discuss his ideas about scientific communication before applying those ideas to scientific images as an explanation of diversity.

3.2.2 Scientific communication – aims and modes

To make use of Fleck's theory to describe and, perhaps also to account for, the diversity of scientific images, the two parts of his explanation of scientific communication have to be considered. The first one is about *communicative purposes or aims*, the second about the different *modes of publication*, namely ways of communicating results and hypotheses to the audience.

Both issues are involved in the circulation of ideas that Fleck describes as taking place within the scientific community and also sometimes within the broader public domain. The term 'circulation' (see Fleck 1986c, 85) thereby indicates that Fleck does not conceive of this as being a unidirectional way of communication, but as a process of mutuality.[57] Furthermore, Fleck explicitly stresses the point that this circulation of ideas also implies their transform- ation in certain ways. The mechanisms involved in this wandering and chan- ging of ideas in science will now be described in more detail.

Starting with the aims of communication, a distinction has to be made between the "intra-collective thought exchange", that is, communication within the boundaries of one's own thought collective, and the "inter-collective thought exchange", that is, communication across the borders. Both kinds result in certain modifications of the original idea (see Sady 2012, ch. 5).

Intra-collective communication means that an idea passes through different stages of modification, depending on the target audience and the envisaged purpose. Accordingly, Fleck distinguishes three aims that are related to the communication of an idea within one's own collective, namely to *legitimise* it, to *popularise* it or to *inform* the group's members (see Fleck 1986c, 86f.). These three communicative aims correspond to the collective's structure. If, on the one hand, a specialist from the esoteric circle speaks to laymen of the exoteric circle, the idea transmitted is made *popular*. On the other hand, the communication between members of the esoteric circle can be aimed at *information* or *legitimisation*. It is not surprising, then, that the same idea is expressed differently if it is communicated to different target audiences. Fleck, however, not only emphasises a harmless alteration in expressions, but also points out that the idea itself – its meaning – is changed. He thinks that the latter takes place because he is of the opinion that transmitting ideas always implies a kind of translation that takes into account the recipient's background knowledge. If a specialist speaks to a layperson within her own community, she will have to consider that her recipient will not be able to understand all the details and the technical terms of specialist research. Consequently, the act of popularisation leads to *simplification*. The specialist does not communicate the subtleties of the discussion, leaves out divergent opinions "[...] and stresses some aspects of the problem by way of pictures and comparisons" (ibid., 86). Modern science journalism illustrates Fleck's thesis neatly.

Furthermore, it is not only due to the specialist's attempt to communicate her ideas in a comprehensible manner, it is also the lay recipient's cognitive process of understanding what has been communicated that is responsible for more or less substantial shifts in meaning. The lay hearer will interpret what he thinks he has heard in the light of his own background knowledge. Thus, the fewer the background assumptions shared by speaker and hearer, the more significant the deviation in meaning between what the speaker wants to transmit and what the recipient will understand. Fleck thinks that such changes in meaning *always* happen in the contexts of popularisation and legitimisation, whereas the act of information leaves the idea more or less intact (see ibid.).

Information and legitimisation are purposes of communicative acts between more or less specialised experts. In the first case, a particular scientist wants to transmit her research results to her peers, namely scientists working on the same theory or problem. As they all share the same background knowledge, the informing scientist does not have to modify her transmission in any particular way. No translation process will affect the meaning of the proposition transmitted, thus what is transmitted remains effectively unaltered (see ibid., 87).

Although, in the case of legitimisation, the participants of the communicative act remain in effect the same,[58] its consequences for the transmitted piece of information are radically different.[59] The scientist who transmits her idea asks the collective to legitimise it, that is, to declare its conformity to the prevalent thought style (see ibid., 86). In this sense, the particular scientist

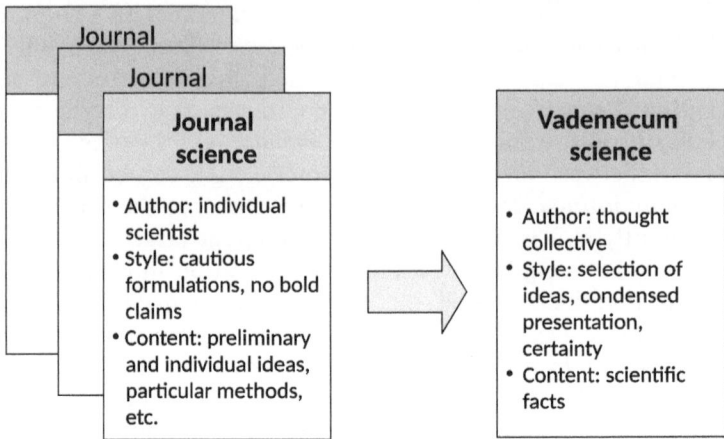

Figure 3.2   Ludwik Fleck: wandering ideas – from journal to vademecum science.
Source: own image.

plays the role of the individual asking for the collective's approval. Moreover, this role also affects the way in which she presents her ideas, namely by using carefully formulated hypotheses and not bold claims. When the process is finished, the stylised, transformed idea is found being integrated in the correlated collective's handbooks (and textbooks). Thus, two different media are involved, namely *journals* to present the initial idea and to raise the discussion and *handbooks*[60] to present the legitimised thought after the debate has closed. Figure 3.2 summarises this relationship between different communicative aims, the style of expression and the media chosen.

The last purpose of communication in science which Fleck mentions, namely the case of *propaganda*, should now be considered. This is the intercollective instance of communication, that is, the attempt to communicate an idea to a different thought collective (see ibid., 85).[61]

Again, we notice a change in meaning of the idea transmitted. Taken in its simplest version, this variation may be comparable to what happens in the case of popularisation. Propaganda, however, can also imply complete misunderstandings, and thereby the destruction of the meaning of what should have been transmitted (see ibid. and Fleck 1979, 109f.). The reason for these difficulties consists in the necessity of translating the idea into the language of the foreign group. The aim is *comprehensibility*, whereby the researcher has to simplify her original statement and to retreat from her technical terms. However, at some point, it might be impossible to find a common language which both communities can understand correctly.[62]

Now that we have an idea of Fleck's conception of the purposes of (scientific) communication and their effects on the ideas transmitted, we can take a look at his considerations about the *modes of scientific communication*. Both

Table 3.2 Ludwik Fleck: four kinds of science

Kind	Target audience	Aim
Popular science	Exoteric circle	Popularisation
Textbook science	Esoteric circle	Information
Vademecum science	Esoteric circle	Information
Journal science	Esoteric circle	Legitimisation / information

aspects will help us to explain the diversity of scientific images as a conse-quence of the social setting of science.

Without doubt, a crucial aspect of Fleck's theory consists in his elaborate description of scientific publishing practices and their influence on transmitted ideas (see Fleck 1979, ch. 4.4). Fleck distinguishes between four different cat-egories of science according to their different media of publication. Table 3.2 summarises these four categories of science and their characteristics. In the following, I will discuss each of them in more detail.

Firstly, there is what he calls *popular science*, that is, "science for nonexperts" (ibid., 112). This is related to the aim of popularisation with the above-mentioned consequences of simplification and apodictic valu-ation. Schoolbooks and science journalism (such as *Scientific American*) offer examples here. They are designed for laymen. Furthermore, an interesting fea-ture of popular science is its tendency to include pictorial elements (see ibid., 114). I shall return to this point in due course.

Additionally, Fleck mentions three further kinds of science: *textbook science*, *vademecum science* and *journal science* (see ibid., 112). He thinks that all of them belong to the collective's esoteric circle and are related to the aims of information and legitimisation. This mapping between aims and kinds of science is a result of their being located within the esoteric circle of a thought collective. However, it should be added critically that it can be questioned whether textbook science is categorised correctly in this way. As it addresses students, that is, laymen, who usually lack the relevant background know-ledge – at least during the first stages of their education – to have them under-stand transmitted ideas correctly, it seems puzzling to regard this as a case of information. Unfortunately, Fleck does not explicitly deal with the case of textbooks needed for education. Thus, there is no solution to this puzzle offered in his work. Nonetheless, this case calls for more detailed investiga-tion and I will refer to this problem in the text below.

With respect to vademecum and journal science, Fleck is much more pre-cise in his analysis. As mentioned above, he explicitly discusses the wandering of ideas between these two kinds of media. In short, an idea starts its journey in journal science, is discussed by the collective, will eventually be added to a handbook in a somewhat modified version and from there it will influence new research in the same field – as it indicates the state-of-the-art now. Fleck describes a *knowledge-cycle* here that, on the one hand, shows the interaction of the different parts of a certain thought collective and, on the other, points out how the act of legitimising an idea proceeds within the community.

The first step is the publishing of particular results in a certain scientific journal, say, *Physical Letters A*. Here, the individual researcher (or a certain group of people) is the author of the idea. However, the aim of this publication is not only to inform others about results, but also to seek legitimisation for those results. The published idea is therefore presented in accordance with a certain formal style corresponding to these aims. Fleck describes the method of presentation of an idea in a journal article as containing especially cautious formulations and no bold claims. And, as individual results are presented, so particular methods, instruments and theories are mentioned. Thus, Fleck describes the style of journal articles as bearing "[...] the imprint of the provisional and the personal" (ibid., 118).

If we now consider a handbook, it becomes clear that the way to create such a condensed work of science cannot consist in just combining all those different journal articles on a certain topic. This would only result in a jumbled mixture of perhaps even incoherent ideas and not in a comprehensive guideline intended for future research. Consequently, Fleck states that:

> [t]he vademecum is [...] not simply the result of either a compilation or a collection of various journal contributions. The former is impossible because such papers often contradict each other. The latter does not yield a closed system, which is the goal of vademecum science. A vademecum is built up from individual contributions *through selection and orderly arrangement* like a mosaic from many colored stones. The plan according to which selection and arrangement are made will then provide the guidelines for future research.
>
> (Fleck 1979, 119f., my italics)

Fleck's description of the handbook's origin seems to be plausible. Surprisingly, however, Fleck claims that the author of the altered idea is not a particular individual anymore, but the collective as a whole (see ibid., 120).

Fleck's explanation of this occurrence hinges on the process of the idea's alteration. It is adjusted to the community's thought style and thereby is deprived of the individual's influence. He explains this by showing that there are always different hypotheses available concerning a particular topic. They are presented in the diverse journal articles which all claim to be the next developmental step of the theory at hand. Vademecum science takes them all into account, assesses their values, combines them with the research tradition of the field and, finally, follows perhaps none of them (see ibid., 124).

One might object that the handbook has nonetheless an identifiable author, namely the one who wrote the book, but this is not Fleck's point of concern. He does not deny that there is an author in this sense; what he denies is that the presented idea is fashioned solely in the mind of a particular researcher without taking into account the set of background knowledge provided by the collective.

Fleck's account is easier to understand when it is taken into consideration that the described adjustment of the idea can also take place within the mind of one particular scientist (see ibid., 120). Although a discussion and an alteration of ideas is usually undertaken by a group of people, it is by no means precluded that a particular researcher carries out the described evaluation and adjustment alone. As Zittel states, "[a]n individual person can also form a collective when he or she discusses with him/herself" (Zittel 2012, 63). And, as Fleck claims, in the end it will not be her idea that she finally presents in the handbook, but an idea which is adequate to the thought style of her collective, taking into account its traditions, its style of writing, its interests and biases.

The *social mechanisms* at work in this context are described by Sady:

> [e]ach member of the group reads different texts (both popular and professional), participates in different experiments, and belongs to more or less different thought collectives (both scientific and non-scientific). So, when they start speaking to each other and reading each other's papers, a series of misunderstandings arises. Ideas circulate within a collective and are enriched by new associations, and therefore the words that are used change their meanings. [...] Having conducted countless studies and conversations and embarked upon a long journey, the scholars finally create a thought style which nobody intended. And after this has happened, nobody post factum knows when and how that style started to operate and who, specifically, created it.
>
> (Sady 2012, ch. 5)

Furthermore, Fleck points out that, as the idea is no longer the product of a particular individual but the product of the community, it seems to be approved by the collective. This is also supported by the presentation of the information, the formal style which declares the published idea to be a common fact. He claims that "[a] statement appears ipso facto more certain and more soundly established in the organized system of a discipline as presented in a vademecum than it does in any fragmentary description found in a journal. It becomes a definite thought constraint" (Fleck 1979, 121).

I agree with Fleck that an idea published in a handbook seems to possess more certainty than a discussion of thoughts in journal articles – at least in the humanities. In the natural sciences, textbooks might apply here. Yet the consequences derived seem to be too strong. Although such publications may work as guidelines for future research, this does not imply that scientists are somehow forced to follow this line of reasoning – as Fleck suggests by using the term "thought constraint". On the contrary, I think that 'being a guideline' can be understood in two different ways, namely either as a guide to the scientist on how to elaborate the theory presented, or as a guide for her on how to criticise it. If this were not the case, it would be incomprehensible how further developments in science could be explained that explicitly contradict the state-of-the-art of a particular academic field.

With this theoretical background in mind, a return will be made to the question of the diversity of scientific images. Fleck developed his theory of scientific communication first and foremost with respect to linguistic representations. Yet many scientific publications comprise visual as well as linguistic representations and, in quite a few instances, also numerical ones. Taking this larger unit of communication here as the relevant subject of research, it comes as no surprise that Fleck's theses apply to them as well. In the following, I will show how the circulation of ideas, described by Fleck, affects the visual elements in scientific communication processes.

3.2.3 Visual representations as proper parts of scientific communication

This chapter started with the empirically supported thesis that there is a broad diversity of visualisations in science – both with respect to their functional roles and their appearances. Downes argues that it would be a mistake to try to explain this phenomenon by pointing to an assumed special representational relation of images to entities in the world (see Downes 2012). He thinks that at least some images can be regarded as models. He uses this thesis to support his rejection of representation-based accounts as suitable candidates of a rationale in this context. By functioning as models, images do not refer to any entity in the world and, consequently, Downes claims that "the epistemic work images do cannot be accounted for via an account of representation that characterizes relations between images and objects in the world" (ibid., 129). Although some might object that it is at least unclear whether images can really be identified with models (see e.g. Perini 2010, 146ff.), Downes's statement brings back to mind one particular functional role of visual representations already discussed above. Amongst others, they can be an *essential part of visual thinking* – in design, engineering sciences, but also in other scientific contexts – and thus Downes is correct in stating that they do not necessarily represent something pre-existent in the world. If this important functional role is not to be excluded, the question why images are epistemically useful in science cannot be answered by pointing to an account that attributes all possible epistemic capacities to representational relations (of whatever kind).

Furthermore, it has already been noticed that social mechanisms play a relevant role when an adequate explanation of the highlighted diversity is sought. As many of these mechanisms are closely related to communication, Fleck's theory of scientific discourse, namely the wandering of ideas, offers a fruitful alternative to those representation-based accounts criticised by Downes. In the following, I will suggest that we can take advantage of Fleck's considerations in two alternative ways.

1 The *developmental stages of a depiction* can be conceived as related to the development of the respective idea. Design and content are simultaneously elaborated. Such a development might either consist in a stepwise refinement of a single image, starting from a sketch and elaborating it into a detailed drawing or painting, or by using different types of depiction, for example, drawing, diagram and computer graphic, to represent different levels of cognition.

2 We can regard scientific images as *certain modes of presentation*, namely as the scientist's attempt to aim at information, legitimisation, popularisation or propaganda by communicating a hypothesis to a particular target audience.

Whereas in the former case modifications of the idea (the content) and the image go hand in hand, the (basic) idea stays fixed in the latter case and the scientist only looks for an appropriate way to communicate it.[63] I will discuss both alternatives in more detail below.

The following analysis is meant to show how visual representations vary with respect to the different social contexts of their usage. The alterations of design will be obvious; however, there might be difficulties in understanding the alteration of the functional roles of images. As my approach aims at an application of Fleck's theory of scientific communication, it might be objected that there is only one function of images being discussed at this point, namely that of communication.[64] As a rejoinder, I want to emphasise the fact that communication serves different purposes. Communication is not a monolithic but a multifaceted phenomenon. This is also what Fleck demonstrates for science, that communicating ideas often implies more than simply transmitting previously fixed data. *Information* might be only one intended aim. Likewise, just as linguistic expressions serve more than one purpose in the context of communication, so do visual representations. It would be better to say, then, that, *within the context of communication*, words and images serve their different functional roles to support the speaker's intended aims. Returning to the above distinction, further investigation into utilising Fleck's theory is now in order at this point.

Visualisations (1) can be regarded as an *integral part of the cognitive process*. Here, use can be made of Fleck's explanations concerning the development of scientific ideas. Such a process starts with rough hypotheses and ends with stating facts. Fleck shows how this works with regard to linguistic expressions. Cautious claims develop into bold statements during the publication process – first in journals and afterwards in handbooks. With regard to visual representations, the starting point of this process might be the scribbling of initial ideas and testing different ways of representing them visually. The scientist's first intuitive ideas about her research object (entity or process) might be accompanied by a rough conception in a visual form, also expressing possible insecurities and questions such as: *does it look like this? Does it work like this? Are these the relevant connecting parts?* After that, the testing of hypotheses will proceed with the designing and rejecting of different images (or at least parts of them). Thus, *trial and error* affect both – the idea and its visual design.

Evidently, such a process of reasoning including visualisations, namely *visual thinking*, will very seldom be accessible to an external observer. Normally, ideas are not communicated until the scientist has worked out the initial results she holds worthwhile to present, which is where journal science enters the picture. In having access to the remnants of the scientists' visual thinking, only historians of science might be an exception when analysing the formers' notebooks.

Edward R. Tufte[65] presents some interesting case studies. Tufte's example of the *cholera map* (see Figure 3.1) has already been discussed as an instance of visual thinking (see Tufte 2010, 27–37). In 1854, when Dr John Snow discovered the cholera transmission route, a couple of years ahead of the detection of the bacterium *Vibrio cholera*, different hypotheses were suggested to explain how the disease might spread among its victims (see ibid., 29). Tufte describes how Snow developed his idea that polluted water from an official town pump was the source of the epidemic by displaying death rates and locations of water supplies on a map (see Figure 3.1). "This map reveals a strong association between cholera and proximity to the Broad Street pump, in a context of simultaneous comparison with other local water sources and the surrounding neighbourhoods without cholera" (ibid., 30). Taking into account that this was but one map that Snow plotted to find an explanation, it becomes clear that working with the maps and visually arranging the data also helped him to develop his thesis on the origin and transmission route of the disease. In this sense, Tufte's example illustrates the intertwined development of an idea and its visual representation.

This first alternative to invoke Fleck's theory to account for the diversity of scientific images is also discussed by Alexander Vögtli and Beat Ernst (see Vögtli and Ernst 2007, 107ff.). However, by agreeing with Uwe Pörksen's view, their proposal includes more steps to the cognitive process than hitherto discussed. They also make use of Fleck's concepts of vademecum and popular science.

Pörksen points out that often a developmental history of visual representations in science can be detected, starting with *sketches* via *schemas* to *depictions* in handbooks and textbooks and, finally, to *canonical depictions* in popular science. Examples of the latter kind are, for instance, depictions of the DNA double helix or the mushroom cloud (see e.g. Bigg 2009). It has to be added that those labels are not meant as categories of depiction, that is, *schemas* can be diagrams as well as photographs. Although neither Pörksen nor Vögtli and Ernst explain their choice of wording, I think that what motivates those labels is the level of certainty that the image producer ascribes to her idea expressed in the visual representation. Pörksen therefore claims that these visual developments can be regarded as analogous to the developmental steps of linguistic representations in science described by Fleck (see Pörksen 1997, 105ff.).

Vögtli and Ernst elaborate on Pörksen's text. According to them, *sketches* are of a purely hypothetical nature. They think that these images function as thought experiments and are not meant to be published (see Vögtli and Ernst 2007, 108). Mind maps might be common tools for enhancing creativity, for example. By putting the concept of interest at the centre, the scientist tries to ascertain how other ideas are related to this central claim. Such sketches might, for instance, contain question marks or dashed lines where the author is still not sure whether certain items are related to each other or not. These images express their producer's insecurities in this case.

Schemas are more elaborated than sketches. The drawing is clearer and explanations or a caption are added to guide the observer. Nevertheless they are more or less hypothetical in nature. What does 'hypothetical' mean in this context? The best way to grasp this aspect seems to be by considering the most common case, namely the use of diagrams. Often they show measurement results, error bars included. These visualisations are meant to be distributed among peers for discussion. In this sense, they are part of the hypothesis presented in the paper which will be checked with regard to its plausibility by the community of peers reading and discussing the article. Thus, the term 'hypothetical image' means that this visual representation is as fallible as the verbal thesis in the text and is therefore also open to reinterpretation, which is emphasised by certain graphical elements such as error bars, dashed lines, etc. In this sense, images belong to the circulation of thoughts within the esoteric circle and are parts of journal articles (see ibid.).

Beyond that, Vögtli and Ernst think that *depictions* in a handbook or textbook[66] are normally modified copies of schemas (see ibid., 108f.). Two aspects are relevant to these modifications. Firstly, depictions which appear in textbooks are prepared for education with didactic elements such as arrows, etc. added to them (see ibid., 109). Secondly, the type of depiction may change. Vögtli and Ernst claim that, whereas diagrams seem to be the main kind of visual representations in journal articles, more *naturalistic pictures* will be found in textbooks (see ibid.). Figure 3.3 illustrates this point. It is part of a textbook of biology, admittedly meant to address the broader public, but nonetheless the main features of visualisation that it reveals can also be found in textbooks for students (see e.g. illustrations in *Purves Biologie* (Purves et al. 2011)). What strikes the eye here is the naturalistic background that is added to the frogs at the bottom. Adding such a background is not necessary, but it greatly contributes to a lively presentation.

Vögtli and Ernst claim that what is typical of depictions in textbooks is the elimination of pictorial aspects expressing insecurities such as dashed lines. Consequently, the schematic nature of the image is turned into a depiction that creates the impression of regarding a three-dimensional, solid body. Drawings in perspective are used for this purpose. Shading[67] is added, whereby the producer of the image tacitly accepts that the idea presented appears as one that is being corroborated. With regard to Figure 3.3, this would mean

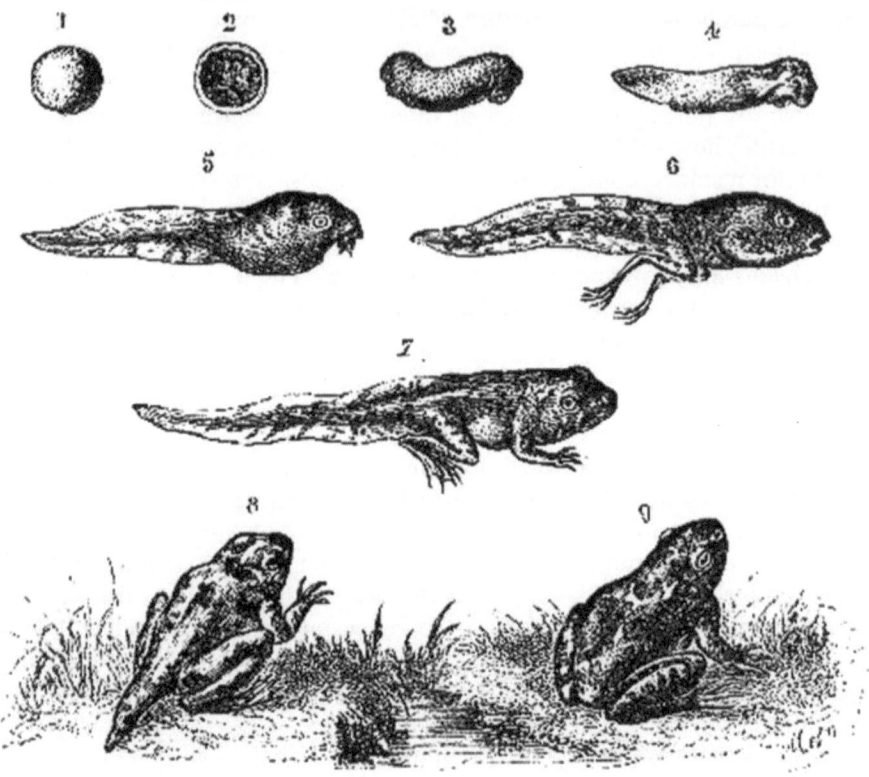

Figure 3.3 Life cycle of a frog.
Source: Figuier, Louis and Gillmore, Parker (eds.): *Reptiles and birds: a popular account of their various orders, with a description of the habits and economy of the most interesting.* London: Cassell and Co., 1883, p. 21. Online edition available at: https://archive.org/details/reptilesbirds00figu

that the designer thinks that the hypothesis concerning the life cycle of frogs expressed in the depiction is true, that the depicted entities exist and that they really look like this in the wild.

Finally, *canonical pictures* used in popular science appear even more naturalistic than depictions in textbooks (see ibid., 110).[68] Sometimes they even enter the field of entertainment, for example, in advertisements or movies, and most of them are not taken seriously by scientists (see ibid., 111). A convincing example in this context that Vögtli and Ernst discuss are depictions (or animations and movies) of dinosaurs. We have all been familiar with such depictions since childhood. Some might even claim that they know what dinosaurs looked like – actually no one does. All that is left of this ancient world are fossil skeletons and footprints; claims about texture and colour of skin are mere speculation. This might also be a reason why many scientists are sceptical, if not worried, about science documentaries that mix science-fiction elements with scientific results (see for instance *Walking with Dinosaurs* produced by the BBC in 1999). Although a lot of speculation is involved in such movies, their producers present their contents as scientific facts and the images become canonical even though they are speculative in nature.

Obviously, Vögtli and Ernst take this developmental process of visualisations to be *unidirectional*, from sketches to canonical depictions. They do not consider the feedback loops between the esoteric and the exoteric circle triggered by these popular depictions. We should, however, add this important aspect in Fleck's sense, as he conclusively points out the crucial influence of these popular visual representations on the expert's work.

What should have become clear is how developing an idea and its visual representation can proceed hand in hand. This is not to claim that it always has to be so, although I think that Vögtli and Ernst present a convincing description based on Fleck's theory of scientific communication. By mentioning the different kinds of visual representations involved in the developmental steps of a scientific hypotheses, they show how and why the appearance of those visual means of communication vary.

Furthermore, the different contexts in which those visual representations appear also point to potential changes of functional roles. Sketches are used for visual thinking, that is, during genuine research the scientist makes use of them to elaborate her hypothesis. Schemas are published in journals. They are meant to present ideas to scientific peers and transmit genuine data that are intended to be discussed within the community. Depictions in handbooks or textbooks present the state-of-the-art. They show what scientists think is the result of their research. They are means employed to teach students how to perceive objects of research within a particular community, are prepared for this purpose by means of didactic elements especially in the context of education, and highlight what is of importance to recognise, what demarcates one species from another. And, finally, canonical depictions in popular science are normally used to persuade and also to entertain.

Although I think that Fleck's theory of scientific communication characterises scientific discourse elaborately, including the alterations to linguistic and visual representations involved, some aspects seem to be exaggerated. I would argue that the developmental process described above does not apply equally well to all visualisations in science. Yet this does not imply that Fleck's theory is of no use here. For those instances which do not fit the above schema presented by Vögtli and Ernst, we can make use of Fleck's ideas in another way to explain the diversity of scientific visualisations.

This way (2) to make use of Fleck's theory is by taking the design process and the development of ideas as separate procedures. Although this approach diverges from Fleck's account, since Fleck defends the thesis that the development of an idea and its representation *always* goes hand in hand, we can nevertheless benefit from his explanatory schema of scientific communication.

In this second context, visual representations are chosen in accordance with what the publisher thinks is the best way – the best mode – of presenting certain information to a particular target audience. Especially, considerations about the assumed level of the audience's background knowledge will determine what kind of visual representation are chosen as an *appropriate mode of presentation*.

Depending on the audience, simplifications might occur. However, the *basic* idea will stay unchanged, as scientific discourse seems to presuppose a certain stock of shared background knowledge even in the domain of non-expertise.[69] The more specialised the idea transmitted, the more background knowledge is to be expected from the audience. Theses on high energy physics will seldom catch the attention of completely uninitiated readers. Thus, although laymen will be addressed by certain scientific publications, they are not totally ignorant of the topic presented.[70] Simplifications will therefore not exceed such a degree that the idea presented will be altered beyond recognition. However, if the presentation of the idea is changed in such a way that its basic content does disappear, then it can be assumed that the testifier in question is simply incompetent in transmitting what is of relevance. This would be an instance of bad science communication, or even of transmitting false ideas.

In addition to reflections on background knowledge, considerations about the intended audience may imply assigning certain functions to the visual representation chosen. Addressing laymen may fall into line with the intention of using visualisations to persuade the audience, or to attract attention, or to trigger an emotional response. In the context of discussions with academic peers, visualisations might be used to offer an overview about a topic, that is, to be a heuristic means, to inform about certain results and perhaps also to attract attention.

As an example, consider the announcement of the discovery of the *most remote galaxy* (called "UDFy-38135539") in October 2010. The details of this short case study show how scientific images are chosen to support two different communicative aims, namely *popularisation* and *information*, as Fleck labelled them. The discovery mentioned was made public in two different ways: on the internet in *Spiegel online*[71] for the interested public and in the journal *Nature* (see Lehnert et al. 2010) addressing the scientific community. How visualisations vary when different readerships are addressed can be mentioned in this instance.

The *Nature* article was supplemented by four diagrams referring, for example, to the spectrographic analysis of the galaxy's near-infrared emissions. The article for laymen, on the other hand, contained three completely different images: the first was a result of a computer simulation which was said to depict the universe at the time when the galaxy's light started its journey. The second one was a collage of two photographs of the galaxy taken by the *Hubble Space Telescope* showing the position of the galaxy in the *Ultra Deep Field*. And the third one was a photograph of the *Very Large Telescope* in Chile, which was the source of the above-mentioned spectrographic analysis.

Apparently these three images were meant to give the reader an impression of the facts explained in the text, i.e. where to look for the galaxy in the night sky, how to imagine the early universe, and what the instrument used for the discovery looked like.

Why did the journalists not just reuse the diagrams that were published in *Nature*? Without doubt, the writer of the *Spiegel* article believed that they would not be appropriate in this context. Such diagrams are not easily comprehensible, they have to be interpreted, and one may reasonably assume that laymen do not have the relevant training to do so correctly. The author therefore chose representations which he thought would be in accordance with the assumed level of background knowledge of the average *Spiegel* reader – and this is exactly what Fleck says about popularisation. Moreover, the audience addressed seemingly lacks the competence to check the presented results critically. They simply accept what the specialist tells or shows, and this is precisely in accordance with Fleck's claims (see Fleck 1986c, 86).

Besides this, the example also illustrates what happens in the case of *information*, that is, in the case of communication between scientific peers. The authors of the *Nature* article provided four diagrams showing the results of their analysis. Obviously, these representations serve here as evidence. Visualisations make research results accessible. They often present the most important information in the article. This thesis – also held by Perini (see Perini 2010, 134, 140) – is supported by the following: some of the world's major research centres in particle physics have implemented a new database called *INSPIRE* that extracts figures from the corresponding articles and makes them independently available and traceable via their captions.[72]

To make visual representations part of a research article means offering the audience the opportunity to see for themselves, to repeat and check the experiment and the correlated data. Sufficient background knowledge on the part of the readership is therefore presupposed by the article's authors. As Fleck claims, "[o]ne can talk about an informing statement only for the case when its recipient has specific confidence in the cognitive abilities of the sender (cognitive confidence) and simultaneously the full possibility of checking the contents" (Fleck 1986c, 86).

Thus, the above example shows how considerations about the intended audience influence the presentation of scientific results in different media. However, the underlying scientific fact, namely the discovery of the galaxy mentioned, is not affected by the different modes of presentation. Thus, in some instances, Fleck's thesis about scientific communicative practices is too strong. He believes that each communicative act also changes the idea (see Fleck 1979, 109f.), whereas the above example makes plain that this is not the case. Obviously, at least in some instances, scientists are able to *keep the content constant while changing its form of presentation*.

I shall now examine an example of *propaganda*. As stated above, it can be doubted that genuine cases of propaganda can actually occur, as the communicating agents will form a new thought collective in Fleck's sense.

Accordingly, within this new context, the participants will try to bridge the gap between their genuinely different communities. This seems to be especially true for the agent who wants to transmit her information. She has to take into account that her audience will share only a minimal set of background assumptions.[73] Although Fleck is right to point out that in such instances attempts to convince recipients are more likely to take the form of persuasion than that of careful argumentation, the former cannot be achieved without aiming at a rudimentary understanding of the agent's assertion by her audience. Thus, she will try to express her thoughts in such a modest way that her audience can grasp at least the basics. I am suggesting an *extreme case of popularisation* here, and not trying to argue for a completely new category, namely propaganda.

Otto Neurath's methodological efforts to transmit statistical data of common concern to uneducated people via visualisations can be regarded as an example of such a communicative attempt (see Neurath 1991). As a means to communicate such information, Neurath tried to develop a system of visual signs, namely *ISOTYPE* (International System of TYpographic Picture Education). *ISOTYPE* was intended as a method to communicate difficult statistical data about economic or social concerns to the members of the working class in an easily understandable way. Consider Figure 2.11, discussed in the previous chapter, as an example.

The audience that Neurath wanted to address with the aid of *ISOTYPE* was not supposed to comprise educated readers or informed laymen (see e.g. Neurath 1927; 1929; 1931). Accordingly, the visual display of data had to be rather simple. Symbols had to be used which could be recognised easily in different settings (see Neurath 2010, ch. 5). Neurath was aware of the fact that this method of data communication also meant omitting certain details and allowing simplification and apodictic valuation to play a role. This is exactly what Fleck claims for the case of propaganda. Nonetheless, from Neurath's point of view, it would be better for his audience to have a rough idea about what was occurring in their environment than to have no knowledge at all (see Neurath 1931, 185). *ISOTYPE* can be regarded as an instance of visual representations used for the purpose of an extreme case of popularisation in the above sense.

Taking stock of the above discussion of examples, it can be stated that the different aims of scientific communication in combination with the different target audiences involved, mentioned by Fleck, can indeed be made accountable for the alteration of the appearances and functional roles of scientific images. Yet the analysis also makes plain that Fleck's theory of scientific communication does not comprise all the relevant aspects in this context. Roughly speaking, they are as follows: teamwork involving public relations officers, etc., different modes (for instance, tables versus diagrams) and contexts of presentation (for example, poster session versus scientific article).

Admittedly, some aspects are the result of more recent developments in methods of circulating scientific ideas. Thus, Fleck was not able to see the whole range of consequences, as he developed his ideas in the first part of the twentieth century. Those aspects omitted by Fleck speak in favour of his thesis that social interactions are most relevant to consider when analysing epistemic processes. In the remaining part of this section, I will discuss these points further.

Firstly, not all ideas are visualised and transferred by the scientist herself. On the contrary, many people co-operate to pursue this task. There are public relations officers employed at universities or research centres engaged in transmitting results in accordance with a target audience. There are science journalists like the ones mentioned in the *Spiegel* example. Graphic designers might also be involved, such as in the *ISOTYPE* example, and many more people might play a role. Their *teamwork* once more emphasises the social nature of science, as it is crucially involved in knowledge-distributing processes.

As the second alternative to make use of Fleck's theory seems to be the more common one, the social influence noted becomes even more relevant to our correct understanding of science. Ignoring its influence would block explanations concerning cognitive processes in this context. It seems, for example, reasonable to assume that, at least in some instances, discussions between scientists and designers also help to shape the idea being presented. By explaining what the essentials are that are to be transmitted, the scientist might become aware of some aspects that are still underdetermined and have to be worked out further. There will be a feedback loop, providing for ongoing developments of the correlated idea.

Moreover, this kind of 'teamwork' can occur, at least partly, within the mind of the particular scientist. There are lots of design manuals and guidebooks (see e.g. Carter 2013; Frankel and DePace 2012; Tufte 2010, 2011) attempting to improve presentations prepared by scientists. Taking their advice seriously also relocates the above-mentioned discussion between scientist and designer within the former's mind, while she becomes an exoteric member of the design thought collective.

Furthermore, the *mode of the presentation* can play a role when considering how to present a particular piece of information. Depending on the kind of information that is supposed to be transmitted, particular modes might be more suitable to meet this aim than others. Matt Carter, for example, highlights the difference between *tables* and *figures* in this context:

> [t]ables display data (numbers or words) organized in rows and columns. Unlike figures, tables usually show data in their unprocessed, rudimentary form. The key to designing a great table is to arrange information clearly and logically so that data are easily accessible and comprehensible to an audience.
>
> (Carter 2013, 85)

Representations in guidebooks for biologists are another example. Although many photographs are available today, most scientists still prefer drawings as they show the relevant details more clearly.

Additionally, the *context of presentation* can be relevant. A simple example concerns the choice of colours. If the target medium is not printed in colour, it can be crucial to change the image respectively. Parts that are coloured differently can, for example, be marked in greyscale which will normally highlight the relevant contrast better than merely printing the original colour graphic in black and white. William Goodwin presents another example in this respect. Referring to the academic domain of organic chemistry, he points out that physical models used in research activities cannot be reproduced in scientific articles. Nonetheless, he thinks that the former can be represented by using a completely different type of representation, that is, not simply a photograph or drawing of the original physical model. He states that:

> [o]f course, it is not possible to reproduce physical models of organic compounds on the printed page, and so structural formulas stand in for physical models in chemical discourse. [...] Names, even systematic descriptive ones, cannot be manipulated to reveal the possible spatial configurations of a compound. Structural formulas – supplemented with some additional conventions – are, on the other hand, objects whose characteristics can be used to infer properties of the compounds they denote.
>
> (Goodwin 2009, 381)

Goodwin thinks that those formulas serve as models within the context of a publication, enabling the recipients to manipulate them for inferring the relevant information.

Another feature that is of relevance with respect to the context of presentation is the scientist's presence or absence during the presentation. Nowadays, poster presentations and slide shows are also important ways of transmitting information. In such instances, however, the scientist normally explains the content of her poster or slides directly to her audience. In case she wants to reuse these contents afterwards as parts of an article, she has to make sure that all the information she gave orally will be included in the text or image, for example, by adding arrows to guide the audience's attention.[74]

All of these aspects – considerations about the target audience, further teamwork, mode and context of presentation – can not only influence the appearance, but also the functional role of visual representation involved in the act of communicating scientific results.

3.2.4 Interim results: what can be learnt from social mechanisms?

Concisely, then, both alternatives to make use of Fleck's theory of scientific communication offer the possibility of explaining the diversity of scientific images as a result of social mechanisms, namely of communicative acts pursued for different aims and involving different target groups.

Table 3.3 Summary: intertwined development of idea and visualisation

Context	Type	Appearance / design characteristics[75]	Communicative aim and characteristic function
Genuine research	Sketch	Incomplete, focus on central ideas, private shortcuts, no clear labelling	Visual thinking: medium to develop hypothesis in a process of trial and error
Journal	Schema	Hypothetical depiction (dashed lines, question marks, error bars), often diagrams, clear caption and labelling added	Information and legitimisation: medium to transmit research results that are supposed to be discussed within the community
Handbook	Depiction	Naturalised, smoothed depiction (3D impression, perspective, shadows), all kinds of visual representations, clear captions	Information: medium to represent the state-of-the-art
Textbook	Depiction	(See handbook) + didactic elements (arrows, explanations, highlighting of important aspects); depending on the context also simplification, explanatory captions adequate for educational purpose	Information and teaching: medium to teach ways of seeing
Popular science	Canonical depiction	Naturalised apodictic depiction, simplifications, colourful, often animated, captions often superficial	Popularisation: medium to gain (and maintain) attention, also via entertainment and to persuade the audience

The first alternative, also suggested by Vögtli and Ernst, shows that sometimes the developmental process of ideas and their visual representations can be deeply intertwined. In this case, the respective level of the cognitive process appeared to be decisive for the function and appearance of its visual companion. Table 3.3 summarises the results of this analysis with regard to the design and function of visual representations within the different developmental phases of such a process. From the top to the bottom row, it shows how the idea, expressed in the visual format, begins as the hypothesis of a scientist pondering alternatives and is afterwards increasingly convincingly and elaborately presented in the different communicative contexts of scientific discourse.

The more common alternative, however, seems to be the second way to make use of Fleck's theory. Here, expanding the idea and developing its visual representation are distinct processes, although feedback loops might link them at certain stages. Here, the main source of visual diversity turned out to be the attempt to adapt the idea to an assumed level of the audience's background knowledge as well as to the mode and context of presentation.

Moreover, it became clear in the analysis of this second alternative that the number of team players interacting in scientific discourse exceeds by far the more simplified model that Fleck seems to suggest. For example, making visualisations part of scientific discourse means – at least – that they have to be produced and printed. In his historical perspective on visual cultures – especially in astronomy – Klaus Hentschel points out that, until recently, scientists were dependent on craftsmen and artists regarding these steps of production (see Hentschel 2000, 31).[76]

Alex Soojung-Kim Pang describes in detail how to conceive of this often difficult teamwork in his case study on astrophotography at the Lick Observatory as it was practised during the late-nineteenth and early-twentieth centuries (see Pang 1997). Firstly, Pang's analysis clearly delineates the problems occurring in this social interaction concomitant with an epistemological analysis of visual representations of celestial objects produced in those social processes. The most important aspect in this respect was the discussion between astronomers and engravers about what kinds of modifications to allow or to prohibit (see ibid., 189ff.). Whereas the former kind was meant to improve the image's visual quality by eliminating artefacts of production processes, the latter would have meant manipulating the essential data that was to be transmitted in the publication. Yet the difference between these two kinds of manipulation were often far from being clear to the craftsmen.

Pang tells the story of one particular event which makes this difficulty especially apparent. Working on an illustration covered with spots for the *Keeler Memorial Volume*, the question arose, which ones were artefacts and which ones belonged to the depiction of the star cluster. Whereas the elimination of the former was epistemically unproblematic, deleting the latter would have meant distorting the scientific result.

> The problem was that some of the spots were obviously artefacts, but some of them looked like stars, and no one was willing to risk accidentally removing the latter along with the former. The engraver could begin by touching out the mistakes, but it was too easy to cross that literal boundary between space that could be altered, and a subject that could not.
>
> (Pang 1997, 197)

It becomes clear that errors might occur that can distort the genuine research result. Yet Pang also shows that both parties involved, i.e. scientists and engravers, did their best to reduce such sources of possible mistakes, for instance by checking proofs of the printed data several times (see ibid., 200). Thus, although Pang's description reminds us of an epistemically careful handling of data published in such a way, it does not undermine its credibility in a particular way. In the end, as science is a human activity, all data should be regarded as *fallible* (see Chalmers 1999, 14ff.). Moreover, fallible data as such do not undermine the search for truth as the aim of science, as Karl Popper points out. As long as we maintain our *critical attitude* towards scientific results, it is possible to learn from mistakes and to approach the truth (see Popper 2002, 309ff.).

Secondly, Pang also points out that scientists and engravers developed a new kind of vocabulary to enhance their communication of what was essential to the visual representation (see Pang 1997, 195f.). This detail of his case study makes it clear how it is conceivable that the exchange of ideas will lead to the formation of a new, common thought collective between speaker and hearer. Moreover, this also supports my thesis that Fleck's conception of *propaganda* as the purpose of inter-collective communication is somehow misconceived. Pang's analysis shows that, contrary to Fleck's thesis, both parties involved had aimed at a mutual understanding in their exchange of ideas, and thereby created not only new expressions but also established new processes of teamwork and even negotiated common standards of picture reproduction. Thus, despite misunderstandings occurring and communication between both groups not being easy,[77] mutual efforts to improve the collaboration seem to have bridged initial gaps between the people involved, and finally united them in a newly established community.

Pang thinks that *aesthetic judgements* played a crucial role in the astronomers' evaluation (of the adequacy) of the engravers' work on the photographs. In this respect, he discusses the formers' emphasis on details, contrast and the colour and depth of the background sky (see ibid., 191ff.). Does this mean, as Anja Zimmermann suggests, that scientific images are *essentially constructed* so that their analysis should, as she claims, *begin* with this constructivist character and not just acknowledge it as being the result (see Zimmermann 2009, 38)?

Although I think that Zimmermann is correct in pointing out the significance of considering the mutual enrichment of the scientific and the artistic domain when theorising about visual representations (see ibid., 194ff.), I refrain here from stating hierarchies in analysis. I am, however, convinced that although aesthetic predicates might be invoked in the assessment of the photographs in Pang's case study, this does not imply that these pictures show mere constructs. Two points in the very same case study support my more realistic reading of his results. On the one hand, Pang stresses that astronomers put their emphasis on "faithful reproduction" (Pang 1997, 190). That is, they wanted to transmit their data correctly, understood in a correspondence sense of truth. This becomes clear when Pang states that engravers were allowed to manipulate the background but not the stars or nebulae – that is, the essential data. On the other hand, Pang draws our attention to the "materiality of images" (ibid., 201). As artefacts of production processes those images are subject to certain sources of imprecision, errors and distortions. The quality of the paper or ink may play a role as well as the reliability of the printing mechanisms and so on. However, although this makes it clear that the recording devices can have an impact on the accuracy of displaying scientific data, none of these ingredients is part of the scientific result to be transmitted. Thus, separating data from artefact might be a hard job, as Pang's case study elaborately shows, but it is often possible nonetheless. Thus, construction might play a role, visualisations are artefacts in the sense of being products of imaging processes, but this does not undermine *per se* the evidential status of such visual representations.

The last point that I want to make in this summary is that the above analysis has also clearly indicated that, because of the steadily growing supply of books and other media supporting self-studies and the availability of software tools to practise newly acquired competences, this connection between the special sciences and design studies can also take place within the mind of a particular researcher. That is, no external collaboration has to occur to connect these domains. A consequence of this way of connecting different communities is that, within the scientific context, people are inclined to expect their colleagues, peers, employees or students to show the relevant competences – although learning to be a competent designer of (visual) presentations of one's research result is not (or only seldom) a proper part of the (traditional) training. Accordingly, Frankel and DePace state that:

> [u]ntil fairly recently it was standard practice for universities and research institutions to hire specialists to help researchers visually communicate their work. Now the research community is primarily responsible for crafting its own graphics – and yet the typical researcher's training rarely includes the developments of such skills and sensibilities.
>
> (Frankel and DePace 2012, 3)

Obviously, this can be another source of potential error. Not being a professional within this domain, people might be tempted to exaggerate their design work. Hentschel expands on this thesis when he states that, as a consequence of this new task of image production in combination with the availability of powerful software tools supporting this work, aesthetic judgements become more relevant

> [...] not only (as you might expect) while preparing 'pretty pictures' in 'enhanced' or 'false' colors for promotional talks and (semi)popular articles, but also during the 'crafting of natural resemblances' in their daily routine, for instance, in the choice of color scale and frame, in the 'tailoring' to bring out a special feature of the picture, and in the selection of the final image from among a whole series of candidates.
>
> (Hentschel 2000, 31)

Thus, from an epistemological point of view, this has to be borne in mind as a possible epistemically distorting factor when analysing visual representations as components of scientific discourse. Again, this should remind us of the fallibility of scientific data in general, but it does not support the thesis that visualisations have to be necessarily interpreted in a constructivist way.

3.3 Summary

In this chapter I have argued for the relevance of maintaining contexts of usage apart when theorising on scientific images. The two main contexts with this respect, analysed in detail above, are the exploratory context – the context of genuine research – and the explanatory context – the context of transmitting information amongst peers and laymen.

Although there is some overlap among the more common distinction between the "context of discovery" and the "context of justification" introduced by Hans Reichenbach into the debates of philosophy of science (see Reichenbach 1961, § 1), I am convinced that the former distinction better suits our purpose of analysing the epistemological roles and status of scientific visualisations than the latter. This is particularly due to the following disadvantage of Reichenbach's distinction. Reichenbach thinks – and many philosophers of science have followed his suggestion – that the context of discovery is *epistemologically inert*, so to speak. It simply does not matter how a particular scientist generates her ideas or how her mind works in constructing theories and so on. This might, at best, be of psychological but not of epistemological interest. This hypothesis also seems to be the reason why Reichenbach introduces this distinction, namely to demarcate those two scientific domains – psychology and epistemology – when analysing cognitive activities such as the acquisition of knowledge.

However, I am not content with this distinction. On the one hand, I am convinced that it is crucial for philosophy to analyse, for example, the nature of creativity, analogical reasoning, or the use of metaphors, to obtain new insights into old domains. That is, phenomena which, following Reichenbach's suggestion, would be altogether irrelevant to philosophical considerations. On the other hand, I part with Philip Kitcher, who suggests that epistemology – and thus also the philosophy of science when dealing with epistemic questions – should be "psychologistic" (Kitcher 1993, 184). Not only should philosophers take into account the logical connections between beliefs, as Reichenbach requires, but also their psychological connections. Mutually informed philosophical and psychological analyses seem to be a more promising approach to comprehending the scientific endeavour of understanding the world epistemically.

In the analysis of scientific images, it has become clear that these visual representations support important epistemic achievements also when the invention and development of scientific hypotheses are at issue. Learning about phenomena by manipulating their visual substitutes or by evaluating their visual measurement-output seem to be essential activities in science. Evaluating whether these processes are performed in a way that justifies knowledge claims based on them seems therefore a valuable task for philosophers of science. For this reason, I think that the above distinction between exploratory and explanatory context which concedes epistemologically relevant activities to both domains suits our purposes better than Reichenbach's proposal which *per se* excludes epistemologically relevant phenomena in one of the domains that he suggested.

The analysis of functional roles played by scientific visualisations in the explanatory and the exploratory context has also made clear that anti-realistic worries exploiting the artificial character of scientific visualisations are often a consequence of confusing those two contexts. It was especially helpful to clarify what kinds of *relata* are connected via scientific visualisations in those contexts in order to see this difficulty. Acknowledging the fact that these *relata* can vary makes it much easier to understand that the information transferred by visual representations can also be of different kinds.

Furthermore, keeping contexts apart has also revealed that there is a variety of reasons that supports the reliance on scientific information inherent to visual representations. Briefly, with regard to the exploratory context, it has been seen that the scientists' background knowledge about the instrument's causal connection to the object of research, about the former's calibration and about the informativeness of the image about its object (in particular a resemblance relation holding between them) established by a relevant mapping can constitute such reasons that can be relied on. In the explanatory context, too, there are many reasons that permit a reliance on visual representations in epistemic endeavours. Most of them are a result of their being embedded in the epistemic practice of testimony, but some are also specific to visual representations. The latter are a result of taking background knowledge from the exploratory context into account in order to check for consistency and hence for reliability.

Finally, this contextual analysis has also opened a way to explain the diversity of scientific visualisations that Downes highlighted and for which a mere representation-based approach obviously cannot account. What has been revealed in the analysis are the social mechanisms involved in scientific cognitive processes making use of visual representations.

My proposal to solve the problem of diversity has been to use Ludwik Fleck's theory of scientific communication processes. Especially, his considerations on the influence of both the different *modes* and *aims of communication* on the content transmitted proved to be fruitful in this context. Regarding visual representations as proper components of communicative acts allows an explanation of their alterations analogous to the one put forward by Fleck regarding the adjustment of linguistic expressions. Yet the analysis has also shown that at least some of Fleck's theses are far too strong. Contrary to his claims, in many instances the idea to be transmitted stays fixed, whereas only the medium of transmission varies. Nonetheless, Fleck's theory makes it clear why functional roles and the appearances of visualisations are altered constantly in such processes of scientific communication.

Notes

1 An overview of attempts focusing on images and pictures in different meta-disciplines of science (history, sociology, philosophy) and Fleck's place therein is offered by Dommann (2004).
2 Fleck worked as a microbiologist, and his philosophy of science was informed by his daily life as a scientist.
3 Collections of case studies on how exactly visualisations play a role in different academic disciplines are presented, for example, by Liebsch and Mößner (2012) and Sachs-Hombach (2005a).
4 It has to be borne in mind that now we are not talking about the photograph of the paintings but of the paintings themselves.
5 Admittedly, the division of epistemic labour has to be taken into account here. That is, in some instances the analysis of the resulting images and their evidential status is not (or not exclusively) performed by the investigating scientists, but by technicians or computer scientists developing the respective instrument or software. Jörg R. J. Schirra describes this more service-orientated approach to visualisations in science from the perspective of computer science (see Schirra 2005; 2012).
6 The topic of visual thinking is discussed in detail by Rudolf Arnheim (1997).
7 Martin Scholz presents a detailed analysis of design images in the context of techno-sciences (see M. Scholz 2000).
8 Letitia Meynell, for example, thinks that they are representations in this way (see Meynell 2008). De Regt, on the other hand, clearly denies this and emphasises the tool character of these diagrams which enables their use for different purposes (see de Regt 2014, 390f.).
9 A web-based simulation of a blink comparator to detect variable stars is offered online by the Nebraska Astronomy Applet Project (NAAP): http://astro.unl.edu/naap/vsp/animations/blinkComparatorSimulator.html, accessed June 27, 2016.
10 Here I make use of the analogue-digital distinction put forward in computer science. I refrain from Goodman's suggestion of a distinction between analogue and digital systems of depiction (see Goodman 1976, ch. 4.8). Defending the idea of characteristics of symbol systems as demarcation criteria of representational modes, Goodman defines a *digital system* in the following way: "[a] digital scheme [...] is discontinuous throughout; and in a digital system the characters of such a scheme are one-one correlated with compliance-classes of a similarly discontinuous set. [...] To be digital a system must be not merely discontinuous but differentiated throughout, syntactically and semantically" (Goodman 1976, 161). Although this definition includes digital images composed of numerical values in a binary code, it seems to be too broad. Goodman's definition would also apply to the children's game "colour by numbers". Moreover, by omitting the fact that one of the corresponding data sets is numerical in kind, Goodman's definition would also allow another system of depiction to play this part, which undermines the difference that is put into focus in my analysis, namely the distinction between two *different kinds* of representations.

11 A detailed description of how computer graphics are produced is presented by Schirra (2012).

12 We will come back to this issue in more detail in section 4.3.3.

13 At least, this is what a proper academic education should aim at.

14 As I have pointed out elsewhere, Ludwik Fleck's theory explains nicely the difference between synchronic and diachronic problems of understanding in science (see Mößner 2011).

15 Within the context of media reports on political topics, Kjørup offers some interesting examples of the epistemic status of photographs. Moreover, he makes us aware of the fact that manipulations are not only due to post-processing, but are often a result of the interpretation offered in the text accompanying the picture. That is, the author tells the audience how to understand the image *correctly* – to take it as evidence for x and not for y, so to speak (see Kjørup 2013).

16 I discuss this difficulty in detail also in Mößner (2017).

17 Most likely, she refers to the analogue mode of picture production here, as it can be argued that few people really understand how pictures are produced by digital cameras.

18 Max Dauthendey offers an interesting insight into the initial difficulties in and obstacles to producing photographic images shortly after the invention of this new technology (see Dauthendey 1981).

19 A detailed discussion of the method of calibration and an analysis of its reliability as a benchmark of the veracity of measurement results is offered by Allan Franklin (1999, ch. 8).

20 Michael A. Bishop (1992) discusses this phenomenon under the label of "expectation argument" as one theoretical approach within the debate of theory-ladenness of observation in Bishop (see ibid., 295ff.).

21 Christian Suhm (2005) offers an elaborate discussion of the different theses of scientific realism in Suhm (see ibid., ch. 2).

22 Goodman's ideas concerning the assumed arbitrariness of visual representations, that is, their ability to stand for any kind of referent that is ascribed to them in the symbol system they belong to, has been discussed in section 2.2.2 of this book.

23 A convention to fix pictorial meaning in comics, for example, states that *lines following a drawn character are to be read as indicators of the character's movement*.

24 'Bandpass' means "the overall sensitivity of an instrument as a function of wavelength: it includes the effects of filters, plus characteristics of the detector, and telescope mirrors" (Michael Richmond, http://spiff.rit.edu/classes/phys445/lectures/colors/301_filters.html, accessed June 27, 2016).

25 The term 'image producer' seems to comprise two different types of people here: the *scientists* using preconfigured imaging technologies and *technicians or engineers* developing such instruments. However, in science, this distinction is not particularly relevant. Many instruments, imaging technologies and software have been invented by scientists. I therefore regard it as rather uncontroversial to take these inventions as proper parts of the scientific realm and not referring to engineers – apparently laymen – who are supposed to make the relevant decisions mentioned above.

26 For a full definition of this term see Mößner (2010b, ch. 2).

27 This approach is critically discussed by Perini (2012c).

28 See http://cms.web.cern.ch/news/observation-new-particle-mass-125-gev, accessed June 27, 2016.

29 'Independent' means that those reasons are not acquired via testimony, neither by the testifier's own words nor by the words of other people. A recipient has to gain them from another epistemic source, such as perception, etc. This requirement of independence is meant to avoid a vicious circle that some philosophers see lurking in invoking testimonial beliefs in the process of justification (see e.g. Fricker 1995). It has to be added that not all philosophers share this worry. Jennifer Lackey, for example, argues that supporting beliefs concerning the speaker's competence and sincerity can be derived from testimony (see Lackey 2008, 185ff.), though she too qualifies her claim by stating that this is permitted "so long as they are not *ultimately and entirely* testimonially grounded" (ibid., 186, her italics). Thus, the question remains whether Lackey's proposal offers a true alternative to claims such as Fricker's.

30 Showing signs of dishonesty or incompetence, such as avoiding eye contact while speaking, are typical examples of such defeating reasons mentioned in the debate (see e.g. Scholz 2001).

31 Meanwhile, philosophers have developed intermediate accounts, combining the two traditional approaches in one way or another. An example is Lackey's "dualism" (see Lackey 2008, ch. 6).

32 For a detailed analysis of both accounts see Mößner (2010b, ch. 3) and Faulkner (2011, ch. 2, 4).

33 The difference between both accounts is due to the fact that anti-reductionists take testimony as having parity with the other epistemic sources mentioned above, whereas reductionists regard the word of others only as a derivative source which has to rely on other sources for justifying beliefs transmitted via testimony.

34 The problem of gullibility is discussed, for example, by Fricker (1994).

35 Gloria Origgi offers an insightful discussion of the interrelation between the epistemological and the moral definition of trust that surfaces here in the context of testimony (see Origgi 2004).

36 On the relevance of the division of cognitive labour in science see e.g. Kitcher (1993, ch. 8).

37 He writes for example: "[w]e feel some obligation to believe speakers, but not all speakers on all occasions. In many cases, our testimonial uptake is based on a careful consideration of the evidence, and not anything like trust; and in some cases we would be culpable if this were not so" (Faulkner 2011, 188), and: "[e]qually when it comes to explaining why it is that we trusted in the Testimony Game [...] we find that both the trust-based and reductive, or game-theoretical, explanations of action have their own spheres of application" (ibid., 198).

38 Graham (2006a) develops his epistemological theory of testimonial beliefs, namely a "liberal fundamentalism".

39 Martin Kemp also suggests such personal experiences as a possible source for justifying reasons to evaluate visual representations in scientific publications (see Kemp 2012, 44).

40 Actually, this is what the notion of an *impact factor* is really about, a quantitative benchmark of how often articles of a particular journal are cited. It is an evaluative concept of journals, but not of individual articles, let alone of their authors.

41 Similar cases can be found in traditional journals, too. An example are the results about cold fusion published by Martin Fleischmann and Stanley Pons in the *Journal of Electroanalytical Chemistry* and in *Nature* in 1989. All attempts to repeat the experiment yielding the data that Fleischmann and Pons claimed to have

obtained failed. Sceptics think that results were published that could not have been obtained and that this is therefore a case of scientific fraud (see Achenbach 2015).

42 A coherent set of beliefs not only shows the feature of consistency, but is also characterised by mutual explanatory relations between particular beliefs (see Bartelborth 1996, 136ff.).

43 This might vary, of course, with respect to different academic disciplines. Theoretical particle physics might be a harder case than media ethics, which latter is much closer to the novices' daily life.

44 As a definition of 'thought style' Fleck claims that "[w]e can therefore define thought style as [the readiness for] directed perception, with corresponding mental and objective assimilation of what has been so perceived. It is characterized by common features in the problems of interest to a thought collective, by the judgements which the thought collective considers evident, and by the methods which it applies as means of cognition. The thought style may also be accompanied by a technical and literary style characteristic of the given system of knowledge" (Fleck 1979, 99).

45 This last aspect is often discussed within the context of a *rhetoric of visual representations* (see e.g. Sachs-Hombach 2012, 39f.).

46 Klaus Hentschel (2014, 61) states a similar concern.

47 Intentionality here refers to the aboutness of scientific representations, that is, their informative character.

48 Chakravartty explicitly discusses the case of *isomorphism* in his paper.

49 An overview on the wide range of topics in social epistemology is offered by Goldman and Whitcomb (2011) as well as Haddock, Millar and Pritchard (2010).

50 CMS stands for Compact Muon Solenoid.

51 As stated on their website: "[t]he CMS experiment is one of the largest international scientific collaborations in history, involving more than 3500 scientists, engineers, and students from 184 institutes in 42 countries" (see http://cms.web. cern.ch/content/cms-collaboration, accessed June 27, 2016). Picking up Hardwig's point, they state that a normal CMS paper will have more than 2,100 contributors mentioned as authors of the article (see http://cms.web.cern.ch/content/people-statistics, accessed June 27, 2016).

52 The following ideas are partly presented in Mößner (2016). In this article I additionally examined Fleck's discussion of scientific images as ideograms.

53 A comprehensive introduction to his philosophy of science is offered by Schäfer and Schnelle (1980). Further reading material – letters and biographical information, etc. – is presented in Fleck (2011). Paweł Jarnicki discusses problems of translations in Fleck's theory, as Fleck published his ideas both in Polish and in German (see Jarnicki 2016).

54 I have discussed Fleck's term in more detail, also in comparison to Kuhn's term "paradigm", in Mößner (2011). Claus Zittel (2012) offers an interesting analysis of Fleck's term from a historical and genealogical point of view.

55 This last aspect is partly explained in Fleck's description of how natural scientists pursue a particular ideal in their community. He writes that "[i]t [the intellectual mood, N.M.] is expressed as a common reverence for an ideal – the ideal of objective truth, clarity, and accuracy. It consists in the belief that what is being revered can be achieved only in the distant, perhaps infinitely distant future; in the glorification of dedicating oneself to its service; in a definite hero worship and a

distinct tradition. This would be the keynote of the common mood in which the thought collective of natural science lives its life" (Fleck 1979, 142). Zittel (2012) offers a detailed discussion of this term.

56 Fleck denies being a relativist (see Fleck 1979, 100). The interested reader may obtain more information on this topic in Markus Seidel's article. He tackles this problem thoughtfully by interpreting Fleck's theory as a *relational stance* comparable to Karl Mannheim's line of reasoning (see Seidel 2011).

57 Claus Zittel describes this wandering of ideas as follows: "[t]houghts may wander within a thinking-collective from the esoteric to the exoteric circle of the lay person, where they become factual knowledge and return to the collective as such" (Zittel 2012, 64).

58 Fleck does not explicitly mention the involved parties here. However, his examples (see e.g. Fleck 1979, 120f.) suggest that it is a discussion between experts.

59 There is a problem of demarcation here. Unfortunately, Fleck does not explain why and how the difference between information and legitimisation occurs. His suggestion of a separate discussion seems to be plausible – although further investigations on the topic are necessary, but remain beyond the scope of the current analysis.

60 Fleck uses the relatively uncommon expression "vademecum" to refer to this media.

61 Admittedly, it can be asked whether a genuine case of propaganda, as Fleck conceives it, can ever take place. Fleck takes it for granted that there are overlaps between the different purposes of communication and that seldom, if ever, do the purposes mentioned appear in their pure mode (see Fleck 1986c, 88). However, there seems to be a greater difficulty at large in his theory. This is why I think that his theses on the topic are flawed: Fleck points out that communication – as performed during an act of propaganda – *always* leads to the construction of a *new* thought collective including the communicating parties. Consequently, they will form a new common collective during their exchange, whereas his concept of propaganda presupposes that speaker and hearer belong to different communities. Now, either speaker and hearer belong to one commonly shared thought collective or to two different collectives regarding the content of their conversation, but not to both at once. Consequently, Fleck's concept of propaganda seems to be defective as it implies this contradiction. I owe this point to Ludger Jansen.

62 I have discussed Fleck's theses on the synchronic and diachronic problems of understanding in more detail elsewhere (see Mößner 2011).

63 Admittedly, overlaps between both methods might occur rather frequently.

64 I would like to thank Helmut Pulte and Ulrich Krohs for making me aware of this difficulty.

65 Tufte is not a historian, but has taught courses in statistical evidence, analytical and interface design.

66 Obviously, they do not make the difference between *handbook* and *textbook* which Fleck suggests. Depending on the scientific discipline, this seems to be correct, as in some instances handbooks are used for educational purposes and no particular textbooks are produced in those contexts.

67 On the epistemic role of shadows in scientific images see Hennig (2009).

68 It has to be added that from Pörksen's point of view some visual representations may also skip one or two steps in this developmental schema. As an example

he discusses the development of genetics in biology and its accompanying visualisations. For instance, the schema of the double helix, first published in a journal article, directly made its way into textbooks and popular science without being significantly modified (see Pörksen 1997, 123f.).

69 At least this is how the system of education in western communities works. Basic ideas are supposed to be transmitted at school, being refined later in university courses.

70 This is also due to the fact that *communication* usually involves at least two people – speaker and hearer. Of course, the speaker should make sure that his assertion can be understood by his potential audience. However, it also depends on the recipient's interests. If the topic being transmitted simply does not fit her interests and needs, it can be assumed that she will simply ignore the information on offer – instead of engaging in endless interpretive efforts. Being interested in a particular topic, on the other hand, also means trying to understand what is communicated, that is, actively acquiring a corresponding stock of background knowledge.

71 See "Geschichte des Universums: Astronomen erspähen Methusalem-Galaxie", www.spiegel.de/wissenschaft/weltall/a-724247.html, published October 20, 2010, accessed June 27, 2016.

72 See www.projectthepinspire.net/, accessed June 27, 2016.

73 Of course, the assumption that at least *some* overlap will be available is a consequence of my realistic intuitions. From my point of view, the scientific realist thesis that all people live in the same world and that nature will thus (at least) provide the necessary obstacles to prevent the free invention of different worlds, receives further support from there being some scientific visualisations that can indeed give us epistemic access to the unobservable part of our world and thus enable the inter-subjective testability of correlated claims.

74 Carter offers an overview concerning the advantages and disadvantages of a variety of different presentation modes (see Carter 2013, 18f.).

75 The following characteristics are typically, but not necessarily, to be found in correlation to the different kinds of visualisations.

76 See also his analysis on collaborations with scientific illustrators and image technicians in Hentschel (2014, ch. 6).

77 Pang states that "[a]stronomers described pictures using terms that were not easy to understand, especially when communication occurred over great distances, and were far harder to satisfy than other clients" (Pang 1997, 183).

Visual Representations in Science
Concept and Epistemology

Nicola Mößner

First published 2018

ISBN: 978-1-138-08993-8 (hbk)
ISBN: 978-1-315-10890-2 (ebk)

Chapter 4
The epistemic status of scientific visualisations

 Routledge
Taylor & Francis Group
LONDON AND NEW YORK

4 The epistemic status of scientific visualisations

As the considerations in the previous chapter made clear, visual representations are, without doubt, part of many epistemic processes in contemporary science. Scientists present diagrams in their publications and talks to communicate their research results. They investigate computer-generated images as substitutes for research objects. Drawings in textbooks are used to educate novices, to introduce them to a new field of knowledge and so on. Moreover, it was pointed out that in quite a few instances images might also be used for non-epistemic purposes, for example to gain the attention of a particular audience.

The preceding contextual analysis helped to clarify and, to some degree, to systematise these diverse functional roles played by visualisations in science. However, although the functions of visual representations in epistemic contexts have been rendered identifiable, it might be argued that this is only a descriptive result. That is, merely noting and assessing scientific activities does not imply that these processes are also the best way to achieve the epistemic aims aspired to. Perhaps scientists are wrong in their decision to rely on visual representations to such a degree in their epistemic processes. Perhaps they should make use of other modes of representations in those instances instead.

These considerations correspond to the two perspectives inherent to the philosophy of science. Thomas S. Kuhn advises paying close attention to what is really going on in the sciences, how practices and theoretical assumptions have developed over time, and in what sense social processes have played a role in epistemic contexts. This is the *descriptive task of the philosophy of science*, the component that is supposed to connect philosophers' hypotheses and theories to real world activities in science. Considering ideal situations might be interesting and sometimes helpful, but clinging to an unrealisable ideal can also turn out to be destructive in the long run – for example, by wrongly eliminating established working practices. Yet Kuhn's advice does not mean that philosophers of science have to confine themselves to a mere unreflected recording of activities. On the contrary, there is also a second, *normative part* of their work. That is, not only are philosophers supposed to record, but also to analyse and evaluate epistemic practices observed in the scientists' strivings for knowledge. These analyses might result in regarding the prevalent practices as being defective and requiring improvements in some way. Thus, philosophers are also asked to make suggestions on how to improve those practices from an epistemological point of view.[1] Hans Poser, for instance, discusses these two aspects of the philosophy of science (see Poser 2001). He points out that, among the different disciplines that are concerned with science on a meta-level – such as the history of science, the sociology of science, etc. – the philosophy of science is unique in that it deals with descriptive as well as with normative questions about its subject (see ibid., 15ff., 36).

At least, parts of our analysis of the epistemic functions of scientific images are in this sense a descriptive endeavour.[2] Now I also want to add a normative dimension. I will investigate the question of whether scientists are right to use visualisations in the way they do. Leaving non-epistemic purposes of image practices aside, issues such as the following will have to be discussed: can people gain knowledge from visual representations? If so, what kind of knowledge do they acquire by those means? How does the process of information transmission work with respect to visualisations? Are there crucial differences in information transmission via numerical or verbal representations?[3] Is there a kind of hierarchy involved when comparing these three modes of representation from an epistemological point of view? Such a comparative task highlights what is at stake here, namely the precise *epistemic status* that visual representations possess in comparison to other representational means in epistemically relevant contexts. It has to be added that, as the functional roles of visual representations are diverse in science, it seems rational to assume that there is not one exclusive status to be ascribed to these representational means, but various ones.

In this chapter, I will focus on the explanatory context of science, the context of information transmission and the role of visual representations therein. More precisely, I am interested in the two following questions: are visual representations a suitable means in this epistemic context at all, and are there epistemic purposes that visual representations are particularly suitable to serve in comparison with competing modes of representation?

To find answers to these questions, I shall proceed as follows: firstly, an examination is needed concerning the basic philosophical problem that underlies the discussion of the epistemic status of scientific images. The question is why philosophers are concerned about visual representations in epistemic processes at all? What do they think is wrong with scientific practices making use of images? Is there anything wrong at all? At this point, there will be a return to Perini's considerations, mentioned at the beginning of this book. The apparent tension, pointed out by her, that appears when comparing actual scientific practices and philosophical reflections about them (see Perini 2005c, 913f.) will be discussed, as well as the nature of arguments, and it will also be explained why philosophers think that visual representations are not suitable means to be used in argumentation. This critical attitude will be contrasted nonetheless with some suggestions on how to conceive *visual arguments*.

These considerations will direct an investigation into the *cognitive content* of visual representations, which will form the second part of the analysis. On the one hand, that information presented in different representational modes can, at least partly, be translated from one mode to another speaks in favour of the thesis that visual representations can contain a cognitively accessible content. On the other, as already mentioned, some philosophers and semioticians (see e.g. Eco 1994, ch. 7) object that communicative acts based on visual representations face serious limitations of expression. I will contrast these rather critical stances towards the epistemic capabilities of visual representations with the more optimistic assessments put forward by Kulvicki as well as by Kitcher and Varzi (see Kulvicki 2010a; Kitcher and Varzi 2000). This discussion will lead to the task of more seriously considering the way cognitive access is obtained to the content of visualisations, namely via perception. It has to be asked *what kind of knowledge* we acquire by making use of this epistemic source. What follows from the fact that we access pictorial information by *vision as the primary sense* of human beings to cognitively access the world?

I will argue, and this is the final part of this investigation into the epistemic status of visual representations in science, that acknowledging the fact that the information process is based on perception in this way makes clear in what sense scientific images can actually be worth, in Kitcher and Varzi's words, "2aleph0" words (see Kitcher and Varzi 2000). Apparently there are at least two different approaches to spelling out these epistemic merits of visual representations. One will lead to a discussion of different *kinds of knowledge* and the question in what sense visual representations might be particularly helpful in their transmission. The other will be about *scientific understanding* and how visualisations can contribute to its achievement.

The discussion will include a comparison of the three different representational modes, namely the visual, the linguistic and the numerical. Whereas the second part strives for the rather moderate aim of showing that these different kinds of representation reveal comparable characteristics in the context of information transmission, the last part of the analysis is more ambitious. Here, the question will be pursued whether there is a kind of epistemic surplus of scientific images not realisable with the other vehicles of communication under consideration.

4.1 Visual arguments?

As seen in the previous chapter, visual representations in the exploratory context might seem problematic for several reasons. Image manipulation, the artificial character of visualisations engendering a constructivist interpretation of their referents and the problem of theory-ladenness of observation have already been discussed. All of these difficulties dwell on the *evidential status* of visual representations in research processes. Thus, the question was whether scientists can be epistemically justified when referring to an image of their object of research to justify knowledge claims about it. The analysis above made clear that no general worry about (digitally) manipulated images, a missing referential relation between visual representation and its object of depiction, or a theoretically induced misinterpretation of visual data can be maintained that would turn visualisations in science into a particularly unreliable source of information. On the contrary, a variety of reasons were identified showing that the scientific epistemic practices are reasonable, albeit fallible.

But even though this result speaks in favour of the epistemic capacities of images in science, philosophical problems remain. Interestingly, these are expressed in particular with respect to the explanatory context, where visual representations are intentionally used to convey information. Although this potential of transmitting information motivates scientists to use images in the exploratory context too, philosophers tend to question this capacity of visual representation. In particular, they are sceptical of the ability of images to fulfil this epistemic task in argumentative contexts.

Before going into the details about the exact nature of the philosophic contention here, let me briefly draw attention to an imbalance of the philosophical discussion. The point is that, although the mode of epistemic access, namely *perception*, to information encoded in visual representations remains the same in the explanatory and the exploratory contexts, its philosophical evaluation varies. Few, if any, philosophers would deny that perception can serve as an *epistemic source*. They do not contest the claim that accessing the information encoded in visual representations will perceptually allow the recipient to acquire at least some knowledge about what was supposed to have been transmitted by their means. However, the argumentative context referring to this way of accessing information does not apparently suffice to make knowledge claims plausible which are based on this source of information. Thus, whereas perception seems to be an admissible source when it comes to scientific observations and experiments, the reasonableness of referring to it in the context of scientific arguments is questioned.

Now, it might be argued that what is at issue is that in many instances there is *a necessity to interpret* scientific visualisations to understand their content correctly. However, two points can be mentioned as a rejoinder.

Firstly, interpretations of what has been perceived also seem to play a role with regard to ordinary instances of perceiving the world. The more precise the observer's background knowledge is, the more she will be able to discern, as Alan F. Chalmers makes clear when comparing a layman's observations with those of a professional botanist travelling through the Australian bush (see Chalmers 1999, 11f.). He points out that, undoubtedly, the botanist will perceive more and also more detailed facts about the native flora than the layman will be able to notice. What explains this is that the botanist "has a more elaborate conceptual scheme to exploit" (ibid.) than the layperson. That is, the botanist has more background knowledge at her disposal than the layperson with which to reason about what exactly she has observed. Thus, the necessity of interpretation in the process of understanding is by no means a unique feature of visual representations.

Secondly, background knowledge about the reliability of the instrument and about the mapping function that defines the informativeness of the resulting image with respect to the object depicted limits the spectrum of possible interpretations of scientific visualisations reasonably. Scientists' reasoning is not driven by guesswork, but by these theoretical and practical restrictions. Hence, highlighting the relevance of interpretations to understand scientific images does not constitute a rationale framed to question their assumed epistemic capabilities.

Actually, the bone of contention consists in something else. It is the question about the *propositionality* of the information transmitted that is critically discussed by philosophers, especially in the explanatory context. Epistemologists still tend to understand 'knowledge' primarily as 'knowing-that', namely propositional knowledge (see Grundmann 2008, 71).[4] From this point of view, the question about the epistemic status of visual representations is essentially about their capacity to transmit propositional knowledge.

In the following, philosophical worries concerning the epistemic status of visualisations in the explanatory context will be scrutinised more closely. Moreover, an alternative approach to the topic resulting from the previous considerations about how best to analyse visual representations in the explanatory context will be discussed. All of these reflections will finally contribute to a better understanding of what kind of epistemic contributions visualisations can make with respect to scientific arguments.

4.1.1 The philosophical challenge

In the previous chapter, it has been argued that scientific images possess the *capacity of information recording, storage and transmission*. Contrary to scientists making use of these capacities in their epistemic processes, philosophers still call them into question in the explanatory context. What exactly are the difficulties that cause this sceptical attitude in philosophy?

Apparently not all branches of philosophy are affected alike by these misgivings. Yet, as pointed out above, epistemologists and philosophers of science have shown crucial shortcomings in dealing with visual representations until rather recently. Two aspects are relevant to explain these difficulties in dealing with visual representations. On the one hand, there is a certain philosophical tradition which can be traced back to Plato and René Descartes, whose proponents question the suitability of images as epistemic means. On the other hand, a particular focus on (natural and formal) language in analytic philosophy and, consequently, in the philosophy of science being based on this tradition, has to be mentioned in this context.

The first aspect has often been pointed out as a difficulty in picture theory. I will therefore only briefly summarise the related line of reasoning.[5] Perception is commonly regarded as a source of knowledge nowadays, but this has not always been the case. In particular, so-called *rationalists* were rather sceptical about the reliability of sense perception as an epistemic source. It is here that Plato's and Descartes's misgivings can be subsumed. Both are convinced that observation is not able to yield knowledge and that we have to rely on reason instead if we are looking for a secure basis for our knowledge claims. Consequently, if perception is not admissible as a source of knowledge, perceiving images cannot fare much better from an epistemological point of view. Plato explicitly discusses the case of images in his *Politeia*. Here he points out that pictures only mimic the entities of the real world, which themselves are only imitations of ideas. Consequently, images only show the appearances of appearances, but not the essence of things. This is why they are not suitable means to transmit knowledge (see Platon 2006, Politeia 597a-598d).

Descartes is similarly sceptical about our perceptual abilities to acquire knowledge of the world. He discusses the unreliability of sense perception in his attempt to point out a secure foundation for our knowledge system (see Descartes 1994). He makes his worries particularly clear when analysing, as an example, what the essence of 'wax' might be (see ibid., 23f., Second Meditation). Here he demonstrates that we cannot rely on our perceptual abilities to answer this question, as apparently none of the characteristics that can be observed, tasted or felt remain unaltered if the wax is heated. Descartes infers from this that none of these properties belong to its essence. To put it differently, perception is not a reliable source of knowledge if we are interested in investigating those essential facts. Yet, if perception in general cannot be regarded as a reliable epistemic source, picture perception cannot be taken as an epistemic means either. Obviously, philosophers following this rationalist tradition will then be rather reluctant to take visual representations into account in their epistemological analyses.

Jakob Steinbrenner addresses the second point mentioned above. He highlights the fact that most of all the *focus on language* of early analytic philosophers still influences philosophic discourse. He describes the resulting difficulties via a comparison of some of Gottlob Frege's main theses about verbal expression with characteristics of images (see Steinbrenner 2009, 285f.). What becomes clear is that none of the main features of verbal expressions – for example, being truth-bearers or being objects of logical operations such as expressing inferences, entailments, negations – show (at first glance) an analogous counterpart in visual representations. Steinbrenner adds that, even though none of these theses has remained uncontested by contemporary analytic picture theorists,[6] this heritage of the philosophical founding fathers, so to speak, has long hindered a thorough tackling of the topic in analytic philosophy (see ibid., 286). Yet what exactly did Frege say that caused these problems?

In *The Thought. A Logical Inquiry* (Frege 1956),[7] Frege deals with the topic of truth. In particular, he is interested in the question concerning what entities can be bearers of truth values. He suggests that it is only "thoughts" that can fulfil this task. But what exactly is it that he calls a "thought"? Frege offers the following description:

> [w]ithout wishing to give a definition, I call a thought something for which the question of truth arises. [...] the thought is the sense of the sentence without wishing to say as well that the sense of every sentence is a thought. The thought, in itself immaterial, clothes itself in the material garment of a sentence and thereby becomes comprehensible to us. We say a sentence expresses a thought.
>
> (Frege 1956, 292)

From his point of view, a thought is then an *abstract object*.

It is already inherent to this short summary of Frege's theses on thoughts what constitutes the basis for our problems with respect to scientific images. The two relevant phrases are 'bearers of truth values' and 'sense of a sentence'. Why are these two characteristics of thoughts so problematic in the current context?

Firstly, with regard to truth, Frege explicitly states that it is truth at which scientists aim. "What are called the humanities are more closely connected with poetry and are therefore less scientific than the exact sciences which are drier the more exact they are, *for exact science is directed toward truth and only the truth*" (ibid., 295, my italics). Consequently, if we are interested in scientific endeavour characterised in this way,[8] we have to focus our attention on those entities that can contribute to this attempt. This introduces the second point of concern derived from Frege's account.

The question is, what kinds of entities are possible candidates for our analysis and, in particular, whether visual representations are amongst them. There are two lines of reasoning related to this question in Frege's text. One is explicitly about images, the other about bearers of truth values. Frege questions whether our common habit of calling pictures true is justified (see ibid., 290ff.). He points out several aspects in this regard. In particular he makes clear that it is not a material object that we call true, but a representational relation intentionally brought about (see ibid., 290). This is why we use the adjective 'true' regarding pictures but not regarding other material objects such as stones. To be concise, Frege suggests that our practice of calling pictures true is a comprehensible, albeit illegitimate, extension of the scope of the term 'true'. He states that "what is *improperly* called the truth of pictures [...] is reduced to the truth of sentences" (ibid., 291f., my italics). It is not pictures that can be called true, but sentences – or, more precisely, the sense of sentences, as Frege claims.

Gombrich reiterates this point in picture theory, when he states that in the same sense that verbal assertions cannot be called red or green, pictures cannot be called true or false (see Gombrich 2004, 59).[9] It seems as if visual representations belong to the wrong category and thus cannot be considered seriously when epistemologically reflecting on the pursuit of truth in science.

From Frege's point of view, they cannot be the bearers of truth values, and thus cannot contribute to the proclaimed epistemic aim of science.

However, this critical stance towards visual representations in epistemic contexts derived from Frege's account did not remain uncontested. As there will be a discussion in detail concerning the possibility of visual arguments, and thus the connected question about the truth-bearing capacities of scientific images, in the following two sections, I will only briefly mention, as a concluding remark, two philosophers who argue for a positive status of scientific images.

Firstly, there is Marcia Eaton, who explicitly tackles the question about truth values of visual representations (see Eaton 1980). Similar to my approach, she suggests regarding images as parts of communicative acts. She also makes use of Kjørup's suggestions (see ibid., 16), mentioned in our analysis above. It is by this embedding of visual representations in communication that they can be regarded as (parts of) assertions. Now, Eaton argues that it is these assertoric acts that are evaluated along the lines of truth and falsehood. In detail, her approach amounts to the following thesis:

> [t]hus pictures lie somewhere between interpreting a set of symbols and interpreting a state of affairs. Judging that a picture is true or false necessitates the viewer's taking an active role in which he or she first formulates a statement which he or she believes to be a possible interpretation of the picture and then relates that statement to his or her beliefs.
>
> (Eaton 1980, 21)

Here, Eaton states her conviction that the content of images has to be translated into linguistic expressions before the question of truth can be tackled. In this sense, she tries to evade the difficulties pointed out above in a twofold sense, namely (a) by claiming that images can be translated into sentences and (b) that this translation, being based on the embedding of images in communicative acts, has the recipient judge what she believes the producer of the image intended to express with its aid.

Eaton's strategy seems to fit nicely with the characteristics of the explanatory context of science. Here, recipients indeed try to figure out what the intended message of a communicative act might have been. Yet, as we will see in section 4.1.3, this is only part of the meaning that can and will be extracted. Moreover, Eaton's approach is somehow at odds with what happens in the exploratory context of science. Investigating the photograph of Olympus Mons (see Figure 2.12), the scientist will not be interested in the intention of the photographer, but what this image can tell her about the height profile of this volcano on Mars.

The second scholar to mention is David C. Gooding (see Gooding 2010). He is not so much concerned with questions about truth as with processes of visual reasoning in science. Nonetheless, he points out that hypotheses based on visual representations can be empirically validated by "checking for correspondence between features derived from models and observed features" (ibid., 28). Assuming that a correspondence theory of truth rules here does not seem far-fetched. Without addressing this issue in any critical way, Gooding spells out how scientists make use of visualisations to derive hypotheses needed for explanations and predictions in their domain of research. In particular, he points out that scientists constantly manipulate their representational means. That is, he argues that it is not a particular image that is relevant to the scientist's reasoning, but a process involving different images (see ibid., fig. 10 on p. 27).

What Gooding explains here suggests an analogy to Vertesi's descriptions of how scientists work with rover images taken on Mars (see Vertesi 2015). She points out how working with those photographs, namely by digitally manipulating them, for example highlighting structures by colouring them or adding numerical or verbal explanations, converts them into maps of Mars (see, for instance, ibid. fig. 4.1. on p. 109), thereby transforming those images into tools not only to manoeuvre the rover geographically on the surface of Mars, but also in accordance with the tasks it is supposed to perform, for example, to analyse certain samples of soil (see ibid., 116ff.). Gooding describes a similar process at a more abstract level. He claims that:

[m]oving from interpreted sources to structural models and on to process models generates visual theories that satisfy the explanatory aims of science. This transformation from simpler to more complex representations increases information content, enabling models to incorporate more domain knowledge.

(Gooding 2010, 25)

Obviously, then, what allows scientists to draw inferences on the basis of visual representations is their embedding into larger processes also containing other kinds of representation.

In a similar way that Kjørup's suggestion draws attention to the wider context of image usage, Gooding's account reasonably broadens the focus of attention by calling to mind that scientific investigations and problem solving is a constant process rather than a single judgement. Helpful as this insight might be, there are nonetheless two critical remarks to be made about his proposal. On the one hand, his approach still leaves us with the question of how exactly, contrary to Frege's suggestion, images can be said to transmit information and thus constitute the basis of those processes of reasoning described by Gooding. On the other, his talk about 'models' in this context brings in another confusing issue, because it suggests the question of whether images can and should be understood as models, whereby the term 'model' itself is highly contested in the philosophy of science (see e.g. Bailer-Jones 2009; Hesse 1970; Morgan and Morrison 1999).

From this point of view, it seems advisable to set our focus back on scientific images and questions concerning their epistemic capacities. This is then the point at which to elaborate on Laura Perini's work on the topic. Contrary to Eaton, she defends the thesis that no prior translation of images into linguistic expressions is necessary to judge them along the lines of truth and falsehood.

4.1.2 Laura Perini on visual representations in scientific arguments

Not least because of Laura Perini's important contributions to the debate (see Perini 2005a; 2005b; 2005c; 2010; 2012a; 2012b; 2012c), philosophers of science have finally started discussing what the ubiquity of visual representations in science means in epistemological terms. Acknowledging this inspiring role that Perini's work has played for the philosophical discussion – and likewise for this book – her suggestions concerning the roles and status of visualisations in epistemic processes of science will be examined more closely now.

With respect to visual representations in the explanatory context, she states that philosophers tend to regard them as " 'mere illustrations' that are redundant expressions of information presented in the text, or convey information inessential to the argument" (Perini 2005c, 913). Yet it seems quite unlikely that scientists would put such an emphasis on visual representations if they were that useless in the epistemic context of argumentation.

Perini not only makes us aware of this tension in the assessment of image-practices by scientists and by philosophers, she also tries to offer a different evaluation of the epistemic capacities of scientific images. Here, Marianne Ina Richter, who critically discusses Perini's attempts in her PhD thesis,[10] makes clear that there are apparently three different strategies that Perini could have chosen to make her point. She could have defended the view (i) that images *are* arguments, or (ii) that they are *functional parts* of arguments, or (iii) that visual representations *contribute* to arguments without being parts of them (see Richter 2014, 159). I agree with Richter that Perini selects the second option (see ibid., 165ff.). Yet it has to be added that Perini presents two different lines of reasoning in this respect. Firstly, she tries to show that images can indeed be proper parts of scientific arguments. Secondly, she also analyses whether there are particular epistemic tasks that can be fulfilled exclusively by visual representations.

Perini's first line of reasoning is deeply intertwined with the question about the *truth-bearing capacity* of images. She states the difficulty concisely: "[p]hilosophers define arguments in terms of sets of statements. This may explain why philosophers of science have paid so little attention to figures" (Perini 2005b, 262f.). Apparently images simply belong to the wrong representational category to play the role of premises or conclusions or, at least, proper parts of such.[11] Moreover, Perini explains why the difference between visual and linguistic representations is commonly regarded as so important in the context of argumentation.

The support that premises provide a conclusion is analyzed in terms of validity or strength, and soundness, so any representation that is an integral part of an argument must be one to which those features could be relevant. Validity, strength, and soundness are understood in terms of the truth conditions of premises and conclusions, so representations that contribute to arguments must have the *capacity to bear truth*. To show that figures are nontrivial parts of scientific arguments, the first question that must be addressed is whether visual representations can be true or false.

(Perini 2005b, 263, my italics)

Assessing arguments in a logical way presupposes that we are able to say something about (a) whether what is expressed in their premises and the conclusion is true or false, and (b) whether what is derived in the conclusion is already entailed in the argument's premises. However, these are exactly the logical characteristics that, following Frege, visual representations apparently do not possess.

Now, Perini is convinced that this last commonly held assumption is false. She tries to show that visualisations actually *can* be truth-bearers (see Perini 2005b; 2012c) and thus can be regarded as (proper parts of) premises or conclusions of scientific arguments. Her line of reasoning works as follows.

Firstly, she discusses some examples from logic and mathematics, in particular from set-theory and geometry (see Perini 2005b, sect. 1). She points out that, in this realm, there are *some* visual representations that can indeed bear truth values. She refers to Venn, Euler and Peirce diagrams and states that they allow "truth-preserving inferences from diagram to diagram" (ibid., 264).[12] To illustrate this point, take the classical example of the syllogism which entails a universal statement as one of its premises. As has been pointed out in section 2.1.4 of this book, such a universal statement can be expressed by Venn diagrams. Figure 4.1 shows how the universal statement 'All men are mortal beings' can be expressed by using a diagram instead of a sentence and, thus, how a diagram can work as a premise of the syllogism.

Furthermore, Perini points to the existence of visual proofs of certain mathematical questions, in particular in Euclidian geometry (see ibid., 265f.).[13] In what sense images can play a role in mathematical explanations is shown by Max J. Kobert, who demonstrates how visual representations can be used to explain algebraic statements (see Kobert 2010, 135f.). For instance, the formula $(a+b)^2 = a^2 + 2ab + b^2$ can be explained by considering Figure 4.2. That is, the algebraic formula is explained by turning it into a geometrical problem which can be solved visually.

Yet despite these initially positive results, Perini sets them aside by claiming that "[b]ecause scientific arguments are significantly different from the deductive diagrammatic systems [...], we cannot extrapolate from their results to the figures that appear in contemporary research journals" (Perini 2005b, 267).

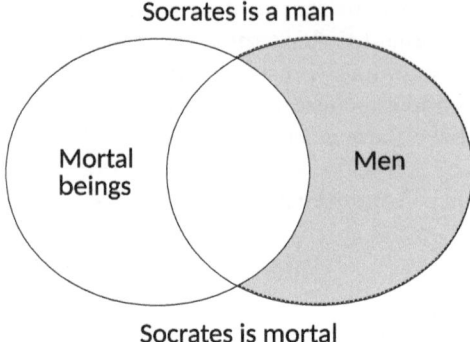

Figure 4.1 Syllogism containing Venn diagram, general statement (All men are mortal beings) expressed by Venn diagram.

Source: own image.

From my point of view, she dismisses too much here because, although Perini is right in claiming that *not all* scientific visualisations are of the kinds used in logics and mathematics, *some* are.

With respect to images used in the arguments of the empirical sciences, some points have to be made clear in advance. (1) Perini thinks of depiction along the lines of Goodman's theory as explained in section 2.2.2 of this book. That is, images are parts of symbol systems whose difference to linguistic symbol systems is characterised on the syntactic level (see ibid., 267ff.). (2) She rejects resemblance relations as the relevant feature to discern what is meant by 'depiction' (see ibid., 268). (3) Yet she admits that similarity might still play a role. She points out that, although it is by conventions that the characters of a given symbol system are ascribed their meaning, it can be the case that some conventions refer to resemblance relations to determine the meaning. "Content can also be determined by conventions that relate symbol and reference through resemblance relations" (ibid., 268).

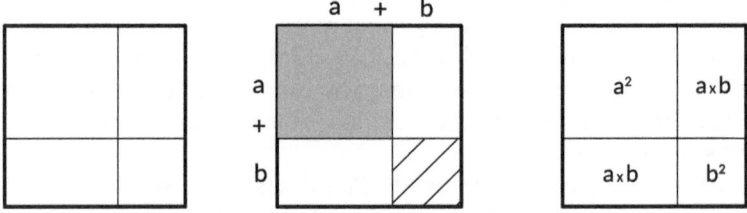

Figure 4.2 Pictorial explanation of algebraic formula $((a+b)^2 = a^2 + 2ab + b^2)$.

Source: own image.

(4) These conventions belong to particular symbol systems. That is, if an individual image is interpreted as depicting y because it resembles y, we do so because this image belongs to a particular symbol system that entails resemblance relations as interpretative rules. (5) It is not by translating images into sentences that they are assigned a meaning. (6) It is not the meaningful content of a sentence that is assessed as being true or false when images are considered (see ibid., 274). (7) It is a feature of the respective system, not of individual representations, that truth values can be ascribed to the latter (see ibid.).

Taking these points as the background to her considerations, Perini then makes use of Alfred Tarski's theory of truth to defend her claim that images can bear truth values. Now, the classical Tarskian schema of truth is: "x is true in L iff p". The standard example is: "'snow is white' is true iff snow is white". Thus, with respect to visual representations, we now have to make clear, what 'x', 'L' and 'p' refer to. It is here that, despite her initial dismissal of linguistic expressions as mediating representations, Perini claims that we are in need of names and sentences to make statements about

the definition of truth for a visual system.

(Perini 2005b, 276)

She points out that such a definition "requires a way to assign a linguistic name to each symbol, based on its structural form, and also a linguistic expression of the content of each representation. A definition of truth for a visual symbol system consists of a statement of this form for every figure f in a system: Name(f) is true IFF statement(f)" (ibid., 276). Thus, to make claims about the truth-bearing relation between 'x' and 'p', that is, between a symbol (for example an image) and its meaning, both have to be expressed linguistically. Moreover, it is by the interpretive rules of the system to which the symbol belongs that the meaning of the respective symbol is fixed. Thus, Figure 4.1 is true, if and only if, all human beings (premise 1) are mortal (premise 2). The conventional rule that helps us to interpret this diagram consists in the following statement: *interpret the sets of two intersecting circles (1, 2) as identical if all elements of 2 belonging to the set of 1, that is, are entailed in the intersecting part of 1 and 2, and no element of 2 is left outside this intersecting part so that the remaining part of 2 is shaded, thus marking an empty set.* Interpretative rules could also be defined at an even more basic level here when the shaded part is stated to mean an empty set or the intersecting part of both circles a shared set.

Interpretative rules such as those last statements then connect an image or a symbol (x), named e.g. by calling it 'Figure 4.1' or 'intersecting part', with its meaning (p) within a particular symbol system (L). 'Meaning' is thereby understood as the reference of the symbol which can be an object, but also a state of affairs as in the example above. Interpretative rules then determine the truth conditions of a particular diagrammatic symbol. It is in this sense that Perini thinks that Tarski's theory is a suitable means by which to discuss the truth values of images.

Tarski's work shows that even though there is no general theory of truth with which we can test individual visual representations for the capacity to have truth value [...], certain symbol systems are characterized by a systematic relation between symbol form and referent, in which the truth-conditions of symbols are a function of their form.

(Perini 2005b, 275)

Admittedly, such an account works very well with respect to highly conventionalised symbol systems such as the diagrammatic. However, as Perini points out herself, many images in science are not of this kind (see ibid., 279). As an example, she discusses electron micrographs (see ibid.). Despite the difficulties that she highlights with respect to images of this kind, she is optimistic that it is still possible to define a concept of truth also holding in pictorial symbol systems.

What x (symbol), p (meaning) and their relating rule are in this context has to be re-analysed in order to discover the cause of the problems here. The mapping of symbol and meaning becomes much more difficult if images are considered such as the micrograph, that is, images of a less conventional style in comparison to diagrams. In particular, there are many characteristics that are not related to any linguistic expression. Consequently, truth conditions mapping symbols to states of affairs cannot be formulated in the way suggested with regard to the diagrams above.

Therefore, Perini suggests that it is relevant to consider "the form of the image as a whole" (ibid., 280). That is, the micrograph represents the scanned sample as being of a particular shape – this is what Perini calls the "form of the image" and which constitutes the x of the Tarskian schema. The meaning (p) is the actual shape of the sample. The truth condition of an electron micro-graph reads as follows, "an electron micrograph is true IFF the shape of the micrograph is a geometric projection of the shape of the sample scanned in producing the micrograph" (ibid., 280).

Yet despite this explanation of how to deal with images of this latter kind, one might object that her strategy presupposes the *translatability* of pictorial symbols into linguistic expressions to check for the relevant rela-tion. In particular, this problem comes to mind when she contemplates the difficulties of naming entities (see ibid., 282). Many visual representations in science display characteristics, such as those depicted by the micrograph, that cannot be translated into linguistic expressions because the latter are simply not available. There are no names, not even linguistic descriptions, at the scientists' disposal either to name the symbol or to describe its meaning. If the content of such visual representations cannot then be translated and thus stated by making use of linguistic expressions, does this imply that the method employed to define truth conditions explained above is not applicable here?

Perini is aware of this objection. She claims that there are two reasons speaking against this negative assumption:

[f]irst of all, this investigation was launched to show whether non-linguistic representations can bear truth or not. This question is begged by invoking the assumption that only representations whose content can be expressed with a linguistic form of representation have the capacity to bear truth. So, the question of whether a micrograph could be true or false cannot be settled in this way. Second, that stance does not fit with the fundamental intuition that truth value depends on reference to states of affairs, since micrographs not only represent states of affairs, but do so without their content being assigned to them through the mediation of another symbol system.

(Perini 2005b, 282f.)

The first rejoinder seems to me to be a rather weak one. One can state vice versa that the necessity of translation indeed shows that pictorial representations cannot bear truth values. The consequence then would be that the project of defining a concept of truth valid for pictorial symbol systems has failed.

Perini's second rejoinder, however, is more substantial. Pointing out that truth values are connected to states of affairs and that micrographs represent such states of affairs apparently suggests itself as a criterion of evaluation. How does the representation of states of affairs work, then, in the pictorial realm? She argues that images represent states of affairs because what is represented by them is related to the latter by conventional rules that guide the interpretation of those images. This also explains why she thinks that images can represent not only visible features of their object of reference, but also invisible ones. She states that:

[v]isual representations are defined by the fact that spatial features of the symbol are *interpreted* to refer to features of the referent – they are not defined in terms of similarity with their referents. For this reason, visual representation is not restricted to visible subject matter.

(Perini 2010, 137)

William Goodwin takes a different track in criticising Perini's account of the truth-bearing capacity of visual representations (see Goodwin 2009). He is also sceptical about the theory of depiction and the theory of truth chosen by Perini. However, his main critique is focused on "the idea that the only way to understand visual representations as having a 'genuine' role in scientific discourse is by finding some way to understand these representations as capable of being true or false" (ibid., 374). Goodwin shares Perini's conviction that visualisations can play essential epistemic roles in scientific discourse. However, contrary to her approach, he thinks that this important task can be explained without struggling with their presumed capacity to bear truth values. How then does Goodwin conceive of the precise epistemic function of visual representations in scientific arguments?

The starting point of his objection is his asking what the expression 'contributes to an argument' might mean. From Perini's point of view, making such contributions implies that the respective entity plays the role of (parts of) premises or conclusions. As Goodwin claims, such a conception means that "a representation 'contributes to an argument' only by being in logical or evidential relationships with the other representations in the argument" (ibid., 376). Postulating such relationships then motivates the search for adequate conceptions of truth in this context, since the other relevant representations are linguistic expressions. However, as Goodwin points out, there is also a second sense in which the statement 'contributes to an argument' can be understood, namely as "simply being important to the expression of the argument" (ibid.).[14]

It is this second sense that Goodwin highlights as relevant when considering particular kinds of representations, for example, the functions of words or numerical data in the context of scientific arguments. This second sense of 'contributes to an argument', however, does not require that the entities analysed in this regard have to be capable of bearing truth values. Thus, Goodwin infers that:

> Perini has provided no general philosophical reason to suppose that the visual representations used in science must be understood to bear truth. Recognizing that such representations contribute to scientific discourses in which they occur leaves open the question of how they contribute.
>
> (Goodwin 2009, 377)

Moreover, Goodwin highlights the fact that, even though scientific images might in some instances play the epistemic role that Perini ascribes to them, this does not preclude the possibility that their roles are different on other occasions.

To support his view, Goodwin discusses the example of structural formulas in organic chemistry. He suggests that they can be regarded as diagrams, and thus as visual representations in Perini's sense. Now, with respect to their functional roles in scientific discourse, Goodwin points out two different tasks that they fulfil.

On the one hand, he claims that they are used as *descriptive names* in scientific papers (see ibid., 378). This he derives from their interchangeable use with IUPAC labels[15] (see ibid., 380), that is, labels that are generated by using the rules provided by the nomenclature developed by this union.

On the other hand, Goodwin argues that structural formulas also serve the function of *models* in scientific texts. He writes:

> [o]f course, it is not possible to reproduce physical models of organic compounds on the printed page, and so structural formulas stand in for physical models in chemical discourse. [...] Structural formulas – supplemented with some additional conventions – are, on the other

hand, objects whose characteristics can be used to infer properties of the compounds they denote.

(Goodwin 2009, 381)

Thus, Goodwin thinks that structural formulas can be manipulated to infer information about the chemical compounds that they represent.

Regarded this way, structural formulas as visual representations provide essential information, but it would be wrong to ascribe truth values to them.

> When used as a model it is facts that are true *of*, or *in*, the structural formula that make it so useful in the discourse of organic chemistry. So when a structural formula is used as a model, it is not true or false. Rather, certain claims are true or false in virtue of the structural formula, and these claims license conclusions, perhaps by way of assumed similarity relations that obtain between the relevant objects, about the chemical compounds that it denotes. The role of structural formulas in this context is therefore that of truth maker, not that of truth bearer.

(Goodwin 2009, 388)

From Goodwin's point of view, his example shows that Perini is wrong in assuming that visual representations can make epistemically relevant contributions to scientific discourse only if they are capable of bearing truth values. Contrary to this, he argues that at least some visualisations function as models and, in this sense, make certain claims put forward in this discourse true or false.

Now, there are several questions to be asked about Goodwin's account. Two aspects, also discussed by Perini in her rejoinder to Goodwin's critique (see Perini 2012c), concern, firstly, the scope of his claim and, secondly, the question of whether his example shows what it is supposed to demonstrate.

I will start by discussing Perini's second rejoinder first. She calls into question whether Goodwin's example, namely whether those structural formulas published in papers in organic chemistry, really function as models. She emphasises Goodwin's thesis that it is the possibility of *manipulating* the parts of the model to gain additional knowledge by using it (see ibid., 144). However, as a 2D representation in a scientific text, those structural formulas cannot be manipulated in the sense intended by Goodwin. "All the diagrams Goodwin discusses are static markings on a flat surface. The positions of their parts can't be manipulated" (ibid., 145). In this sense, the analogy that Goodwin tries to establish with regard to models and scientific visualisations does not hold from Perini's point of view.

Nonetheless, she does acknowledge a certain similarity with regard to their epistemic function, but only via a detour through imagination. She states:

> [t]he viewer can imagine a three-dimensional object after viewing the diagram, and then imagine what that three-dimensional object would look

like if its parts were rotated. Then they could use the same reasoning that was applied to their observations of the actual three-dimensional models after they were manipulated. [...] If this is how the diagrams are used, then some model-based reasoning is involved, but the diagram itself is not functioning in the same way as the model in that reasoning process; instead, the imagined compound functions like the model.

(Perini 2012c, 145)

Now, Perini uses this description of how the analogy between inferences made with the aid of models and by using scientific visualisations can be preserved to support two different claims of hers. (1) She points out that, with regard to the particular example discussed by Goodwin, the analogy does not hold. That is, he has not shown that, in this particular instance, scientists were interested in gaining new knowledge via mentally manipu-lating the presented structural formula. On the contrary, she thinks that the diagrams in question actually work in the sense she suggested earlier, namely "to make a claim about the structures of the isomers" (ibid.). (2) Perini admits that visualisations do have the capacity to function in the way intended by Goodwin and that some of them are also used in this way, namely as tools in visual thinking (see ibid.).

Perini's rejoinder seems convincing. Obviously, visual representations can serve the function of tools of thinking, as has been pointed out in section 3.1.1 of this book. However, it can be questioned (and Perini does so) whether it is this functional role that is intended when distributing the respective visu-alisation in a scientific paper. More often than not, the presentation of results seems to be relevant, namely the *evidential role* that Goodwin also mentions in his text. This topic will be discussed in due course.

Before going into those details, the analysis should be concluded by also examining Perini's first rejoinder to Goodwin's critique more closely. Although I agree with her first reply, I do have some worries concerning the scope of her and also of Goodwin's claim. Both emphasise the functional roles of the scientific images they discuss in their respective articles as the more or less predominant ones. This also seems to be the rationale why Perini claims, for example, that "[...] the fact that *in some circumstances* scientists use images in ways that are best described as using the image as an object (or description-fitter), doesn't imply that *most* scientific images are used this way" (ibid., 146, my italics). What is missing in both accounts is a more precise analysis of contexts of usage as presented in section 3.1 of this book. Such a more detailed perspective not only reveals that an epistemology of scientific visualisations can accommodate both accounts of functional roles without devaluing the relevance of either of them in the scientific context, but also that, even if the discussion is focused on the explanatory context alone, as both Perini's and Goodwin's are, it has to be acknowledged that visual representations can fulfil different epistemic and non-epistemic functions. It depends on the aim pursued and the mode of the communicative act as well as on the audience involved in it that determine the exact nature of their function. I reject, therefore, the predominance of the respectively identified functional roles of visual representations expressed by Perini and by Goodwin.

Furthermore, it might be helpful to mediate between Perini's and Goodwin's account to bring to mind that there is another close connection between models and visualisations that is often overlooked. Vögtli and Ernst remind us of there being many instances of physical models reproduced in visual representations. The fact that, in the processes of copying such images, people pay less attention to what has been the actual object of reference of their template motivates Vögtli and Ernst to urge their readers to be careful in interpreting the content of visualisations as showing real entities (see Vögtli and Ernst 2007, 47ff.).

To return now to the point that visual representations play an evidential role in scientific discourse. Goodwin puts forward an interesting idea *en passant*. In a footnote of his paper he writes:

> [...] consider a piece of physical evidence in a legal trial. Surely it doesn't follow from the fact that this physical evidence plays an important role in the prosecutor's argument that the physical evidence itself must be capable of bearing truth.
>
> (Goodwin 2009, 375, ftn. 2)

He mentions this example to support his initial thesis that certain entities can contribute to arguments, although they are not capable of bearing truth values. From his point of view, such physical pieces of evidence should be regarded as objects that make statements about them true or false. This is the parallel that he wants his readers to draw between this example and his claims about the role of visual representations in the context of scientific arguments.

Although Goodwin's suggestion offers an interesting new perspective on how to conceive of the role of visual representations in science, it still implies that, if scientists want to make use of images in argumentation, they have to formulate statements about them and use the latter in their texts. They still need verbal statements that visual representations can render true or not in order to express their arguments correctly. Yet, contrary to Goodwin, although in accordance with Perini's point of view, the scientific practice of publishing visualisations as parts of scientific papers suggests that scientists do not see the need to translate those visualisations so that they play the roles they are supposed to fulfil. Information given in images is not recorded in the text of the article containing them. Usually there are only references in the text telling the reader to consider the relevant figure in order to get the information relevant to understanding the explanation offered.

This seems to be particularly true with respect to genuine research papers in science. Articles addressing laypeople commonly contain descriptive sections about their visual representations to guide their novice readers in interpreting them correctly. Although scientific images are supposed to play an evidential role in both contexts, their embedding within the written text of the respective article will be different.

Perini is clearly vindicated in pointing out that scientific practices show that visual representations can play an epistemic role in their own right, that is, without prior translation. However, as I am also convinced that Goodman's account does not offer the means suitable to explain the nature of scientific visualisations, I cannot follow Perini's suggestion to regard the latter as truth-bearers as stated in her theoretical framework. Nonetheless, there are alternative ways that might guide our analysis to conclusions similar to those intended by Perini.

Two such strategies will be considered in more detail in the following sections. One is concerned with the *cognitive content* of visual representations. If it can be shown that visualisations do contain such a content, it might be possible to argue that this content works in a way similar to Fregean 'thoughts' when transmitting information. Hence, a subsequent analysis has to focus on the question of whether this kind of content is propositionally structured. If it turns out that it is indeed propositionally structured, a truth-orientated approach, such as the one preferred by Perini, can also be established via this route.

The second strategy, although related to the first one in certain respects, follows a different line of reasoning. Close attention will be paid to the role of *visual perception* when cognitively accessing the information encoded in scientific visualisations. In this context, the epistemic qualities of perception and image perception in particular will be examined. Is knowledge gained via perception and, if so, what kind of knowledge? Is it propositional or non-propositional in kind?

Perini suggests a strategy in a different paper (see Perini 2005c), similar to the idea just sketched, to account for the *evidential role* of scientific images in the explanatory context. Since she also refers to perceiving those visual representations as a relevant step in the process of their cognitive processing, it seems reasonable to combine the analysis of her thesis with a brief review of the discussion of *perception as an epistemic source*.

Yet before going into the details of the two suggested strategies, this section about "visual *arguments*" should be concluded. The level of argumentation as a communicative act will be considered next. Moreover, analysing the topic in a broader perspective such as this is also a consequence of Kjørup's thesis that the object of investigation should be whole communicative acts and not visual representations in isolation if their meaning is to be correctly understood. Taking this suggestion into account, how should visual arguments be conceived?

4.1.3 *Giving reasons, drawing conclusions*

In this section, a couple of ideas will be used to explain how visual representations can contribute to scientific arguments. My perspective on this topic is guided by insights from semiotics, the philosophy of language and argumentation theory. What analytical tools do these approaches provide to analyse the role of images in this context? To begin with, they permit a better understanding of what the terms 'argument' and 'argumentation' mean; in particular, what parts they are composed of. Additionally, they will offer some useful information about the questions to which of these parts visual representations can contribute and how.

We will start (1) with a brief analysis of Scholz's theory of image games. This theory helps us to explain the effects of argumentation on particular images in science. I will call one of these effects a 'layering of meaning'; that is, subsequent uses of one particular image in different contexts can add new dimensions of meaningfulness to the visual representation at hand. Despite this layering, Scholz's theory also allows the conclusion that a kind of basic meaning will still be maintained in the encoded content.

If it can be conceded that such a layering of meanings can take place in argumentation, the question arises how best to analyse this process. This (2) will be examined next by deconstructing the act of argumentation including visual elements. As an auxiliary means employed in the investigation, Kjørup's proposal of pictorial speech acts will be used. This analytic approach will provide the tools to describe the different components of argumentation as a communicative act in science.

Finally (3), this line of reasoning will stress the question of how to conceive of the propositional content inherent in the act of argumentation. To answer this question, I will discuss approaches from argumentation theory dealing with the topic of visual arguments. Two different ways to handle the subject will be considered: on the one hand, some scholars suggest that images do indeed have a propositional content that can account for the task of information transmission in question; on the other, there are philosophers who try to evade the question about propositionality or, respectively, who clearly deny that argumentation presupposes the propositionality of its parts by any means.

Thus, the content-level of visual arguments will finally be returned to. Yet the line of reasoning leading to it points out the possibility of analysing the contributions of visual representations when focusing on the level of argumentation – the communicative act – rather than on the content-level of individual images. To begin with, Scholz's theory of image games demands closer attention. With this theoretical approach, Scholz suggests the term of image games in close connection with Ludwig Wittgenstein's theses about *language games*. This latter approach is meant to explain how linguistic expressions acquire their meanings in conversational settings (see Scholz 2011).

As Scholz adopts Wittgenstein's conception for discussing how meaning is attached to images, it would be reasonable to first point out what he considers to be the essentials of language games. Scholz states that (1) the usage of signs is deeply intertwined with human actions. He points out (2) that words and sentences are not only embedded in contexts of sign usage, but also in cultural contexts. Both contexts are characterised by certain rules that are relevant to the meaning of those linguistic expressions. The thesis is highlighted that (3) to understand linguistic expressions correctly it is necessary to know the rules governing the usage of signs in the different contexts of their usage and within the different cultural contexts.[16] And finally (4) no claims about a presumed *essence* of language games are put forward by Wittgenstein who, on the contrary, defends the thesis that a variety of phenomena subsumed under this label is connected via family resemblance alone (see ibid., 369). These are then the four characteristics of language games on which Scholz relies to transfer Wittgenstein's concept to the pictorial realm and that he uses as a stepping stone to develop his theory of "image games".

As was pointed out in section 2.2.2, Scholz is an adherent of Goodman's theory of depiction. Consequently, he argues that the practice *of using an entity as a picture* implies that this entity fulfils Goodman's definition (see ibid., 371) – that is, images belong to symbol systems that are characterised as syntactically dense and relatively replete. Nothing can be regarded as a picture *per se*. It is due to our treatment of a particular entity that causes it to be thought of subsequently as a picture. However, not any entity can be called an image in that way. In addition to being embedded within the relevant practices, it has to fulfil further conditions, namely the conditions of depiction mentioned by Goodman.

It might be objected that this concept of the term 'image' is somehow in conflict with the anti-essentialist thesis that Scholz wants to transfer from Wittgenstein's account. If all the entities to be called 'images' have to fulfil Goodman's conditions of depiction, their essence not only seems to be clear, but this requirement also considerably restricts what can be called an 'image'. Setting aside this apparently contradictory claim, what else can we learn about Scholz's concept of image games?

Scholz claims that there is a vast variety of different image games (see ibid., 373). Contrary to Wittgenstein, however, he brings in a level of systematising by distinguishing between two different kinds of image games. These two kinds are characterised by their *different directions of fit*. On the one hand, there are image games that imply for their pictorial instances a fit from world to image. Examples are blueprints in architecture or in the engineering sciences (see ibid., 371). On the other hand, there are images that are directed at the world, that is, images that mimic previously present objects of the real world (see ibid., 372).

Two further aspects are of prominent importance with respect to image games. Firstly, Scholz makes clear that one particular image can be used in very different games. For instance, a photograph of a dog can be shown either to inform someone else about what the photographer's pet looks like or it can be meant as a warning, for instance when attached to the fence of the photographer's garden (see ibid., 374). Consequently, *in order to understand an image correctly*, we have to know in which image game it is employed. As Scholz points out (see ibid., 377), particularly regarding a correct historical understanding of certain images, this becomes extremely relevant in addition to background knowledge about the cultural specifics and practices.

Secondly, and as a consequence of this close connection between purposes of usage and meaning, there is (or can be) a distinction between the context of producing an image and of its subsequent usages (see ibid., 372). This distinction implies not only a change in meaning due to the different purposes involved in using the image at hand, but also that completely different individuals might interact in both contexts. Thus, many more agents can be involved than a simplistic focus on the context of image production alone suggests.

Now, Scholz thinks that many image games are related to communicative purposes (see ibid., 373). I would maintain, rather, that *all* of them are somehow related to communication. This point of view is a consequence of how the term 'communication' is understood. I prefer a broad concept here, obviously broader than the one used by Scholz. The concept used here is based on ideas propounded by Paul Watzlawick and his colleagues. They suggest that all kinds of behaviour in social settings can be interpreted as communicative acts and that, as we cannot stop behaving, we will never cease communicating in those contexts (see Watzlawick, Beavin and Jackson 1974, 51).

Although Scholz relies on Goodman's theory of depiction, which has already been pointed out as being unsuitable in the context of scientific images, his concept of image games is nevertheless very useful in the attempt to clarify the role of visualisations in scientific communication and, in particular, in scientific argumentation. The theory of image games makes clear that there are more aspects to consider when trying to understand what is meant by a particular image in a communicative context than its encoded content alone.

Leo Groarke's conceptual theses about argumentation pave the way for further analysis. He claims that " '[a]rgumentation' includes not only arguments, but also the broader dialogue, discussion and disagreement in which arguments are embedded. Acts of arguing attempt to rationally establish some conclusion by providing evidence in its favour" (Groarke 2015, 135). Thus, regarding argumentation as a communicative act[17] permits utilising the theory of image games and, hence, disentangling correctly what has been called a 'layering of meanings'. Such a layering is the result of a multiple usage of one particular image. In such a case, there is the genuine meaning encoded into the visual representation by its producer. This basic meaning, however, will be covered (at least partly) by layers of meaning superimposed on the genuine one by subsequent users. Making use of visual representations in image games can, at least, add a second meaning to what has genuinely been encoded into the visual format, and this second meaning might change due to the different games played with the aid of this particular image. A compilation of meanings, acquired in different image games, might also be possible, but would be rather an exception than an example of common practice.

Nevertheless, adding layers of meaning in image games to what the visual representation was genuinely supposed to express by its producer is the reason that Scholz urges us to reconstruct the genuine image game and cultural surroundings if visual representations are to be correctly understood. Important dimensions of meaningfulness would simply be overlooked if the focus were solely on the context of production. Hence, the theory of image games permits discovering what additional levels of meaning images can acquire by being used in argumentation. The question of how exactly images play a role in argumentation is answered in the second part of the analysis, which is concerned with ways to deconstruct the act of argumentation to see if and how visual representations contribute to its different parts.

Kjørup makes us aware that understanding visual representations correctly has to take into account the whole communicative act, not the image in isolation (see Kjørup 1978, 57). Moreover, by taking *speech act theory* as the relevant paradigm, he makes clear that pictorial speech acts are composed of different parts, namely an illocutionary, a locutionary and the propositional act (or content, see ibid., 61). Taking these levels into account and adding the perlocutionary act offers us the means to show at what levels visual representations can make contributions to scientific arguments.

To begin with, the significance of these sub-acts in speech act theory should be briefly summarised: the locutionary act concerns the uttering of words or, with regard to visual representations, the showing of images in conversational contexts. The illocutionary act is about the kind of act performed, such as warning somebody by showing the image or simply informing her of the appearance of what is depicted, etc. The propositional act is about encoding the basic information, for instance by drawing a certain figure or taking a photograph (see ibid.). And finally, the perlocutionary act is about the effect intended on the recipient, for example, evoking a particular emotion or convincing the recipient of a certain proposition.

Kjørup tries to spell out rules for all of these sub-acts when images are utilised in communicative contexts (see ibid., 61ff.). He points out that the *context of usage* is important both to fix the reference of a particular visual representation (see ibid., 58) and to determine what particular illocutionary act is performed by using a certain image (see ibid., 65). Even a photograph seems to presuppose some background knowledge about what is depicted in order to discern exactly to what it refers. For instance, it is only by reading the caption accompanying Figure 2.15 that we come to know that it shows lava flows on Olympus Mons on Mars. Without this additional information, we would most probably be unable to interpret to what this photograph refers.

Moreover, Kjørup discusses particular illocutionary acts that can be performed with the aid of visual representations. For example, he tries to establish the rules governing the illocutionary act of "illustration" (see ibid., 66ff.), which is particularly interesting with respect to the intertwining of text and image that plays a major role in this case in order for it to be successfully performed.

Applying these ideas to the broader context of scientific argumentation, it seems reasonable to assume that *argumentation* can be similarly subdivided into particular acts, as suggested above with respect to speech acts – although one might object that arguments are composed of *assertions* and that it is this kind of act that should be analysed when discussing the nature of arguments. Correct though this might be, the relevant level of analysis still seems to be about arguments and not about assertions. Moreover, to a certain degree it is also possible to transfer the subunits of speech act theory to arguments. How does this work and why is this of relevance when theorising about visual representations in scientific arguments?

I shall start by investigating the possibility of transferring the units of analysis to the context of argumentation. When arguments have to be uttered or otherwise employed, a locutionary act occurs. Arguments entail certain information that the speaker will encode them in a certain way – the propositional act. They are meant to transmit information by offering reasons to believe in a certain conclusion supported by them that could be called the illocutionary act. This transmission of information can then be connected to the various intentional aims of the speaker. She might want to transmit knowledge, she might want to ponder on the consequences of certain hypotheses herself, or she might want to solve a particular dispute. This last option, for example, is put forward by Groarke as the aim of argumentation. He attempts to clarify the nature of 'arguments' by writing that:

> I define an argument as a standpoint (a conclusion) backed by reasons (premises) offered in support of it. In a typical case of arguing, arguments are an attempt to resolve disagreement […], though they may also function as an attempt to avoid disagreement by securely establishing some belief. I will understand an act of arguing as an attempt to use premises and conclusions to resolve some disagreement or potential disagreement.
> (Groarke 2015, 134f.)

Finally, arguments are supposed to convince the recipient about what has been transmitted by argumentative means; in this case, this could be regarded as the perlocutionary act. This subdivision of arguments is relevant in the current context because quandaries about cases of using visual representations as *parts* of scientific arguments can apparently best be understood as a *consequence of their contributing to different sub-acts of argumentation*. Making use of visualisations in the context of scientific argumentation does not necessarily imply that they are meant to entail information relevant to the argument at hand, that is, they do not have to be parts of the propositional act. They can also be used as a means to support the perlocutionary act. Clearly, visualisations are used to support the persuasive power of arguments in many instances. This functional role *can* be correlated with transmitting additional information such as evidential data. Visual representations might also be used to present data in an organised way that makes correlations salient. However, particularly in popular science, visualisations simply might also be used because of their persuasive and emotional power and, in this sense, have to be evaluated cautiously if the epistemic dimension is of interest.

Finally, it can be asked whether visualisations can also be used as means to perform the locutionary act. Is it possible to 'utter' an argument simply by showing an image? Clarifying this question seems to be particularly relevant in contexts where the contending parties introduce images in an ongoing debate in which, say, party Y is urged to counter an argumentative move made by party X by also making use of images to support her position. A telling example of this kind is discussed by Randall A. Lake and Barbara A. Pickering. Their case study is about the debate between anti-abortionists and pro-abortionists in the United States (see Lake and Pickering 1998) and will be discussed in some detail below. Lake and Pickering's account will be discussed in some detail below. For the moment it will suffice to note that their example makes it particularly clear that there are instances where contending parties seemingly have no other choice than to make use of visual representations in their argumentation.

Let us now come to the third part of my analysis, namely to the question of whether visual representations can also play a part with respect to transmitting propositional contents. Groarke, for example, approaches this topic rather carefully by claiming that:

> [t]he kinds of examples that motivate me are acts of arguing that involve pictures, maps, sounds, diagrams, smells, video clips, and other non-verbal phenomena which are not propositional in the way that verbal statements are (though it bears noting that the relationship between sentences and propositions is itself a matter of much controversy).
>
> (Groarke 2015, 135)

Obviously, he thinks that non-verbal elements can also function as premises and conclusions in argumentation. However, he does not try to pin down the claim that these non-verbal elements are propositional in the sense claimed for sentences.

A different approach is suggested by Lake and Pickering. They examine the possibility of visual arguments by taking a look at films. They do not want to analyse arguments of mixed media, but images as exclusively visual arguments. At least, this is what they attempt to investigate, although in the end they admit that their case study actually shows results only for a "mixed-media environment" (Lake and Pickering 1998, 91) – which is not surprising since they focus on films as their object of research. However, what Lake and Pickering might regard as a shortcoming in their analysis becomes a virtue in the context of our current discussion.

In their analysis, they try to show in what sense images can be said *to refute* one another. The authors start with the assumption that visual representations do not have a propositionally structured content (see ibid., 81). Moreover, they regard the ability to refute as essential to whatever kind of representation is used in argumentative contexts. In contrast to Perini, Lake and Pickering deny that images can be bearers of truth values, but they nonetheless think that images can be used in argumentative contexts since they can *refute* each other.

They point out three different ways how such a refutation can be achieved by visual means, namely:

> (1) through *dissection*, in which an image is 'broken down' discursively, its component parts named and its relations analyzed, thereby opening the image to refutation via traditional (discursive) argumentative means; (2) through *substitution*, in which one image is replaced within a larger visual frame by a different image with an opposing polarity; and (3) through *transformation*, in which an image is recontextualized in a new visual frame, such that its polarity is modified or reversed through association with different images.
>
> (Lake and Pickering 1998, 81f.)

Lake and Pickering discuss these assumptions by analysing films used by opposing parties in the dispute about abortion in the US. They show how anti-abortionists and people defending women's rights regarding abortion make use of different films that become part of the debate and are used in such a way that these films can be said to refute one another.

In particular, they discuss how the anti-abortionist film *The Silent Scream (SS)* is dissected within another film (*A Planned Parenthood Response to 'Silent Scream' (PPR)*) produced by the contending party. This is meant as an example of the first strategy of refutation. For instance, the plastic model used in *SS* to demonstrate the developmental state of the foetus is criticised as inappropriate in scale. Furthermore, the slow-motion technique used in *SS* to create the impression of a very calm and secure environment for the foetus in the womb is highlighted as a manipulative means in *PPR*. They make clear that the *SS* film-makers used slow motion to manipulate the recipients' impressions by visually demonstrating the frame-by-frame changes visible in their film (see ibid., 86). Beyond that, *PPR* also makes use of the strategy of substitution. Whereas *SS* focuses on the foetus as a victim and presents pregnant women suffering passively after abortions, *PPR* shows women as rational and active decision-makers. "It replaces images of women as passive and pregnant with images of women as active professionals" (ibid., 87). Finally, *PPR* also offers an example of the last strategy of refutation, namely transformation. Key images of *SS* are presented here in a different visual frame. For example, a specialist is shown watching *SS* on a TV screen, commenting critically on what is shown (see ibid.).

What becomes clear in Lake and Pickering's analysis is that participants in a controversy not only have to pay attention to their opponents' verbal statements, but also have to operate in a visual way if their rivals make use of visual representations in the course of the debate. One reason for this obvious necessity might be found in the commonly stated *persuasive power* of images.[18] Lake and Pickering offer a much stronger reason when indirectly referring to the *evidential role* of visual representations. Interpreting the anti-abortionists' communicative intention to use films in the debate, they conclude that for these people films even have the status of *proofs*.[19] At the end of their paper, Lake and Pickering state that:

> [i]nterestingly, SS (and anti-abortion rhetoric generally) implicitly adopts the view [...] that pictures *cannot* argue, and adapts this view to ideological ends; that is, it contends that images of the fetus are *beyond* argument and constitute *irrefutable* proof of the fetus' humanity.
>
> (Lake and Pickering 1998, 91)

More precisely, the thesis that is supposed to be proven visually by *SS* is that "the fetus is an unborn 'child' and that abortion therefore must be murder" (ibid., 84). Obviously, Lake and Pickering regard 'argumentation' as an exchange of reasons in this context. If one side achieves for its thesis the status of a proof, a continuing exchange would make no sense, as apparently no question is left for further discussion.

Lake and Pickering contrast this claim about what anti-abortionists attempt to achieve by showing their film with what the authors think is the underlying assumption of the second party's way to make use of films in this context.

> In contrast, PPR implicitly adopts the view [...] that pictures *can* argue, and adapts this view to contrary ideological purposes; that is, it contends that images of the fetus are *only* argumentative, are *susceptible* to refutation, and constitute *misleading* evidence of the fetus' humanity.
>
> (Lake and Pickering 1998, 91)

Lake and Pickering elaborately show different strategies of utilising films – and thus of visual representations – in an already highly charged emotional debate. They point out the dynamics of the debate that played a role in producing and making certain visual representations popular. In particular, it becomes clear that, because of the persuasive power of images, the contending parties were simply not able to ignore this shift in representational means in the debate and stick to an exchange of linguistic arguments. They had to bring into play something that could deal with this persuasive capability of images. Doing so, however, does not necessarily imply that in such instances *pathos* is given priority to *logos*, as critics usually object. Contrary to this, Lake and Pickering point out that refuting the visual representation of *The Silent Scream* also means presenting a detailed and careful analysis of what is shown in this film and making clear how and why those images were produced. These analytical steps are accompanied by visualisations that are meant not only to support the verbally but also the visually presented explanations of experts (see ibid., 86). For example, the expert's claim that it is neurologically impossible that the foetus screams inside its mother's womb is supported by images showing the development of the human brain. At the relevant stage, it is neurologically impossible that the brain of the foetus is able to register alarm (see ibid.). This illustrates the point that making use of images in argumentation does not necessarily imply dramatisation and appeals to emotional stimuli. They can also be used in an explanatory and/or evidential way.

Beyond that Lake and Pickering's analysis also shows how images can make contributions to the different sub-acts of argumentation.

- Visualisations were necessary to utter the argument (locutionary act). The examples in this context are films. Moreover, visual representations – such as the ultrasound images of the foetus, or images of the developmental stages of the foetus's brain – are displayed within this audiovisual medium.
- Images were used as persuasive means (perlocutionary act). For example, the ultrasound images of the apparently screaming foetus and its apparent attempts to avoid the abortion are supposed to provoke feelings of sorrow and compassion amongst the recipients.
- Visual representations entail certain information (propositional act). The *PPR* film-makers, for example, demonstrate visually that their opponents produced a propaganda-film making secret use of slow-motion techniques to provoke certain emotions.
- Finally, images were brought into play to *securely establish some belief* – to use Groarke's words (see Groarke 2015, 135) – (illocutionary act). By showing their films, both parties try to convince an audience of their respective opinion, namely that abortion is either murder or that it is not.

What remains to be shown is how to deal with the thesis that visual representations can provide propositional knowledge, although it is disputed by traditionalists that images are propositionally structured. Two further proposals of how to conceive of visualisations in the argumentative context should therefore be examined more closely at this point.

To begin with, (1) Axel Arturo Barceló Aspeitia suggests taking them as instances of what he calls "heterogenous arguments, i.e. arguments that are not conveyed through a single medium, but instead make use of both verbal and visual resources" (Barceló Aspeitia 2012, 356). He thinks that, in these contexts, visualisations can indeed make proper contributions which "can be either sub-propositional (i.e. properties and functions that, properly combined with information conveyed through other means, like words, or available in the context, can yield full propositions) or fully propositional" (ibid.). Thus, Barceló Aspeitia is convinced, like Perini, that visual representations can entail propositionally structured information. Contrary to her conviction, however, he does not claim that images can be bearers of truth values. By analogical reasoning, he compares heterogenous arguments with sub-sentential ones. The latter contain fragmentary statements, either as premises or as conclusions, and are nonetheless understood as transmitting full propositions (see ibid., 357). In such cases, the context yields the missing information. This contextual information enables the hearer to understand what was meant by the sub-sentential argument. Barceló Aspeitia thinks that, just as language can be used in a fragmentary way to transmit complete propositions, images can also play this functional role. Thus, he tries to show how, in certain instances, image and text work together to transmit the information intended by the speaker or writer.

Barceló Aspeitia points out that it is common practice to use a combination of different modes of representation – or media, as he calls them. The examples he mentions are wanted signs, store catalogues and advertisements. We do not usually have any difficulty understanding that the combination of an image and a numerically expressed price tells us that a certain item is meant to be sold for a certain amount of money (see ibid., 360). In this way, image and text interact to complete the message.

Yet it might be objected, as Barceló Aspeitia points out, that this information transmission is only possible if the proposition in question is reconstructed verbally by the respective recipient (see ibid.). Contrary to such an assumption, however, he defends the thesis that paraphrasing is not necessary. As a rationale for this claim, he refers to the difficulty of translating the non-verbal elements in a precise and unambiguous manner. He concludes that "[p]recisely because it is not always possible to translate heterogenous arguments into verbal ones, it is very unlikely that that is what happens every time we interpret heterogenous arguments" (ibid., 361). Nonetheless, the mere possibility of highlighting certain translations as more adequate than others and, furthermore, of ruling out certain sentences as completely inadequate as translation demonstrate, from his point of view, that visual representations convey propositional content – that is, a content being used as a benchmark to assess the adequacy of certain translations.

An example is provided by the successful detection of gravitational waves mentioned in the introduction to this book. A crucial part of the argument that a successful detection had been achieved was that two measurement devices registered the signal independently and thereby raised the statistical significance of the evidential status of the event registered. This claim is supported by the fact that the two diagrams (see Figure 1.2) showing the registered data match to a significant degree. This seems to be the usual way to translate visual representations, such as the diagrams in this example, in the context of a heterogenous argument. However, the fit between both diagrams claimed here is doubtless only a rough approximation to what is visually conveyed. Yet the conflation of the two diagrams also shows that a claim such as "The LIGO project at Hanford registered completely different data from the LIGO project at Livingston" would not be an admissible translation of these diagrams.

Barceló Aspeitia comes to a positive result when considering the question of whether visual representations can transmit propositional content.

> Thus I conclude that the contribution images make to this kind of argumentation is substantial and direct: exploiting information from the context, they provide information necessary for the communication of the propositions that play the roles of premises and conclusion. Furthermore, they achieve this directly, without the need of verbalization.
>
> (Barceló Aspeitia 2012, 365)

Yet contrary to some conceptions of 'proposition', he also defends the thesis that propositions are not linguistic entities (see ibid.). They *can* be transmitted linguistically, but this can also be achieved by using other kinds of representations. Thus, what alters the scales here is the question of how exactly to conceive of 'propositions' – and Barceló Aspeitia at least partly removes this concept from an assumed close connection to verbal expressions. That is, his considerations undermine the assumption of a need to express propositions in verbal statements right from the start, or at least the possibility of translating the content transmitted by other representational means completely into verbal statements. This interesting move will be scrutinised in more detail in due course when analysing the nature of the *cognitive content* of visual representations below.

The second strategy (2) to deal with visualisations in argumentative contexts that will be discussed here is Groarke's approach to the topic. His proposal about "multimodal arguments" offers a different perspective with regard to the relevance of propositionality in the context of argumentation (see Groarke 2015). He agrees with Perini and Barceló Aspeitia that visual representations can be regarded as proper parts of arguments. Furthermore, Groarke defends the claim that these parts do not have to be propositionally structured to play the epistemic roles they are supposed to fulfil in those contexts (see ibid., 135). He attempts to broaden the concept of argumentation in an even more wide-ranging sense than, for example, Barceló Aspeitia. He writes: "[t]he classical model of argument understands arguments to be entities made up of words and sentences. In using other modes of arguing, arguers build arguments from other kinds of ingredients – pictures, diagrams, non-verbal sounds, tastes, and so on" (ibid., 140). Thus, Groarke regards a broader class of "ingredients", as he calls them, as possible candidates in argumentative structures. As a result, his conception by far exceeds my attempt to show how the three modes of representation – linguistic, visual and numerical – interact in the context of scientific argumentation. Groarke's broader perspective, moreover, also permits the detection of an even closer connection between visual representation, observations and sense perception as an epistemic source.

What exactly does Groarke suggest? Having pointed out that there are a lot of different "ingredients" a speaker might utilise in her argument, he proceeds by defining different modes and sub-modes of an argument along the lines of these ingredients. In order to analyse these multimodal arguments, he suggests the usage of a "Key Component" (KC) table (see ibid., 135). This table summarises the different components used in a particular argument by displaying in a first column "acts of arguing" (actions such as the utterance of claims, the display of visual data, etc.), the structural parts of the argument (premises and conclusion) in a second, and, finally, in a third column the mode used to express the different parts (visual, verbal, etc.). Although Groarke admits that using such a table to analyse multimodal arguments also means interpreting the different components to some extent (see ibid., 139), this method nonetheless permits a more faithful reconstruction than does a mere verbal translation of the relevant parts. Groarke's suggestion seems to be a useful addition to the traditional methods of argument-reconstruction in verbal forms employed to analyse their validity and conclusiveness.

Even more importantly, however, he points out that not only visual representations but also actual visual demonstrations can be parts of argumentation (see ibid., 148f.) – respectively, components that we perceive by our other senses such as smells or sounds (see ibid., 149f.). All of them are instances of using *perception* as an epistemic source. In order to better understand in what sense it can be said that these instances contribute to argumentation by adding information to premises and conclusions, the epistemic capacities of perception should be examined more closely. How do they work and what kind of knowledge (propositional or non-propositional) do they yield?

To sum this up, both, Barceló Aspeitia and Groarke try to pave the way for a more moderate conception of argumentation which would justify Perini's observation that it seems to be common practice to make use of other (in particular visual) components in the context of scientific arguments. Nonetheless, both accounts also raise the question of how (and what kind of) information is extracted from these different components to be employed in argumentation. This leads to a discussion of perception as an epistemic source and is closely connected to the question about what kind of content we cognitively access when deciphering information encoded in visual representations. Both questions will be considered in section 4.2.

4.1.4 Interim results: what can be learnt from argumentation theory?

In the previous sections, a closer examination was made of the possibility of visual arguments. This term is used in the debate to refer both to arguments that contain visual representations as (parts of) premises or conclusions, and to images that are supposed to express a complete argument. Perini has pointed out the tension between, on the one hand, scientists' practices of making extensive usage of visual representations in epistemic contexts of this sort and their obvious conviction that visualisations can undoubtedly fulfil the epistemic tasks they are supposed to perform and, on the other, the tendency of philosophers to ignore them or to hold them to be epistemically impotent.

It was argued above that this ignorance is at least partly a result of the linguistic turn at the beginning of the developmental history of analytical philosophy. This focus on language still dominates studies of this discipline and hence approaches connected to this field, such as the philosophy of science. In particular, arguments – that serve, amongst others, as means to transmit and defend scientific hypotheses – are typically regarded as linguistic entities. Moreover, as the discussion of this traditional point of view has shown, the emphasis on verbal expressions is also related to the question of what kinds of representations are suitable candidates in logical operations. In particular, entailment relations and the capacity to bear truth values are of relevance when it comes to (scientific) arguments. Both features, however, are associated with linguistic expressions in the Fregean tradition. Thus, visual representations are dismissed in this context as apparently belonging to the wrong kind of representation.

At best, people defending the traditional point of view in philosophy would call for a translation of information potentially contributed by visual representations to linguistic arguments. Thus, they think that proper contributions are only possible via a paraphrasing – a method that is supposed to take place in the recipient's mind – if it is not made explicit in the relevant publication. It is interesting to note that the demand of such a paraphrasing into linguistic expressions is not restricted to the realm of scientific argumentation.

Exactly the same sort of discussion can be found, for example, with respect to the epistemic functions of *thought experiments* (see e.g. Brown and Fehige 2011, sect. 3.2).[20] Here, too, the necessity to paraphrase the content presented visually is discussed to make use of it within the epistemic process in which it appears. With regard to thought experiments, this similarity to the above discussion comes as no surprise because the approach defended by John D. Norton in this respect is simply that "thought experiments in science are merely picturesque *argumentation*" (Norton 2004, 1142, my italics). Thus, Norton not only draws an analogy between the epistemic function of thought experiments and scientific arguments, he actually affords them parity. From his point of view, arguments can replace thought experiments without losing essential information (see Norton 1996, 336). In developing this thesis, he tries to make a case against James Robert Brown's ideas on the topic. Contrary to Norton, Brown thinks that the picturesque element of thought experiments is not simply reducible to (verbal) arguments. He states that thought experiments work epistemically *because* they are visual in certain respects. In particular, he emphasises the relevance of perceptually accessing the content of thought experiments. Of course, such a conception presupposes certain hypotheses about the nature of phenomena, their correlations (laws of nature) and the way these essential facts are understood.

To start with, from an ontological point of view, Brown defends a Platonist approach. That is, he thinks that abstract objects, such as mathematical objects, exist independently of human beings and do not belong to the realm of time and space. Moreover, he assumes that laws of nature are relations between properties. These relations are regarded as abstract objects as defined above (see Brown 2004, 1130f.). But how can we grasp these laws of nature? This is the point where it is of relevance that thought experiments are *visual* entities. Brown is convinced that we are able to *perceive* the laws of nature when visually engaging in thought experiments. He writes:

> [a]ccording to mathematical Platonism we can perceive the abstract entities of mathematics. Not all, of course, but we do have intuitions of some of them. So, it's possible to perceive abstract entities, at least some. Usually we learn laws empirically, by seeing instances. But laws are abstract entities, so they could be perceivable, too. How could we have an intuition of a law of nature? The obvious thing to conjecture is that we grasp them via thought experiments. Laws and numbers are both outside of space and time. If we can see one, then we should be able to see the other. Thought experiments are telescopes into the abstract realm.
>
> (Brown 2004, 1131)

Translating thought experiments into verbal arguments would then imply cutting off the ability to epistemically access the relevant content that they reveal to the recipient. At least our understanding of the relevant correlations would be hampered to a certain degree.

Brown's ideas on the topic undoubtedly deserve closer attention in order to understand correctly what is implied by his different hypotheses. In particular, his approach to regarding thought experiments as perceptual tools to grasp laws of nature as abstract objects is obviously not an easily handled topic. Consequently, the discussion of this account here is admittedly rather brief, as I will not go into the details of Platonism; nonetheless, the controversy between Brown and Norton is worth mentioning in the context of the current discussion. It makes explicit an important question already present above in the discussion of visual arguments in science, namely what a translation of possible contents of thought experiments into (verbal) arguments would imply.

The question about the consequences of a translation reveals different dimensions of the topic that should be kept apart. (1) If a translation is possible, then there is a cognitive content of some kind present in both instances. Are there particular difficulties with regard to such a translation? If there are, this might mean, in a negative sense, (2) that one of these representational means is inferior to the other in performing a certain epistemic task. Ambiguities, for example, might cause problems in translations. If precision is required in the situation at hand, representations that do not meet this standard seem to be inappropriate tools to serve the respective purpose. This also shows that (3) evaluations of inferiority and superiority are not absolute. They are relative, depending on the context of usage and the task to be performed with the aid of representational means. Nonetheless, there might be contexts (4) in which our perceptual sense is of particular use in grasping certain information, as evolution has emphasised its development in human beings. Vision is our primary sense to perceive, to cognitively access the world. It should not come as a surprise that images, which allow us to make use of our visual sense to decode them, are easier to handle in the epistemic process than other modes of representation. Finally (5), there might be certain aspects that can be accessed via perception only (this is what seems to lie at the heart of Brown's ideas about thought experiments) and thus undermine the suggestion of translating visually presented information into verbal information in any possible context.

These five theses constitute the stepping stones to the next questions. Do visual representations contain a cognitive content? And, if so, is it similar to the propositional content of verbal expressions? If it is similar, this might provide a clue to rethinking the concept of propositions as currently in use in the philosophical community. The tight linkage to linguistic expressions may have to be considered anew because propositional content can also be transmitted by visual means. However, if it turns out that, although the content entailed in visual representations is not similar to the nature of the alleged propositional content in the Fregean tradition, it can nevertheless be shown that at least some visual representations are epistemically potent, this might then be taken as a clue to reconsider the thesis that epistemic usefulness is exclusively tied to propositional content.

The discussion of these two alternative approaches on how best to explain the epistemic potentiality of images in the context of scientific arguments naturally takes its course by analysing perception as an epistemic source. It is through perception that the content of visual representations is cognitively accessed. Clarifying this route of access will thus also contribute to an understanding of how visual representations may be epistemically effective. This, then, is what will be analysed in more detail in the following section.

4.2 The cognitive content of visual representations

Making use of visual representations in scientific arguments in the way Perini explains this process, namely that scientists "describe figures as supporting, or as expressing, the conclusion of an article" (Perini 2010, 134), implies a particular attitude towards those representational means. Scientists seem to assume that those images entail certain information and, additionally, that they are able to transmit this information in an appropriate way. These attitudes towards visual representations in science are one of the main reasons supporting the initial claim that scientific images can best be regarded as signs.

Perini takes this observation about communicative practices in science and the place of visual means therein to demand that "[w]e [i.e. philosophers of science, N.M.] need to take a more fundamental approach to understand what kind of contents visual representations convey, and how they do so" (ibid., 135). Thus, there is a need for "a more specific account of visual representations as communication tools – a theory of signs or a semiotics of visual representations" (ibid.), as she expresses it. In this section, an approach to account for this capability of visual representations to serve as communicative tools and will proceed as follows.

Firstly, a start can be made by focusing on the capacity of visual representations as *content providers* in scientific arguments. It is necessary to analyse in more detail in what sense this capacity can be accounted for and what exactly the nature of the content is that can be transmitted via visual representations. This topic is connected with considerations about translatability and the limits to what is expressible by certain representational means. Certain theses of this sort have already appeared in section 2.2 of this book. The consideration now must be why certain characteristics of visual representations do not apparently preclude the latter from being used in scientific communication, although philosophers tend to regard them as shortcomings.

Moreover, taking seriously the possibility of *translations* from one representational means into another raises the question of whether this transfer of information only works in some directions but not in others. Taking into account the initial distinction made earlier between numerical, linguistic and visual representations, it might be argued that certain translations will imply a loss of information, which might lead us to the assumption that one of those representational means is more ontologically and epistemologically basic than the others. That is, although all information can be encoded in a (visual) representational means when transferred into another medium, the content provided might lose particular pieces of information in that process. In section 3.1.1 of this book, we have already come across the puzzle that digital images present when discussing the epistemic status of scientific visualisations. With regard to digital images, the question arose of whether numerical data should be considered as the more basic kind of representation, as, obviously, digital images are rendered in such data sets. In the following, such an apparent *reducibility thesis* will be examined more thoroughly.

Closely connected to this issue of reducibility is the question of how exactly to conceive of the content of visual representations in science, that is, the question of whether it is propositionally structured or not. What links this to the discussion of reducibility is the following consideration: although there are many different accounts of what the term 'proposition' exactly means,[21] one aspect that is commonly accepted is that propositions are expressible by linguistic sentences. At least in this way, propositions are tied to language, although they themselves do not necessarily have to be linguistic in kind. A corollary of this is that if the content contained in and transmitted via visual representations is non-propositional in certain respects, it cannot be translated and thus is not reducible to linguistic expressions without a loss of information.

Secondly, the topic of *perception as an epistemic source* will be discussed in order to come to tackle this question about the assumed propositional or non-propositional content of scientific images. Obviously, the information that is presented via visual representations has to be perceived, that is, their content has to be cognitively accessed by vision – the primary human sense. It does not therefore seem far-fetched to assume that the epistemic mechanisms of perception will also disclose something about the special case of image perception, i.e. the decoding process of information presented visually. A closer look at the epistemological debate about perception in general thereby reveals an interesting fact concerning the question of propositions. On the one hand, contemporary philosophers seem to have no particular problem in regarding perception as an epistemic source, i.e. as a source of propositional knowledge. On the other hand, there is no consensus about the content of perception and how exactly it features in processes of justification concerning perceptual beliefs.[22] This presents a difficulty similar to the case of visual representations in science. Although knowledge claims based on visually transmitted information (either via perceiving images or by perceptual experiences of other entities in the world) are commonly accepted, there is a dispute over whether the content of images is propositionally structured or not. Contrary to the case of everyday perception, this latter question is employed to undermine claims concerning the epistemic capabilities of visual representations in communicative contexts, whereas a potential non-propositional structure of perceptual experience in general is not usually indicated as a particularly serious obstacle to gaining knowledge via perception. This discrepancy between the two cases seems to be worth closer investigation.

It can be assumed that this difference is a consequence of the initially defended thesis that scientific images can best be regarded as *signs*. Taking this claim seriously, one might object that, particularly in the context of scientific communication, those signs cannot simply be perceived in order to decipher their content, but that background knowledge is necessary to interpret them correctly. It might be argued that permitting the thesis that non-propositional content plays a significant part in visual discourse would clash with the communicative aims of transmitting precise information and clear messages. The compatibility of the claims that scientific images are signs and that perception is the primary way to decipher their content will therefore be analysed thirdly. Before discussing these difficulties, a start can be made by scrutinising the content characteristics of visual representations in science in more detail.

4.2.1 *Content translatability and the reducibility thesis*

In this section, the question will be investigated whether it can be stated that visual representations possess a content that works similar to Fregean

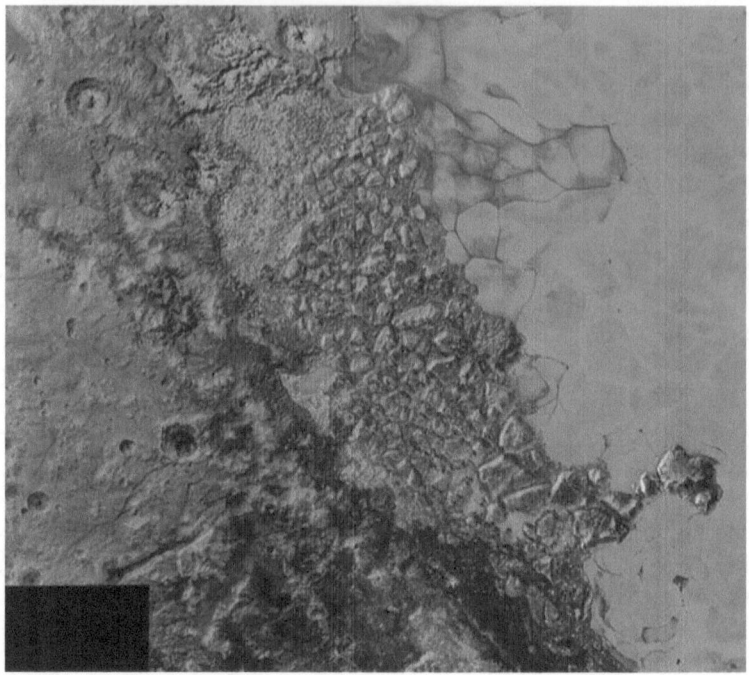

Figure 4.3 Image of Pluto's vast, icy surface. This part of its surface is informally
called Sputnik Planum. The image was taken from a distance of 50,000
miles (80,000 kilometres) as *New Horizons* flew past Pluto on July 14, 2015.

Source: Southwest Research Institute, www.nasa.gov/image-feature/image-of-plutos-vast-
icy-plain-informally-called-sputnik-planum

'thoughts' in linguistic expressions, that is, whether they have a content that
human observers can cognitively access and process to gain knowledge or not.
My approach to this topic is via an analysis of content translatability from
one representational means to another. An example will illustrate what is at
issue here.

In 2015, after a nine-year voyage, the NASA spacecraft *New Horizons*[23]
finally reached its first mission destination, the dwarf planet Pluto and its
system of satellites. Since its arrival, the space probe has started taking and
transmitting photographs of Pluto (see Figure 4.3) and its satellites. Of
course, these are digital images, not analogue photographs. What is transmitted to Earth are digital data that allow the corresponding image to be
rendered when the data have been received. In this case, a translation between
the numerical and visual mode of representation apparently works without a
loss of information.

This kind of translatability is not very surprising, because digital photography works by storing information detected by a charge-coupled device

(CCD) sensor in binary units that allow the reconstruction of visual information as a picture later on. Interestingly, this process also works in reverse. That is, starting with an analogue image, we can transfer it into bits and bytes by simply scanning it, and it is not only images that we digitalise in this way but texts, that is, written language, as well. Moreover, oral language can also be recorded and digitally stored. Thus, switching back and forth between numerical data on one side and all other modes of representations on the other seems to be fairly unproblematic. Again, this is not very surprising as this is how information-technology devices work.

Regarded in this way, then, no representational means seems to be more basic than numbers, or rather 'digital units', as even numbers are reduced to bits in IT contexts, that is, to 'basic units of information'. These units are expressed by binary representations such as the commonly used 0–1 schema. However, they could also be represented quite differently.

Now, what does this translatability of all three kinds of representation to digital units mean with regard to the question of the content provided by visual representations and its usability in scientific epistemic discourse? Firstly, it shows that, at least in a trivial sense, *there is information* contained in all representational means that can be transferred from one medium to another, namely information about the respective medium itself. However, it is the representational relation involved that is of interest here, namely the question of whether what can be transferred in such translation processes can go beyond this trivial kind of information. That this is at least sometimes the case becomes clear when regarding the process of digitalisation in reverse, as Zachary C. Irving does. He states that:

> [c]omputer simulations and other computerized instruments are interesting case studies because they output data sets that can, in principle, be represented in many different styles (e.g. numerically or visually). Since each of these stylistically distinct outputs represents the same data set, they should have the same content and thus, *prima facie*, be intersubstitutable in whatever scientific argument they are used without a change in soundness.
>
> (Irving 2011, 775)

Does this mean that these different representational variants of the basic content can then be used interchangeably in communicative contexts?

Such an assumption seems to be immediately blocked when human beings are considered an essential part of the epistemic process under discussion. This is also the objection that Irving raises. Although it might be possible to transfer all kinds of information into digital units, it would be senseless to present these sequences of bits and bytes in scientific arguments because we would simply not be able to *understand* them.[24]

Moreover, in section 3.2 of this book, it was pointed out that not all kinds of visual representation are equally suitable for diverse communicative purposes. The background knowledge of the audience, the goal and the context of communication were highlighted as conditions essential to influencing the decision on how best to present the information. A similar case can be made with respect to the applicability of digital units. They might be useful in the context of information technology, but not as a basis for human epistemic processes. As such they have to be (re-)converted into a format cognitively accessible to human investigators.

The usefulness of digital information processing in epistemic practices is indubitable, but digital information itself should not be regarded as an *epistemically more basic* category of representing information. It is only a tool, a technology that works very well, the design of which, however, could also have been totally different. Nonetheless, we can infer the positive result that the possibility of digitalising images (as well as texts) shows that they contain information which can be extracted and decoded by using other representational means. Yet it might still be objected that the kind of information disclosed in its processing by digital devices is more or less trivial in nature. For example, it does not allow us to differentiate between images that are only used for decorative purposes and ones that are essential to the scientific argument. Thus, the focus of the question should be shifted slightly: are visual representations capable of bearing *meaningful information* instead of information in general (whereby 'meaningful' here refers to information relevant to the cogency of the arguments presented)? An instance of meaningfulness that plays a major role in this context regarding scientific images is the presentation of measurement data. As our example of the Higgs image (see Figure 2.4) has made clear, in many instances these measurement processes are connected to information-technology devices. This means that the results obtained can be presented by more or less all kinds of representational means. However, a translation between different modes of representation is also possible if no digital information is available. In many instances, there is, for example, the choice of presenting correlations between numerical data values in a table or visualised in a diagram. Accordingly, such visual representations can indeed be *bearers of essential information*, which was one presupposition for their possible role in scientific discourse.

Apart from this basic translatability, is there a kind of epistemic hierarchy involved when choosing between the different modes of data presentation? There are two points to be mentioned as a reply. Firstly, in section 2.1 of this book, it has already been shown that, depending on the context, some ways of presenting the data are more suitable to the task at hand than others. Secondly, such a task-orientated approach allows for the possibility that in some instances visual representations are epistemically superior to other kinds of representational means because they allow for an evaluation of data that otherwise would not be possible. Jutta Schickore offers an example in her discussion of a case study in astronomy (see Schickore 1999). Here, diagrams of a spectrographic analysis were used for the purpose of clarifying whether the data received showed two different interstellar objects or just one object, doubled by a gravitational lens (see ibid, ch. 3). Schickore points out that, in the end, it was a matter of a *qualitative visual evaluation* of the diagrams that guided the scientists in their decision. The *resemblance* between the two diagrams supported their judgement that only one object had been observed and not two different ones (see ibid., 282). It can be assumed that this judgement would not have been possible if the scientists had only the numerical data at their disposal.

In any case, it is not just resemblance and other relations between data that visual representations may easily reveal, but measurement errors can also be detected in this way. Sometimes the results have first to be depicted in a visual form before it can be decided what the real output is and what is just background noise. The Higgs image (see Figure 2.4) is a good example. One might argue that if inconsistencies or anomalies were detected, scientists would turn to the numerical database for clarification. This assumed procedure is then taken as supporting the claim that numerical data are epistemically more basic than their visual presentation. However, there are two points which call this reasoning into question. Firstly, scientists detect anomalies because they evaluate those very images and base their judgements on their own visual experiences with those representational means. That is, detecting anomalies would not be possible without visual representations. Secondly, not only does the epistemic process start with an image, but it also ends with one: the detection of an anomaly will surely result in the rendering of another image that can be used for comparative tasks. Thus, the whole process refers fundamentally to visual means. Of course, this line of reasoning should not be mistaken for claiming that processes of error detection and correction *always* work this way. However, it would be equally wrong to state that visual representations can in principle play only epistemically inferior roles.

Summarising the results so far, the analysis shows that numerical measurement data can be translated relatively easily into visual representations and vice versa. In addition it has been pointed out that there are particular tasks and contexts that make the use of certain representational means more reasonable than others. Consequently, there seems to be no fundamental hierarchy in place with respect to epistemic processes when comparing these two forms of information presentation. Despite this initial comparability between both means of representation, there is another way that might provoke claims of epistemic difference here, namely their usefulness in the context of scientific arguments.

In the previous section, it was highlighted that there are certain requirements that constitutive parts of arguments have to fulfil. They have to be bearers of truth values and they have to allow for certain entailment relations that make deductive reasoning possible. These characteristics are usually attributed to propositions. Moreover, it was stated that, despite there being rival conceptions of what is meant by the term 'proposition', one predominantly uncontested characteristic is that they are supposed to be *linguistically expressible*. Georges Roque gives the following synopsis of the dispute about visual representations in (scientific) argumentation that follows from accepting this condition of propositionality. "One underlying assumption is that arguments should be propositional. Consequently, if images are not propositional, therefore visual arguments are impossible unless the recipient extracts and builds propositions from them, which is a problematic process" (Roque 2015, 180).

Of course, the degree of the difficulty involved in translating the content of visual representations in a propositional way is greatly dependent on how we conceive of the concept of propositions itself. A more moderate stance towards this concept can be used to more or less circumvent the whole quarrel. J. Anthony Blair, for example, an early defender of the thesis that visual arguments are possible, chooses such a strategy. He claims that there is no particular problem related to translating visual content into propositions, as the latter can be expressed in a variety of different ways. He states that "[p]ropositions can be expressed in any number of ways, including silence [...], but also by signs or signals [...], or by facial or other body-language expressions [...]" (Blair 1996, 26). As a consequence of this moderate conception of propositions, he infers that "[t]he visual expression of propositions, then, is familiar and relatively unproblematic" (ibid.). Whereas Roque argues that visual arguments are impossible as images are not propositional, Blair shifts the perspective by claiming that propositions can be expressed by very different means. That is, a claim about characteristics or essence is turned into a question about capacities of expressibility. This shift in perspective then makes it relatively easy for Blair to claim that "[...] visual arguments are not distinct in essence from verbal arguments. The argument is always a propositional entity, merely expressed differently in the two cases" (ibid., 38).

Roque, on the other hand, emphasises the relation between propositions and linguistic expressions, particularly when denying that images have the capacity to bear propositions. Thus, by mentioning the process of *extracting and creating propositions*, he refers to the question of the translatability of visual representations into linguistic expressions. This clearly more difficult approach to the topic can be obviated, however, by simply adopting Blair's suggestion that propositions are in principle expressible in a diversity of ways, which would then render the task of translation redundant.

Perini's approach attempts to deal with the persistent problem of the capacity to bear truth values. This becomes clear when Blair's later works (see Blair 2004) are examined, for it is here that the objection based on the truth-bearing capacity of propositions is taken up in earnest. He offers two rejoinders to the objection that, as images cannot be bearers of truth values, they cannot express propositions (see ibid., 47). His first rejoinder is simply to offer an example of what he considers to be a paradigmatic case of a visual argument (see ibid., 47f.), namely a political, pre-World War II cartoon showing an Englishman sitting beneath a pile of boulders. These boulders are labelled with the names of European countries.

The bottom boulder, sticking out but wedged under and holding up the rest, is marked 'Czecho'. [...] A thick rope is attached to the out-thrust end of the 'Czecho' boulder and pulled up overhead and out of sight. Clearly a strong pull on that rope would dislodge the 'Czecho' boulder, causing the rest to come crashing down on the Englishman below. The cartoon's caption reads, 'What's Czechoslovakia to me anyway?'

(Blair 2004, 48)

The message, i.e. the proposition expressed by this cartoon, is that the political fate of Czechoslovakia will indeed have a major impact on England. Moreover, Blair points out that this proposition was either true or false at the time of its publication (see ibid.).

His second rejoinder consists in highlighting that arguments are not only used to change the recipient's beliefs, but also her attitudes, intentions, or behaviour (see ibid., 48). These latter instances do not have truth values, as he points out (see ibid.). Apparently, this concession implies a softening of Blair's initial thesis. Yet it might be objected that, although Blair is right in claiming that "attitudes, intentions and conduct do not have truth value", this does not imply that argumentative attempts to change them do not rely on truth-bearing propositions. Blair's intention to ascribe a different important role to images in the context of argumentation by drawing on the psychological effects of visual means instead of their assumed epistemic contributions does not therefore undermine the thesis that they express a propositionally structured content.

Be that as it may, Blair concludes by pointing to the more moderate achievement that he thinks he has obtained, namely that his rejoinders to this objection "[...] at least [...] shift the burden of proof" (ibid., 49). Alternatives have now to be contemplated, if his rejoinders do not convince, that create a way to keep visual representations nonetheless involved in the argument.

This brings us back to Roque's suggestion that although images cannot represent propositions themselves, their content can perhaps be translated into representational means capable of fulfilling this task. This comes down to questioning whether the content presented visually can be paraphrased, i.e. translated into linguistic expressions. Perhaps a mental reconstruction of arguments that is performed has to be taken into account of which the contemplating scholar is more or less unaware. But is the cognitive content of visual representations translatable into linguistic expressions in the way required for scientific arguments?

Blair thinks that, in principle, such a translation is possible (see Blair 1996, 25). Actually, he argues that visual representations pose no particular problem of translation. This optimistic attitude is a result of his more moderate conception of propositions. Yet he also seems to implicitly admit that there are limitations to the scope of such translations when he says that "[w]hether such descriptions or translations can be complete or fully adequate is a separate question" (ibid., 25). The difficulty has already been mentioned that, if visual representations are to be employed in communicative acts, certain limitations of expressibility have to be acknowledged. A related lack of image precision is highlighted in Elliott Sober's account. Similar to Blair, Sober starts in an optimistic way by expressing his conviction "[...] that every representational picture[25] has a sentential counterpart of a certain form which captures the content of the picture with which it is identified" (Sober 1976, 111). Consequently, Sober defends the thesis that a translation of visual representations is indeed possible.

Interestingly, his conception of how to translate visually presented information into linguistic expressions also allows him to explain in what sense images can be called true.

> A representational picture gives it to be understood that a certain state of affairs obtains. If a picture is *true*, then the state of affairs *does* obtain. For this reason, we will identify each representational picture with an existential hypothesis which posits the existence of certain specified objects. A picture will thus be true if and only if the world contains the kinds of objects demanded by the sentential counterpart. True pictures are thus ones that have *verifying instances*.
>
> (Sober 1976, 111f., his italics)

Sober's suggestion is much more precise than the proposal made by Tim Crane, who discusses the question of whether images can have a propositional content (see Crane 2009). He assumes that "[f]or any picture P, there is a sentence which gives the content of P" (ibid., 460). At first glance, this seems to be quite similar to Sober's approach. However, Crane thinks of the sentential counterpart of images as mere descriptions of their contents, not as an "existential hypothesis which posits the existence of certain specified objects" (Sober 1976, 112), as Sober does. Crane uses this line of reasoning to conclude that there is still an important difference between the image's content and its description (see Crane 2009, 460). He infers from this that such sentential counterparts to the content of pictures do not support the thesis that images have propositional contents (see ibid., 461). Obviously, this reasoning is a consequence of Crane's identification of "a sentence which gives the content of P" with "a description of P" which allows the latter to be rather loosely connected to the visually encoded information. Here, Sober's thesis is much more precise in pointing out that the sentential counterpart has to entail an existential hypothesis of a certain kind. From my point of view, this precision is ample reason for preferring Sober's account to Crane's approach.

Yet this preference does not imply that Sober's approach presents no difficulties. Whereas the idea that there is a kind of test to check whether images that are supposed to represent certain entities of the world fulfil their task successfully seems to be more or less uncontroversial, the moot point in this conception is, of course, Sober's claim that it is the picture and not the correlated sentence that is true. Suggesting such a way of evaluating the content of visual representations apparently calls for precision when it comes to the task of translating its content. If we want to check whether a particular state of affairs that a certain picture shows actually obtained, as Sober puts it, we not only need a vague description of what is going on in the picture, as Crane e.g. suggests, but a precise translation. And although Sober's account is preferable to Crane's, as the former makes clear what kind of description is called for, for two reasons Sober's approach unfortunately does not fulfil requirements either. Firstly, there is Sober's own discussion of the limits to what is visually expressible. Secondly, there is his concession that analogue pictures cannot be completely translated in the way suggested above (see Sober 1976, 141).

Concerning the first aspect, Sober points out that visual representations lack the capacity to express negations[26] and disjunctions. With regard to the former, he notes that:

> [...] the absence of negation has two parts: On the one hand there is no operation on a picture that produces a picture whose interpretation is the negation of that of the one operated upon. On the other, predicates occu[r]ring in the interpretation of any representational picture are such that their negations never so occur.
>
> (Sober 1976, 128)

Adhering to his idea of translatability, he points out that there is no logical way to produce two images whose sentential counterparts are contradictory, and consequently he claims that there is no such thing as pictorial negation.

One might object that there are, for example, drawings in field guides for biologists that show a particular characteristic that dragonflies, say, belonging to the species *Emperor* always show a black mid-line stripe on the abdomen. Of course it is possible that this characteristic is omitted in another drawing. But although in such a case there is a picture showing p (black stripe), and the other one does not *not show p*, i.e. an example of an Emperor without a black stripe, this is not a negation of the former image but simply an image of a different species.

Sober presents a similar argument concerning the possibility of disjunctive claims via visual representations. "There is no operation on pairs of pictures which effects their disjunction, and if two predicates occur in an interpretation, their disjunction never so occurs" (ibid.). No doubt, there is a broad variety of different pictures, but how could a relation between two of them be produced to express an alternative to what is shown?

Although Sober has highlighted serious shortcomings of visual representations as means of scientific communication, he does not regard this as proving the impossibility of translations between images and verbal expressions. On the one hand, he does not defend the thesis that a presumed translatability works both ways – from the visual to the linguistic and vice versa. He only emphasises that the first route – from visual to linguistic representations – is always possible in the case of representational images.

On the other, he circumvents the problem by diminishing the relevance of those logical relations, discussed above, in human sense perception: human vision is developed by evolution to allow for a well-informed orientation in our natural environment. "For this reason, it [human perception, N.M.] is supposed to pick out properties of the environment that are significant in terms of prediction and explanation" (ibid., 131). Sober thinks that it is only natural properties that can fulfil the relevant task. Consequently, he claims that human vision is adapted to perceiving predominantly natural properties. He transfers this line of reasoning via an analogy to picture perception. By claiming that the basic cognitive processes in both instances are quite similar, he takes it that "[h]ence one should expect pictorial interpretations to fail to include irreducible disjunctions and negations, since these fail to pick out properties" (ibid., 131). What Sober apparently hints at is that, as we decode images via perception, the presentation of visual information has to obey the same limitations (or environmental accommodations, to put it in a positive way) as everyday perception does. In short, if we do not observe negations in the world, it should come as no surprise that they do not belong to the repertoire of what is expressible via visual representations.

Of course, this is quite a controversial claim that Sober inserts into his argumentation at this point. Although he might be right that human sense perception is particularly adapted to noticing certain entities in the world better than others, this does not constitute a reason that the expressibility of pictorial representation should be limited. Why should images not be able to present information that we usually do not perceive in our natural environment? Why should it not be possible to learn with this respect? Many scientific visualisations probably exhibit phenomena unfamiliar to everyday perception: the Higgs image (see Figure 2.4), for example.

Anyway, what should have become clear is that there are certain relational properties, such as negations, that cannot be expressed via visual representations.[27] Now, for the initial concern about questions of translatability from one representational mode into another, this means that there are certain constraints on transferring information *from the linguistic to the visual domain*. Although this does not pose a problem for Sober's own thesis that all representational images can be translated into linguistic expressions, a difficulty arises that does affect his project, namely translations *from the visual to the linguistic domain*. Here, complications arise when we want to translate what he calls *analogue pictures*.

What are analogue pictures in Sober's sense?

> In saying that a representational system is analog, we are claiming that there is a property of representations which the system treats as significant, and which is such that if there were a continuum of values of the property, there would be a continuum of significantly different representations.
>
> (Sober 1976, 139)

Because of these possible continua of property values, an analogue system of representation is usually richer than a digital one regarding the information it presents. Although translation into a digital system – such as natural languages – is possible, it implies a loss of information and, thus, "a loss in precision", as Sober points out (ibid., 140). A translation of the content of analogue pictures into linguistic expressions can therefore be achieved only approximately, that is, with a considerable lack of precision (see ibid., 141). This lack of precision is the result of there being a number of very different possible sentences derivable from an analogue picture via translation, which also limits Sober's initial thesis that the truth of pictures can be proven via verifying instances expressed in the sentential counterparts of images. If there is a variety of different sentential counterparts, no definite verifying instances can be stated.

To sum this up, Sober's account of the relation between linguistically and visually presented information seems to be correct. We are obviously able to describe what we see in representational images to a certain degree, and can thus take this as supporting the thesis that *there is* a cognitive content of visual representations that can be used to transmit certain information. Moreover, the discussion of Sober's ideas also makes plain that there are certain limitations to what is expressible in the visual format, but also highlights the fact that these limitations hold vice versa, i.e. that the informativeness of the visual content can outstrip the capacity to represent information in linguistic expressions. As a result, *a reduction* of visual information to the latter *is not possible*. Or, as Sober puts it:

> [h]ence, not all pictorial systems are reducible to impoverished linguistic systems of a certain kind. The relation of analog pictorial systems to language is more complicated: With respect to logical operations on representations, linguistic systems are more powerful, but with respect to expressing specifically visual relations between posited objects, linguistic systems can be more impoverished.
>
> (Sober 1976, 139)

It might be objected that images not only contain more information, but also raise the issue of ambiguity by this. Hence their richness of information might turn out as a particular shortcoming, especially in the context of scientific arguments. The soundness and validity of an argument depend on the precision of its premises and its conclusion. If visual representations cannot guarantee this, they might be inappropriate as representational means in this context.

A rejoinder to this objection has been worked out by Blair. He highlights several aspects diminishing the sceptical attitude towards visual representations in scientific arguments (see Blair 2004, 46f.). Blair states that ambiguity and vagueness are not attributes exclusively attached to the visual domain, but also to linguistic expressions. Nonetheless, difficulties of understanding are mostly ruled out by information provided by the context of uttering an argument. Moreover, visual arguments are usually mixtures of verbal and visual representations so that the former can (and are) used to disambiguate the latter.

This fits neatly with what was stated earlier about taking the whole communicative act into account when considering the meaning of particular visual representations. It is advisable here to follow Kjørup's account of the topic. Similar to Blair's idea, Kjørup points out that the context and the combination of visual and linguistic expressions will enhance our understanding of visually transmitted information, as the former will help to disambiguate the latter. Thus, we can conclude that ambiguity and vagueness pose no particular problems with respect to visual arguments.

What would be more interesting to know is whether the comprehensive informativeness of visual representations, pointed out by Sober and others, also entails *information that in principle cannot be transmitted by other representational means*. If this were the case, not only would the quantity of information of certain images be different from other representational means, but also their quality, i.e. the kind of information that can be transmitted via visual representations, would differ.

Taking a look at concrete examples suggests that this might indeed be the case. For instance, Perini discusses the question of whether a micrograph can be translated into linguistic expressions. Pointing to the same continuum of property values that Sober highlighted above, she comes to the conclusion that a reduction to verbal language is not possible. She offers the following reason as a rationale for her inference:

> [b]ecause the form of pictorial symbols is correlated exactly with features of their referents, a dense range of extremely complicated properties can be expressed by a pictorial system. This kind of system is extremely useful in science because it allows for the representation of properties, whether or not the vocabulary exists to refer to those properties with linguistic representations. As a result, a verbal description of the shape of the figure will convey less specific information about the shape of the sample than the figure itself.
>
> (Perini 2005c, 923)

Here, she highlights the fact that visual representations can transmit information that have no linguistic counterpart, i.e. no linguistic expressions available to adequately describe, let alone translate, what is occurring in the image. It is this characteristic feature of visual representations, namely that they enable communication about entities and states of affairs without the appropriate words available for them, which makes them particularly useful in scientific arguments (see Perini 2010, 144).

However, if we take it that propositions are connected to linguistic expressions, at least in the sense of being expressible by them, then Perini's example suggests the interesting conclusion that some scientific images can transmit *non-propositional content*. Moreover, if images contain such a kind of cognitive content, then the strategy of translating their content in order to adhere to the traditional conception of proposition-based arguments will be blocked. It is also this line of reasoning that motivates Blair's claim that "[f]or visual argument to represent a radically different kind of argument, it would have to be non-propositional" (Blair 1996, 34).

Moreover, as Perini's example makes clear, it would be wrong to dismiss visual representations as devoid of substantial information. On the contrary, they can contribute a kind of information that apparently cannot be transferred by linguistic means. Yet how is this information perceived, how is it cognitively processed so that it can play a role even in scientific arguments, and can knowledge be acquired in this way if the latter is usually considered to be propositional in kind?

Obviously the main process of cognitively accessing the content of visual representations is by visual perception. Thus, the next question to answer must therefore concern how perception is connected to propositional content or to non-propositional content respectively. Analysing these relations will offer an explanation of the possibility that the non-propositional content of images can play a role in scientific arguments.

4.2.2 Perception and non-propositional content

This section will be concerned with the nature of the content of perception. In particular, the question will be discussed whether this content is propositionally structured. The answer to this question will facilitate greater precision about the kind of content perceivable when regarding scientific images. Two aspects of the following line of reasoning are to be highlighted at this initial stage: firstly, it is particularly the theoretical assumptions put forward by Christopher Peacocke that will afford a connection between the debates in the philosophy of perception and in picture theory. Secondly, although the main focus of the analysis is on perception as an epistemic source, it will turn out that the interplay between different sources of knowledge is of relevant consideration in order to explain sufficiently what kinds of content can be transmitted in image perception as well as in perception in general.

Philosophical considerations of perception are related to different branches of this academic discipline. There are phenomenological approaches as well as epistemological ones, but the topic is also connected to questions of the philosophy of mind.[28] Regarding the current topic, two constraints need mentioning that will restrict the set of theoretical approaches that have to be taken into account: epistemological questions on the one hand and the cognitive mechanisms of vision on the other – although other sense modalities might play a role as well. In the following, only those parts of the debate about perceptual knowledge that are closely related to epistemological questions of picture perception in science will be examined.

It is not usually called into question that knowledge ascriptions are made, and are also often justifiably allowed to be made, based on the *epistemic source* of perception. It seems to be perfectly admissible to reply "Because I saw *that p*" to the question "How do you know *that p*?". This becomes particularly apparent when considering the debate about knowledge by testimony. Here, perception is commonly regarded as a more or less uncontroversial exemplar of an epistemic source, whereas philosophers still dispute whether testimony can have parity with perception in this respect. Participants of this debate seemingly have no qualms about regarding perception as an *epistemic source*. What is meant, then, when *sources of knowledge* are discussed?[29]

Robert Audi specifies the epistemic achievements of those sources in the following way:

> [a]s I am understanding sources of *knowledge*, and as they are generally conceived in philosophical literature, they are not just where knowledge comes from; they also provide the knower with grounds of knowledge. Grounds are what it is in virtue of which (roughly, on the basis of which) one knows or justifiedly believes. [...] sources indicate the kinds of grounds to expect a person to have when a person has knowledge through that source.
>
> (Audi 2002, 82, his italics)

Thus, epistemic sources not only tell us how certain beliefs are acquired, i.e. which cognitive resources are used in in this process, but also yield reasons that can justify the corresponding beliefs. Mentioning an epistemic source as a reply to the question "How do you know *that p*?" is commonly regarded as a *prima facie justification* of the corresponding belief. Hence, from an internalist's point of view, epistemic sources differ from mere causes of beliefs in that they provide us with reasons to think that the belief acquired is true, as Thomas Grundmann points out (see Grundmann 2008, 453). From an externalist's point of view, epistemic sources specify *reliable ways (or methods)* of belief formation. Of course, these cognitive processes are not infallible, but philosophers acknowledge them as usually leading to true beliefs. They also agree that epistemic sources *can* yield knowledge, i.e. *propositional knowledge*. Grundmann makes clear that calling x an epistemic source implies a positive epistemic evaluation of x. Making use of x in one's cognitive processes will usually imply that one will gain knowledge from x (see ibid., 455).

Perception is usually regarded as belonging to this set of epistemically distinguished sources. In the debate about knowledge by testimony, perception is even regarded as a *paradigmatic example* of such a source. Acknowledging this means accepting that perception enables epistemic subjects to gain propositional knowledge.[30]

Despite the agreement of many philosophers on this positive epistemic assessment of perception, and also their acknowledgement of the thesis that it is propositional knowledge that is at stake when contemplating the achievements of epistemic sources in epistemology, there is no consensus on the nature of perceptual content (see e.g. Siegel 2015). A primary consideration here is whether this content is propositional or non-propositional in kind. A propositional content would provide accuracy conditions with respect to claims concerning what has been perceived by a particular speaker (see ibid., sect. 2). As was pointed out above, considered in the Fregean tradition, a proposition is commonly regarded as an abstract object which is either true or false. Scholars also agree that propositions can be expressed by linguistic statements. Beyond this, however, there are also *concepts* that are related to both propositions and to language.[31] Now, some think of concepts as the "constituents of propositions" (Margolis and Laurence 2014, sect. 1.3). Regarded this way, the question of whether the contents of perceptions are propositional or non-propositional in kind also implies the question of whether their contents are conceptual or non-conceptual. Saying that those contents are propositional would then also mean to claim that they are conceptual, and, following Frege, these contents can then be expressed by verbal statements.

However, the precise relation between concepts and propositions is far from clear and the discussion about the contents of perception therefore becomes even more complex when taking this relational aspect into account. Grundmann, for example, questions the cognitive process involved in perceptual activities in this respect. He wonders how perceptual content is turned into propositional knowledge if perception transmits non-conceptual contents, an assumption which he defends (see Grundmann 2008, ch. 7.1.2). As a rationale for his thesis that the content of perception is non-conceptual in kind, Grundmann states, amongst other things, that we are able to perceive entities, for instance, different hues of a colour, for which we have no concepts (see ibid., 492).[32]

In particular, philosophers ask how sense experience and perceptual beliefs are related. Perceptual beliefs are regarded as propositional attitudes here. From a Fregean point of view, fulfilling their task as accuracy conditions also implies being expressible by verbal statements, which seems to presuppose that those contents are conceptualised. Grundmann makes clear that we should not simply identify perceptual content and perceptual beliefs (see ibid., 488f.). In particular, cases of optical illusions show that sense perceptions are quite robust even when challenged by background knowledge. That is, although the epistemic subject is aware of being misled by her perceptual apparatus, because she knows how the optical illusion at hand works, she is not able to change her perceptual experience. Consequently, she will have a particular visual impression, but will not believe what she sees because she knows about the illusionistic mechanism at work. Interrupting the process of belief formation in such a way, however, seems to be rather exceptional because we do not usually question what our senses tell us about the world but just believe what we perceive. More often than not, perceptual experiences are then turned into perceptual beliefs. For the moment, the answer to Grundmann's question about how exactly this process of transformation works will be left open. It will be explained below when we consider Peacocke's theory of non-conceptual perceptual content. What is of importance to the current investigation, however, is noticing the fact that this process usually works reliably. That is, perceptual contents are turned into perceptual beliefs, even though the former can be non-conceptual in kind, whereas the latter are commonly regarded as propositional attitudes that presuppose a conceptualised content.

In the context of science, Hentschel offers an illustrating example of observations that yield non-conceptual content and of scientists' subsequent attempts to conceptualise what they have observed. He describes the first observations of the surface of the sun in the nineteenth century (see Hentschel 2000, 23ff.). The solar structures, now known as granulation, visible with contemporary telescopes, were completely unknown to astronomers of that time. Thus, they not only lacked the words to describe what they had observed, but also had no conceptual background knowledge about the mechanism producing this phenomenon. As a strategy to circumvent the descriptive problem, astronomers made use of different metaphors, e.g. "willow leaves in an ocean of fire", "rice grains", etc. (ibid., 23) to report their observational results. Moreover, these metaphors were also used to pictorially illustrate their observations in their publications. Here, Hentschel points out that those images were particularly important to subsequent research activities about the related phenomenon, as they influenced significantly the way scientists observed it later on:

> [w]hat is at issue here is far more than terminology: Behind these competing 'descriptive labels' are mutually exclusive options on how best to see a new feature, that is, which Gestalt is assimilated to their visual impressions. [...] And these published illustrations in turn strongly influenced the perception of other observers [...] about what they anticipated seeing, that is, recognizing, in their telescopes.
>
> (Hentschel 2000, 23ff.)

By making use of metaphors (linguistically and visually), those scientists started conceptualising the phenomenon of granulation that they were observing. Those metaphors allowed them to refine their observations by offering a point of comparison to already-known phenomena, to exchange ideas about their observational results and, later, to develop hypotheses to explain the mechanism in the background. This example makes plain that human observers are apparently able to receive perceptual information, although they lack the concepts to explain and to describe in concrete terms what they have seen.

Here we find an explanation for why it is possible that some scientific images can transmit information we cannot describe by using linguistic expressions as we do not have the relevant words to do so – i.e. the thesis defended by Perini above. On the one hand, images can be used to transmit information about phenomena during the process of their conceptualisation, as Hentschel's example demonstrates, and, on the other, scientific images – in particular those working in a causal way, i.e. that are causally related to the object under investigation – can also contain non-conceptual information about the entity they represent; and, as it is through perception that we grasp such non-conceptual information, so human observers can learn about those entities by regarding the respective images.

Philosophical analyses of perceptual content also allow us to learn what *non-conceptual contents* might be like. Postulating such a kind of content is the result of acknowledging the above-mentioned thesis that the content of perceptual experiences and corresponding beliefs are not identical. Susanna Siegel calls a corresponding thesis of identity which explicitly refers to concepts in this context "experience conceptualism" and defines it as follows: "[f]or any object *x* and any property *F*, a subject has an experience as of *x* being *F* only if she has concepts of *x* and *F*, and deploys those concepts in the experience" (Siegel 2015, sect. 6). The relation between beliefs and experiences expressed in this statement can be grasped more quickly by identifying the contents of beliefs with the contents of experience. Siegel calls this the "same-content" thesis: "[f]or any experience as of an object *x* having a property *F*, if the experience has content *p*, then it is possible to have a belief with content *p*" (ibid).

Both of these theses are challenged by the fact that perceptual experiences can apparently be informatively richer than the concepts people are able to deploy, as Siegel points out (see ibid., sect. 6.1). Consequently, it can be argued that if non-conceptual information is acquired via perception, then (at least) some perceptual contents are not bound to concepts in the first place. The example of the first observations of the granulation of the solar surface illustrates this point nicely. Scientists at that time had no concept of the phenomenon at their disposal. That is, they could neither verbally describe what they had observed nor explain what was occurring, namely what caused the phenomenon they were observing, what mechanisms were responsible for causing it, nor why it looked the way it did. Nonetheless, they were able to see those particular structural aspects of the solar surface.

A first approximation to the concept of 'non-conceptual content', then, consists in acknowledging the fact that we are dealing with a contrastive concept here, as José Bermúdez and Arnon Cahen point out (see Bermúdez and Cahen 2015, sect. 2). That is, our understanding of the term 'non-conceptual content' depends crucially on how we conceive of the term 'conceptual content'. Consequently, there is a variety of hypotheses available about what exactly the non-conceptual content of perception might be like.

The contrastive nature of the term 'non-conceptual content' is also highlighted by Christopher Peacocke (see Peacocke 2001a, 243). He briefly addresses the relation assumed between concepts and language in his discussion of this topic.

> If someone holds that a concept user must have a language in which he can express at least some of his concepts, that is a substantive, nondefinitional thesis that needs to be established. [...] It should, however, be uncontroversial that any content that can be expressed in language by the use of an indicative sentence, including sentences containing indexicals and demonstratives, will be a conceptual content.
>
> (Peacocke 2001a, 243)

Thus, having concepts at one's disposal does not necessarily urge us to admit that we possess a language to express those concepts. Consequently, the inability to speak, either temporarily or permanently, is no necessary indication that the corresponding person does not have concepts at her disposal. It is only the weaker thesis, namely that what has already been formulated in language is automatically related to concepts, that Peacocke holds to be uncontroversial when considering the question in what sense concepts and language are related to each other. Thus, Peacocke's ideas caution against claiming too close a relation between language and concepts.

Can the nature of non-conceptual content of perception be described in a more substantial way? Here, Peacocke's approach is particularly interesting, as his considerations bridge the gap between the discussions in the philosophy of perception and in picture theory. With respect to the latter, he thinks that philosophers should take into account results from cognitive science, in particular from theories of perception (see Peacocke 1987, 383, 404). In the following summary, together with a discussion of some related examples, a recapitulation of what he suggests in the context of his theory of depiction will help to clarify why the transmission of non-conceptual content is important in epistemological terms and what kinds of cognitive processes are at work.

As was explained in section 2.2.1 of this book, Peacocke uses the concept of the "visual field", borrowed from theories of perceptual experience, to elucidate where exactly to look for resemblance relations between an object and its depiction. According to this assumption, it is not the direct properties of an object and its depiction that are compared, but properties belonging to the perception of image and object. That is, what is compared are experiences of the object and of the image in the visual field of the perceiver. The visual field can be regarded as a kind of mental intermediate plain between perceiver and the object of perception. It is then in experiences of the image and the object in this visual field that resemblance relations are detected (see ibid., 386). Now, if the perceiver possesses the relevant concept to identify and label the object correctly, i.e. if she makes use of this concept as a comparative instance to her experience of the image at hand, Peacocke suggests that what is presented to the perceiver in the visual field is then "F-related" ('F' refers to 'field') to that very concept (see ibid., 387). The Altamira cave painting is a good example (see Figure 1.1). According to Peacocke, the viewer will experience a particular similarity in the shape of the painted animal and of a living example of the same species in her visual field. Moreover, if the viewer possesses the concept of an aurochs, she can cognitively proceed by calling the cave painting *an image of an aurochs.*

But what happens if the corresponding concept is not available to the viewer? Here, Peacocke brings in the idea of a non-conceptual content in his theory of depiction. He argues that there are images – most of all, abstract paintings – that are experienced as having representational qualities,[33] that is, they will be experienced as showing particular shapes, although the viewer does not possess the concepts that would allow her to describe her visual field experience linguistically. Moreover, Peacocke thinks that the content of the viewer's experiences can nonetheless be "assessed as veridical or non-veridical" (ibid., 395).

To explain why it is that the viewer's experiences of the respective image represent the latter as showing certain shapes, although the relevant concepts are not available, Peacocke brings in what he calls the "analogue content" of pictures (see ibid.), which he considers to be non-conceptual in kind. Moreover, he is convinced that assuming the possibility of such a kind of content being transmitted in at least some processes of picture perception is crucial as:

> [...] it explains how there is room in the space of logical possibilities for something that actually occurs: the acquisition of a recognitional capacity for a kind of object (or an individual) by seeing a depiction of a member of that kind (or of that individual).
>
> (Peacocke 1987, 395)

An illustrative example of this is provided by images in field guides for biologists. Consider the following case: I observe a particular bird of prey circling Lake Geneva. As I do not know what this species is called, I want to identify it with the aid of a bird guide. Before consulting this book, all I know about the animal is that it is a bird of prey hunting for fish and has a certain appearance during its flight, that is, I know how its tail is formed, perhaps its colour, and I have a rough estimation of its size. So, my observation and my interest in the species have already triggered my attempt to conceptualise my observation. Nonetheless I still will not be able to classify the bird with the aid of a verbal description, as there are too many uncertainties correlated with my brief observation. The bird guide, however, provides a series of drawings of flying birds of prey that I can use as a means of comparison. That is, I can now compare these pictures with my mental image, derived from my memory, of what the bird looked like during its flight. Obviously, what finally makes a classification possible is this comparative task when considering similarities and dissimilarities between the appearance of the bird that I remember and the respective image printed in the book. Hence, although resemblance relations might not be the *non plus ultra* explanation of the phenomenon of depiction in general, they nevertheless prove to be an important aspect in our handling of images. In particular, they help us to conceptualise a phenomenon by using images as teaching tools. What is made plain by this example is that, like perception in general, picture perception can transmit non-conceptual contents, for example about similarities and dissimilarities between an object observed and a drawing in a field guide. They can contribute to our acquisition of concepts and be turned into linguistically expressible beliefs later on. Therefore, Peacocke's concept of "analogue content" connects picture perception with our broader visual capabilities.

Peacocke suggests a similar line of reasoning. He thinks that, in general, images can help us to recognise later on what they depict. That is, by showing you an image of a flying swift,[34] you will be able to discern one in the wild. You will know, for example, how to distinguish a flying swift from a flying swallow, although you have not had the relevant concepts to explain and describe this difference beforehand. In this sense, images can contribute to the conceptualisation of phenomena. It is the non-conceptual content of perception that explains how we are able to acquire observational concepts in the first place (see Peacocke 2001a, 252ff.). That we finally possess the relevant concept is indicated by our ability to correctly recognise the respective entity afterwards.

This, then, is how analogue content works and why it is important. But what exactly does it consist of? Siegel claims that non-conceptual content in Peacocke's sense can be subdivided into "scenario content" and "proto-propositional content" (Siegel 2015, sect. 6.2). For both of them, it is essential to note that Peacocke distinguishes clearly between the properties of the content of an experience and the properties of an object, or event, that is perceived. These properties can show relevant similarities, but need not do so.

An example of this phenomenon concerns distorted experiences. That is, perceiving an entity x from an oblique angle – or from a non-optimal point of view – will present x with a distorted shape in our experience. Its shape might, for instance, appear squeezed or stretched – although x's actual properties of shape have not been altered in any way.[35]

The content of experience can thus diverge from what has been perceived by the viewer. There is an additional dimension, so to speak, entailed in the content of our experience, namely *the way we perceive* those entities.

> So, in characterizing the fine-grained content of experience, we need the notion of experience representing things or events or places or times, given in a certain way, as having certain properties or as standing in certain relations, also as given in a certain way. Henceforth, I use the phrase *the content of experience* to cover not only which objects, properties, and relations are perceived, but also the ways in which they are perceived.
>
> (Peacocke 2001a, 241, his italics)

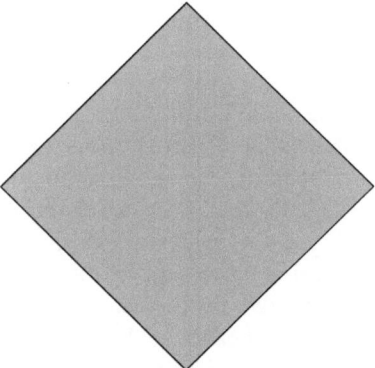

Figure 4.4   Non-conceptual content of perception: the same shape can be experienced
either as a regular diamond or a square.
Source: own image.

To illustrate what kind of information is added when our perceptual content contains the ways we perceive things, Peacocke discusses an example
presented by Ernst Mach. According to Mach, Figure 4.4 can be perceived
either as a regular diamond or as a square, without the actual properties of
the figure being altered. Consequently, the way that we perceive the figure is
either as a regular diamond or a square. This information is added to our
content of experience of Figure 4.4, although the properties are not changed.

Now, although the content of our experience and what has been perceived
by us can diverge in the way just explained, the latter is nonetheless the cause
of the former (see Peacocke 2001a, 248). This is also why Peacocke thinks
that the content of our experience can indeed be assessed with respect to its
correctness (see ibid., 240f.) which implies that we cannot just make up our
minds. What can be perceived is restricted by this causal connection between
the entity and our content of its experience.

Within this framework, Peacocke highlights two different kinds of non-
conceptual content of experience. The first one is what Siegel calls "scenario
content". She explains that "[s]cenario content is a set of ways filling out the
space around the perceiver, relative to an origin and axes marking directions,
that is consistent with the perceiver's experience being veridical" (Siegel 2015,
sect. 6.2). Or in Peacocke's words: "[s]cenario content is essentially the notion
of a spatial type, a type an experience can represent as instantiated in the
volume of space around the subject of the experience" (Peacocke 1994, 420).
Thus, the scenario content is related to the location of the perceiver in her
environment during a concrete perceptual episode. Usually the scenario content of an observer is related to her conceptual states. Yet this does not imply
that the scenario content itself is (completely) conceptualised (see ibid., 422).
Think, for example, of the concepts of being left or right. Apparently we are

able to experience a cat crossing our path from the left to the right although we might not possess the concepts of 'left' and 'right'. We would not say that, before acquiring them in our childhood, we were not able to entertain the relevant experience.

The second kind of non-conceptual content is called *proto-propositional content*. This type of experiential content is needed as there are instances when the scenario content of a perceptual experience remains the same, although these experiences differ phenomenally (see Siegel 2015, sect. 6.2). An example that Peacocke discusses with this respect is Figure 4.4. Although both the shape and the scenario content remain the same, this figure can be regarded either as a square or as a regular diamond. What changes is "the way in which some property of relation is given in the experience" (Peacocke 2001a, 240). Yet, despite this last explanation, it remains difficult to understand what kind of information is added by this way of givenness in the perceiver's experience. Apparently it is not about the perceptual setting – i.e. directions, relations, lighting conditions, etc.– as this is what the scenario content is supposed to transmit.

Nonetheless, the ways things are given in our experience are essential, as Peacocke makes plain, because they provide the foundation to conceptualise what we have perceived. He points out that concepts presuppose some kind of raw material or information, so to speak, from which they can be formed. Thus, the non-conceptual content of our experience precedes the process of conceptualisation (see ibid., 245f.). Moreover, those ways of givenness in experience correspond to certain kinds. "On reflection, it should not be a surprise that for each way, there is a specific kind such that the way is intrinsically a way for something of that kind to be perceived" (ibid., 246). Concerning Figure 4.4, this means for example that the way this figure is given in our experience can correspond either to the perception of a square or a regular diamond, but not to the perception of a circle. Experiencing the figure as having four edges is the condition that allows us to subsume it in a certain category and this way of experiencing Figure 4.4 excludes it from belonging to the set of phenomena that can be called a circle. It is the way Figure 4.4 is given in our experience that allows us to judge that this is a square if we do possess the concept of squareness, i.e. if we know that a condition of being a square is 'having four edges' (see ibid., 253). This is why the way of givenness in our experience is proto-propositional. It is the foundation for respective judgements which allow us to formulate corresponding that-clauses, but it is not conceptualised as such.

Peacocke explains what is going on in such instances when discussing how the transition from non-conceptual content to observational concepts is achieved. He suggests that perceivers:

> [...] will be displaying a sensitivity to a particular boundary. This is the boundary between those states which are, and those states which are not, mentioned in the possession condition for a given concept such as *square*.

[...] One, perhaps the basic, way to make this transition is to ask questions that are in the first instance not about concepts, but about the world. [...] One can answer such questions by drawing on one's ordinary, ground-level abilities to react rationally to one's perceptual states in coming to make judgements about the world.

(Peacocke 2001a, 258)

Thus, the thesis is that we do need the non-conceptual content of experience to acquire the concepts that allow us to make judgements about the world, to describe and communicate about it.

Here, we can add an important point related to this nexus between the non-conceptual content of experience and the acquisition of concepts. Peacocke points out very clearly that it is because of this kind of *content* that this latter cognitive achievement can be accomplished. He thereby rules out an important rival to his theoretical approach, namely discussing *states of perception* instead of *contents of perception* in this debate. Tim Crane suggests that Peacocke's theory of scenario contents can be used also as a rationale to defend such a state-orientated view of perceptual experience.

Peacocke's theory might seem to be a version of the content view, but properly understood, it is a version of the state view. The reason why a state with scenario content is non-conceptual is because S is not required to possess any of the concepts that canonically characterize the scenario in order for S's state to be canonically characterized in terms of it.

(Crane 2009, 468)

Peacocke, however, highlights the possibility of making transitions from non-conceptual to conceptual contents of experience. Apparently this is what we experience in the processes of learning, and it seems to be much more plausible to assume that our concepts are formed – or carved out – of a *raw content*, namely a non-conceptual content, than that they appear because of their relation to a non-conceptual state.

In the following, we will discuss some clarifying remarks that will bolster Peacocke's claim concerning how to conceive of the cognitive process of concept acquisition. Two supplementing ideas referring to human developmental history are worth considering here. Those ideas will shed some light on what Peacocke might have had in mind when writing about proto-propositional content.

Firstly, it can be assumed that our perceptual apparatus has been crucially shaped by evolutionary processes. That is, only those of our species survived who were able to react reasonably adequately to what they had perceived. Forming proto-propositional contents about one's environment, about food, predators, members of and threats to one's community apparently seem to have been evolutionarily advantageous. Such a line of reasoning appears to lie at the heart of Peacocke's thesis that human beings are sensitive to certain boundaries concerning the conditions of the possession of concepts.

Secondly, as mankind has always been a social species, it is strongly advisable to take into account that the process of learning has not to be mastered by the individual alone. On the contrary, we can, and usually do, rely on members of the family, relatives and other human beings in general. That is, we rely on the division of epistemic labour. Thus, it is rational to assume that not only perception and testimony often work hand in hand as epistemic sources when it comes to language acquisition (see Bermúdez and Cahen 2015, sect. 4.1), but that they also co-operate when we learn about concepts in the broader sense. Parents show their children the entity they name for them (see Harris 2012, 8). By repeating this show-and-tell event and by the children's listening to subsequent parental explanations about what is characteristic of those entities, children will come to be aware of the boundaries of certain concepts and, later, will know how to deploy them in language. Requiring that concepts always come first would apparently undermine the whole cognitive process just described.

However, it would be equally wrong to deny that the ability to deploy concepts can contribute much to what we are able to *observe*, as Chalmers has highlighted (see Chalmers 1999, 10ff.). His concern is that observational facts have to be formulated as statements to be of use in scientific reasoning – a claim that reminds us strongly of the controversy about the role of images in scientific discourse. As a result of this assumption, Chalmers thinks that we are in need of an "appropriate conceptual framework and a knowledge of how to appropriately apply it" (ibid., 11) to formulate those sentences. Moreover, he shows that the better – i.e. the more appropriate to the task at hand – our observational concepts are, the more detailed our scientific observations can become.

This interplay between concepts and observation also appears in Hentschel's example of the first observations of the solar surface. Not only does he demonstrate the difficulties that those first observers had to face in trying to understand the visual information their telescopes conveyed to them, he also explains how their strategy of analogical reasoning, i.e. the application of metaphors to describe their observational results, influenced later observations of the respective phenomenon. Hentschel describes this process as a search for and application of different *Gestalt concepts* (see Hentschel 2000, 23). Moreover, he notes that published illustrations influenced what other scientists expected to see and actually observed (see ibid., 25). We have already become acquainted with what Hentschel describes here through Fleck's ideas about scientific communication processes and the place of visual representations therein (see section 3.2.3 of this book).

Fleck makes plain that often there is a feedback loop between popular science and expert knowledge (see Fleck 1979, 115ff.). In particular, he shows how images, produced as means to communicate results in a simplified manner to laypeople, influence the experts' cognitive processes. "But what was initially meant as a means to an end acquires the role of a cognitive end. The image prevails over the specific proofs and often returns to the expert in this new role" (ibid., 117). Thus, to use Hentschel's terminology, once a concrete pictorial *gestalt* – putatively formed by deploying certain concepts – has been chosen by the expert to present her results, this *gestalt* will determine what she will henceforth observe and investigate.

Moreover, the above discussion of the non-conceptual content of perception allows us to explain why this initial choice of a pictorial or descriptive *gestalt* does not lead into an epistemic dead end and, thus, to a cognitive standstill if those choices are not adequate. Fleck describes what might have been expected, namely a particular tenacity of thought styles and what he calls a "harmony of illusion", i.e. the tendency to explain phenomena in accordance with the prevalent thought style or even to ignore divergent phenomena which cannot be accommodated in the theory at hand (see ibid., 27ff.). Consequently, the question arises how a change in thought styles is possible if this is necessary. How can we perceive something in accordance to a different *gestalt*-concept if the one initially chosen does not fit?

Although Fleck discusses in detail the effects of the harmony of illusion, he does not deny that persistent anomalies can surface in those systems and urge scientists to deal with them (see ibid., 27ff.). On the contrary, he explicitly describes two different stages of scientific observation as a result of such difficulties. Firstly, he mentions a "vague, initial visual perception" which he calls "unstyled" (ibid., 92) and, secondly, the "direct perception of form [Gestaltsehen]" which "requires being experienced in the relevant field of thought" (ibid.). Thus, concrete *Gestaltsehen*, as explained above, is preceded by a weighing of the pros and cons of the various alternatives in order to conceptualise the phenomenon at hand. This balancing of alternatives is made possible by the fact that the individual is simultaneously a member of different thought collectives providing her with divergent suggestions of how best to explain the phenomenon, i.e. what concepts to choose that can provide an explanation expressed in words or pictures known to the scientist.

Here, we can grasp how scientists deal with what is non-conceptually provided by perception to human cognition in their observational processes. It is this epistemic strategy, i.e. the application of concepts borrowed from different thought styles as explanatory means to persistent anomalies, that turns metaphors into the seeds of new theories, and thus offers an explanation of how changes and developments of thought styles can take place (see ibid., 117; Fleck 1986c, 103).

Furthermore, although Fleck seems to think that all that is needed to account for changes in thought styles are communication processes in science and elsewhere (see Fleck 1979, 109f.) and, despite his obvious reluctance to admit the possibility that something can be perceived prior to one's being introduced to the relevant thought style that allows a correct interpretation of the percept,[36] his mentioning of those anomalies[37] suggests that perception provides us with information that cannot arbitrarily be subsumed under concepts suggested by the thought style at hand. The relevant resource to set the stage for anomalies seems to lie in the non-conceptual content of perception that becomes a part of the scientists' observational results.

Taking the above considerations into account, it seems reasonable to assume that both non-conceptual and conceptual contents can be obtained via perception, whereas the former seem to come first. They will partly be substituted by the conceptual content acquired later. Nonetheless, some information will remain non-conceptual in kind even in our later developmental stages. These contents can constitute anomalies within explanatory theories and, as such, motivate further testing which might lead to changes in our theoretical approaches. They can therefore bring about progress in science and are thus of particular importance in this context.

Moreover, as Grundmann points out, adopting an externalist position of justification, i.e. regarding perception as a reliable process of information gathering, also allows us to make knowledge claims based on those non-conceptual perceptual contents (see Grundmann 2008, 497). Consequently, the thesis can be defended that perception not only fulfils its function as an epistemic source if concepts are available for what has been perceived so that observation statements can be formulated as reasons in support of beliefs, but also if those concepts are missing. In this latter instance, it seems reasonable to assume that there is a presumptive right similar to the one highlighted by anti-reductionists in the debate about knowledge by testimony that can be used as a justificatory basis of our perceptual experiences. As Scholz points out, such a presumptive right allows the perceiver to trust her perceptual abilities as knowledge suppliers as long as there are no reasons undermining and defeating this trust, such as indicators of odd lighting conditions or of the previous consumption of hallucinatory drugs, etc. (see Scholz 2009c, 199f.).

But it is not only in this sense that the processes of justification of testimonial and of perceptual beliefs show similarities. As I have argued elsewhere (see Mößner 2010b, ch. 5.5.3), participants in the activity of testimony become increasingly experienced with individuals as testifiers or testimonial contexts during their lifetime. Moreover, it is an important part of our educational system to acquaint students with the relevant practices in their fields of expertise.[38] A similar developmental process can be assumed concerning perceptual experiences. During their professional training, people will usually acquire many of the relevant concepts to verify their perceptual experiences in scientific observations. This becoming acquainted with the relevant concepts as part of the practical training in science is elaborately described by Fleck (see Fleck 1986b). In the next section, we will come back to the process of learning that Fleck explains. For now, it suffices to point out the analogy of development in becoming (a) a trained recipient of testimony in a certain domain and (b) a trained observer in a particular context. It is this developmental progress of the epistemic subject that also allows us to bring in internalist conceptions of justification when concepts are acquired. The observer can then be asked to name the reasons that she thinks support her perceptual beliefs precisely – and they can also be checked for adequacy by others if necessary.

In a nutshell, then, if perceptual experience provides non-conceptual *and* conceptual content in the way argued above, and if it is agreed that the primary way to cognitively access the information transmitted via visual representations is by perception, it does not seem far-fetched to claim that the same mechanisms are also at work in picture perception. Apparently it is these mechanisms that are relevant to consider when enquiring how knowledge might be gained via images in science and how visual representations can make epistemic contributions to scientific arguments. Furthermore, as our line of reasoning moves from the more general case, that is, perception in the broader sense, to the special case, that is, image perception, these considerations also rule out Crane's suggestion, namely that perception is not a propositional attitude, because apparently, picture perception is not propositionally structured either (see Crane 2009). He thus argues conversely, moving from the special case to the general concept of perception. Yet this seems to be an inappropriate way to defend his thesis. In particular, by regarding the issue in converse manner, Crane cannot and does not accommodate the artificial sign character of images in his theory. Yet our ability to decode visually presented information has undergone certain developmental phases because visual representations have become increasingly artificial due to newly developed styles of representation, which has already been explained by making use of the theory of image games.

How to consolidate the theses concerning image perception and those concerning the apparent artificial character of visual representations will be discussed in detail in the next section. The exact implications entailed in pointing to the evolutionarily shaped resource of perceptual experience when analysing potential epistemic contributions of visual representations in science will also be more closely examined.

4.2.3 Evolutionary merits of perception

The question under discussion in this section is whether evolutionary processes that have shaped the development of human vision can be employed in the argument to explain why it might be advantageous to make use of visual representations in scientific discourse. The discussion of this topic will bring into focus our abilities of pattern detection and our discriminatory capacities with regard to colours. However, to mention one constraint to the attempted reasoning by analogy right at the beginning, it has to be taken into account that scientific images have been characterised as signs, i.e. as artefacts. Thinking in this way about the nature of visual representations in science seems to suggest that image perception and everyday perception do not have parity. This apparent difficulty will carefully be considered in the final part of this section.

The capacity of human perception to transmit non-conceptual content is explicitly related to evolutionary processes by Peacocke. He states that "[n]onconceptual representational content is part of our animal heritage" (Peacocke 2001b, 615). Such a line of reasoning arises somewhat naturally from his considerations concerning the perceptual abilities of non-human animals. Here, Peacocke claims that "[I]f the lower animals do not have states with conceptual content, but some of their perceptual states have contents in common with human perceptions, it follows that some perceptual representational content is nonconceptual" (ibid., 614). Admittedly, even lower animals can have perceptual experiences, although they might be different from ours. Moreover, we would usually also assume that, in particular, lower animals do not possess concepts in the way we do or even not at all. Nonetheless, observing the behaviour of those animals suggests that they indeed receive relevant information via their perceptual apparatus to react in accordance to certain stimuli. Thus, it seems reasonable to assume that those animals receive a kind of non-conceptual content via perception (see also Bermúdez and Cahen 2015, sect. 4.3).

Yet a qualifying remark has to be added. Our main concern is with visual perception, i.e. vision, as the predominant human sense modality. Of course, the predominance of vision cannot be so stated regarding other species in general. Some perceptual experiences of animals might be completely different in comparison with human sense perceptions. This is due to some animals being equipped with sense modalities that human beings do not possess, such as a cat's whiskers or the capability of sharks, hammerheads and sawfish to detect the electrical field emitted by their prey.

Here, some interesting questions arise concerning the connection between sense modalities and concepts on the one hand, and concepts and communication on the other. How are our concepts influenced by our predominant sense modalities? Do our concepts evince more details regarding visually perceivable properties (or do we possess more of those concepts?) than regarding properties related to audible, tactile and other forms of perception? Is it because it was more relevant to our ancestors to communicate, for example, visual clues about prey and predators in their environment that our concepts are richer in information regarding visual details than other kinds of information? The archaeologist Steven J. Mithen explains, for example, how a sensitivity to visual information might have been an evolutionary advantage for our ancestors in the context of specialising in certain food supplies.

> Knowledge of carnivore behaviour and distribution would therefore appear to have been critical to early Homo: competing carnivores may have provided both a threat and an indication of a possible scavenging opportunity. In this light it would seem improbable that H. habilis [This ancestor of the human race existed around two million years ago. N.M.] could have exploited the carcass niche *if it had not mastered the art of using inanimate visual clues, such as animal footprints and tracks.*
>
> (Mithen 1998, 115, my italics)

Putting these empirical questions aside, Peacocke definitely has a point when relating the thesis that perception can yield non-conceptual content to the evolutionary development of the human species as visually perceptual beings. In the following, the focus will be on our visual abilities and the discussion of some consequences for images in science as bearers of information resulting from our evolutionarily grounded predisposition to visually access the world.

The opportunity to gain non-conceptual content via perception, and thus to draw on a very rich source of information, may allow us to discover anomalies or new phenomena and even to communicate our findings by integrating images of them in scientific publications. Beyond that, accessing our world visually also enables us to make use of certain strategies of information gathering and processing that are particularly shaped by evolution. Thus, empirical results from the different fields of research about perception should be taken into account. Regarding the analysis of visual representations as bearers and providers of information in scientific discourse, I would like to suggest two further capacities of the human visual sense modality which might alter the balance of the argument when considering the relevance of visual representations in comparison with other representational modes.

Firstly, human observers are obviously good at *pattern detection*. As becomes clear in Mithen's quotation above, the correct reading of – i.e. the correct recognition of behavioural patterns in – animal footprints was of such evolutionary advantage to our ancestors that we can assume it was one of the causal factors in the development of a certain visual expertise in this regard. That, on average, people are indeed particularly skilled in pattern detection makes a more or less current trend in science quite reasonable, namely the initiatives of the so-called *citizen sciences*.[39] It seems worthwhile to select and examine just one of these, namely the project called "Galaxy Zoo" (see www.galaxyzoo.org/, accessed January 12, 2016). Recent sky surveys have yielded huge amounts of data that have to be somehow analysed. "Galaxy Zoo" is part of this analytical project. Here, volunteers – particularly laypeople without special training – participate in classifying galaxies, photographed by space telescopes such as Hubble, according to their shapes. The data are made available online, as well as descriptions of the task, including suggested patterns of what the objects might look like.[40] The user is guided through the classifying task via online dialogues. She can accomplish the required classification without great effort. All that is needed is an *attentive eye* to compare the photograph with the patterns suggested. The project organisers obviously rely on their volunteers' ability to detect patterns. This task cannot be performed at the level of numerical data, i.e. as computational operations by IT devices. There is a certain vagueness involved in classifying these objects that cannot be removed by adhering to numerical data. It can only be reduced by consistent results obtained by various human classifiers. That is, the same object is categorised by several people to minimise the possibility of distorting biases and errors resulting in misclassifications.

It seems to be a consequence of evolutionary processes that human observers are particularly skilled in tasks like this. Thus, making information available in a way that activates these skills, i.e. as visual data that can be investigated by vision alone, can enhance our understanding by connecting these abilities to the cognitive processing of the respective information. In particular, the huge amount of data yielded by today's IT-based research processes, such as measurements recorded as visual data at the LHC in Geneva, the digital photographs taken by the Hubble space telescope, or the data collected in recent sky surveys undertaken in regard to the mapping of the sky by the SLOAN Digital Sky Survey (see www.sdss.org/, accessed January 12, 2016), challenge scientists to think of new ways to evaluate these large data sets. Maynard thinks that the best new ways to achieve this are sometimes rather ordinary old-fashioned ones. "Confronted with the increasingly large data sets from its latest cognitive technologies, our species has fallen back upon its greatest natural information-processing resources, inherited over millions of years: notably those for vision [...]" (Maynard, unpublished paper).[41]

Yet, as the practice of multiple visual classification of the same object by different observers shows, scientists are well aware that blind reliance on the capabilities of human observers would be problematic, as the virtues of our visual skills might sometimes turn out to be vices where science is concerned. Hentschel explains this when discussing the capacities of human observers (see Hentschel 2014, ch. 9). His main example are the pitfalls that led scientists in astronomy seriously astray in the nineteenth century when investigating the surface of Mars with their telescopes. Here, he refers to what has become known as the 'Martian canals'. In 1879 the Italian astronomer Giovanni Virginio Schiaparelli published a map of the surface of Mars, showing "thin lines, apparently interconnecting the various oceans (*mares*), lakes (*lagos*) and huge rivers (*fluvius*) [...] In later maps, he incorporated a profusion of these thin lines, interconnected in a strange maze-like network mostly originating from one of the *mares*" (ibid., 299). Furthermore, by calling these lines '*canali*', which can be translated either (in the more neutral way) as 'channels' or, in the way most of his contemporaries did as 'canals' (which brings the artificial character of those structures to the fore), Schiaparelli created an illusion which trapped human observers till the end of the twentieth century, namely that these observational results show intentionally brought-about artificial structures on Mars indicating the presence of some sort of intelligent inhabitants on the planet.

However, as Hentschel makes clear, those apparent canals were just the result of the poor resolution of the telescopes used at that time and a certain tendency in human vision to connect isolated dots into lines (see ibid., 307). Moreover, people also seem to have expected to see those patterns after Schiaparelli had published his results. Thus, our ability to detect patterns relatively easily can also result in unintended interpretive impositions of expectations of what is actually visible. Accordingly, Hentschel concludes that:

[t]he moral of this little tale is that pattern recognition is not an easy business. It can easily go astray, especially if the all-too-human observers are swayed too much by suggestive analogies to well-known terrestrial phenomena. The foregoing is also a lesson about how finicky human vision is. It is optimized to resolve potential threats quickly and thus evolutionarily programmed to over-interpret rather than under-interpret sensory perception. That is why humans are still quicker than modern computer programs in detecting visual patterns. But this speed comes at a price. Our vision occasionally plays tricks on us.

(Hentschel 2014, 309)

Thus, mere reliance on our perceptual abilities alone seems to be not epistemically advisable.

Bearing this in mind, let us turn to the second merit of visual perception that evolution has been working on. Belonging to a diurnal species, the human eye is particularly sensitive to the spectrum of visible light, that is, to a broad range of colours. It has already been noted above that human observers seem to be particularly skilled in detecting even subtle differences in colour hue – even though we often do not have words to term our observations correctly.

With respect to this phenomenon, Karl Schawelka offers an explanation referring to evolutionary processes in human perceptual capabilities (see Schawelka 2007, 64ff.). He draws our attention to the fact that our visual apparatus is adapted to natural sunlight. Although it might be objected that the spectrum of sunlight is very broad and vast parts of it are simply invisible to the human eye, Schawelka makes plain that our visual apparatus is nonetheless perfectly adapted to its tasks. Because of the filtering effect of the atmosphere, not all kinds of solar radiation reach the surface of Earth to the same degree. There is a notable maximum within the range of visible light that is exactly the part of the spectrum that the human eye is particularly skilled in detecting and discerning, i.e. wavelengths between approximately 400 and 700 nm (see ibid., 65).

Moreover, Schawelka highlights an interesting biological finding, namely that the different visual abilities of our perceptual apparatus cannot be simultaneously developed equally well. Apparently spatial perception, colour perception and visual acuity cannot be optimal at the same time. Being particularly good at one of them also implies a loss in those other visual abilities (see ibid., 68). This takes us to the question of what the evolutionary benefits were that stimulated the emphasis on colour vision in the evolutionary development of our ancestors.

A full-blown colour perception appeared as a characteristic of our early ancestors around 35 million years ago (see ibid., 74). Schawelka points out that being good at discerning different colour hues allowed our ancestors to spot, for example, ripe fruits quite easily (see ibid., 72ff.). As omnivores, our ancestors certainly enjoyed the evolutionary benefit derived from this capability of their visual apparatus. It has to be added, however, that further experiences also seem to be of relevance, as colour alone does not indicate digestible food *per se*. Ripe apples and fly agarics, for example, share the same colour, namely red – a particularly salient colour to the human eye (see ibid., 72) – but this does not imply that we are advised to taste the latter. On the contrary, colours can also indicate poison – not only regarding mushrooms and plants, but also insects, fish, snakes and amphibians, such as South American tree frogs. Apparently we also profit from our visual abilities to discern colours easily in this case if the relevant background knowledge is available to us. The colourful skin of those animals will make us deal with them carefully or simply avoid them as soon as we spot them.[42]

John Campbell connects these more general ideas about the evolutionary merits of colour perception to human capabilities of reasoning when pointing out that "[i]t is often observed that colours of objects have predictive value. The particular colours of various foods are predictive of their nutritional value. The exact colours of particular people and plants are good predictors of their health" (Campbell 2009, 35). Campbell here not only highlights the role of colours as indicators of certain properties, such as ripeness or toxicity, but also draws our attention to the way in which our capacity to visually discern colours can be useful in science. In medicine, the colour of human skin can, for instance, be an indicator of certain diseases such as yellow fever or tuberculosis.

Sabine Müller and Dominik Groß discuss the usage of false colour images in medical science and in astronomy (see Müller and Groß 2006). They point out that this kind of visualisation is used especially for two purposes in those domains: firstly, to make visible what is otherwise not accessible to the human eye, but can be detected by other means such as ultraviolet radiation (see Figure 2.2 as an illustration of this point) and, secondly, for supporting the interpretation of images by enhancing their contrast (see ibid., 94). This latter purpose draws explicitly on the ability of human observers to discern between different colours – for example, the photograph of Olympus Mons (see Figure 2.12), in which false colours allow us to identify the different altitudes of the volcano in its different regions without great effort. In comparison with this, extracting the same information from a plain photograph or a greyscale image would be much more difficult. Consequently, as colours are used to transmit information and apparently support recipients in grasping it relatively easily, Müller and Groß defend the thesis that colours can be regarded as *epistemic tools* (see ibid., 114; Groß and Müller 2006, 79).

Despite these advantages of colour vision, it has to be added that, without going into all the details of the debate, there is a particular difficulty related to suggesting colours as scientifically valid indicators of certain states of affairs. Although human observers are usually very good at distinguishing between different colours, the latter cannot be regarded as *objective properties* of the entity to which they seem to be attached. This problem is also addressed in Campbell's text:

[m]any philosophers – the classical sources are Galileo and Locke – have said that science shows that there is a mistake embodied in our ordinary understanding of colour concepts. We commonsensically take colours to be categorical properties of objects, whose nature is apparent to us in vision, but in fact there are only complex microphysical structures and the consequent tendencies of objects to produce ideas in us.

(Campbell 2009, 40)

What is highlighted by those scholars mentioned by Campbell is that colour is not an intrinsic property of objects, but a phenomenon that is solely present in our perception of those objects. Yet in science we want to find out what the entity under investigation is like, and are not so much interested in the dispositions of human vision to interpret its percepts in certain ways.

What makes matters of colour perception so complex requires two brief explanatory remarks. Firstly, scientists can detect wavelengths of radiation and so measure which parts of the spectrum have been absorbed and which have been reflected by an illuminated entity.[43] Wavelengths are commonly associated with certain colours. Spectrographic analyses in astrophysics, for example, make use of this nexus by assessing the absorption lines of certain kinds of gas in a given spectrum emitted by a particular celestial object to determine the ingredients of a particular nebula or star, yet physicists are reluctant to regard colours inferred from wavelengths of (reflected) radiation as characteristics inherent in those objects. Glenn Elert, for example, states bluntly that:

[c]olor is a function of the human visual system, and is not an intrinsic property. Objects don't 'have' color, they give off light that 'appears' to be a color. Spectral power distributions exist in the physical world, but color exists only in the mind of the beholder.

(Elert 2014, ch. 3.3.7)

The second aspect that makes matters even more complex is that our perception of colours is highly influenced by a variety of factors outside the assumed bivalent relation between our visual system and the object perceived. What kind of light illuminates the object in question – a warm candlelight or a cold light emitted by a fluorescent tube? What are the surrounding colours like that the object is embedded in? Dark context-colours let us perceive a brighter hue of the object's colour than light context-colours do. Schawelka explains these puzzling phenomena of colour perception in detail. He highlights the fact that context-colours can either contribute to our experience of colour consistency or, to the contrary, trigger the phenomenon known as 'simultaneous contrast', that is, two physically identical stimuli that can result in two totally different colour experiences depending on the context-colours of these stimuli (see Schawelka 2007, 49). It therefore seems rather difficult to refer to our ordinary perceptual abilities to detect certain colours as a reliable tool of scientific analysis. In order to achieve reliable results, what seems to be required is the use of instruments unaffected by such distorting visual phenomena; instruments that can deal with these problems by abstracting the objective value from those additional factors.[44]

To sum up my results so far, the above considerations suggest that evolutionary processes have provided human beings with particular visual skills. On average, we are able to gather a wide range of information via visual perception, even if this is non-conceptual in kind. Moreover, nature has endowed us with certain detective abilities, especially with respect to grasping patterns and colours, that have enabled us to quickly focus our attention on aspects of vital interest in this wealth of information. Those abilities were apparently advantageous to the survival of our ancestors – and can still be useful in the everyday context as well as in the sciences.

The usefulness of those skills in epistemic contexts is defended by educational psychologists. In particular, Eleanor Gibson's work on *perceptual learning* is based on the discriminatory abilities of human sense perception. Benedict Carey explains how learning is considered in this context. He writes:

> [p]erceptual learning is active. Our eyes (or ears, or other senses) are searching for the right clues. Automatically, no external reinforcement or help required. We have to pay attention, of course, but we don't need to turn it on or tune it in. It's self-correcting – it tunes itself. The system works to find the most critical perceptual signatures and filter out the rest.
>
> (Carey 2014, 184)

Carey points out that quite a few computer-based learning programs take advantage of perceptual learning. Such tools are, for example, used in flight training to teach trainees "perceptual intuition" (ibid., 188). The dials of six main instruments in the cockpit of a small aircraft are displayed on a computer screen, and the trainee is asked to choose between seven different options of what these instruments tell her in sum about the position and activity of her plane. She is thus trained to make her choices just by looking at her instruments and not by considering each of them in detail, trained to develop a gut feeling, so to speak, about what those instruments in sum tell her, as flying means multitasking, especially during landing manoeuvres. That is, the pilot not only has to read her instruments, but also talk to the tower, consider visual information, perform the necessary tasks – and all of this more or less simultaneously. Thus, the pilot simply does not have the time to read and interpret each instrument separately, but needs to base her decisions on a quick look at them.

This example of training tools drawing on perceptual learning makes clear that human observers are particularly good not only at discriminating between relevant perceptual clues, but can also become even better at this task through regular visual training. Carey's examples thereby support the suggestion that, because picture perception draws on these evolutionary resources, using visual representations as a means to transmit information in scientific discourse allows us to profit from these virtues of our visual apparatus.

Moreover, this line of reasoning offers a rationale for the epistemic qualities of images of various kinds, as pointed out in Perini's work, for example. She shows that, for instance, micrographs can inform us about spatial features of the specimen, even if we do not have concepts to verbally describe what we have observed by inspecting the image. She claims that these images

> can represent unfamiliar phenomena, without the need to articulate hypotheses about results prior to the experiment. As in this case, figures that support the existence of unexpected structural features can be produced. This system can also represent very complicated structural properties, even when there are no linguistic terms for the same features.
>
> (Perini 2005c, 921)

In a similar fashion, Kulvicki's claim that images can transmit vast amounts of information across various levels of abstraction more or less immediately (see Kulvicki 2010b) can be explained with recourse to the epistemic merits of the perceptual abilities of human observers.

Caution is definitely advisable, however, that optimism about similarities between picture perception and perceptual processes in general is not misleading us when analysing the epistemic capabilities of scientific images. Maynard stresses that there are important differences between both perceptual processes (see Maynard 2011). He highlights the fact that images[45] are *artefacts*. Acknowledging this implies that it has to be taken into account that people are accustomed to handle artefacts in certain ways. This means that they regard visual representations as being produced for certain purposes, which they then try to identify when looking at images. This is why Maynard thinks that picture perception is, at least to a certain degree, different from visual perception in everyday life. Moreover, this crucial distinction is also his rationale for rejecting an unqualified application of theories of visual perception to picture perception. He writes that:

> '[v]isual array' treatments of pictures dominant in philosophy and perceptual psychology ignore these basic features about perception. Overlooking that depictions are familiar artifacts and used as such, they neglect the great differences between looking at things and looking at depictions of them. Notably, they hide the fact that we look at depictions, not real scenes, for their depictive 'for' affordances, and in terms of 'on purpose': why they were put or left there, what they are *doing* there – including whether they should be there.
>
> (Maynard 2011, 21)

Thus, although the deciphering of visual representations in science might benefit greatly from being based on perceptual processes that allow the investigator to rely on cognitive mechanisms especially shaped by evolution, Maynard is obviously right in reminding us of this difference between ordinary and picture perception. Following his advice seems to be most crucial with regard to the initial thesis, namely that visual representations in science can best be understood as *signs*. However, acknowledging this artificial character of images in science does not undermine the possibility of drawing on the cognitive resources provided by visual perception. An explanation is due at this point. In section 4.1.3, and with reference to Scholz's concept of *image games*, we already discussed the possibility of a layering of meanings in visual representations; that is, the addition of new layers of meaning to its initial encoding by making use of images in communicative contexts. It is here that interpreters of scientific images will be most attentive to their object's artificial character and question the intentions of those people who either produce or distribute those images. It was suggested that strategies of justification developed in the debate about knowledge by testimony concerning the epistemological role of visual representations in scientific discourse be employed to deal with this situation. Those models of justification are primarily based on assumptions about the speaker and only secondarily on considerations about the content transmitted, such as its coherence with background beliefs. Thus, Maynard's reminder does not pose a counter-example to the above line of reasoning.

As a concluding remark, I want to highlight another finding of Maynard's analysis concerning the cognitive content of visual representations. He emphasises that adhering to the results of perceptual sciences too strictly might lead to an unwanted consequence concerning the epistemic status of (scientific) images, namely that they are regarded as external to the human mind, i.e. as objects of visual investigation similar to natural entities (see Maynard 2011, 11). Such an externalist conception apparently speaks against the possibility of using images as signs that can be purposefully produced to transmit information. Contrary to this, no problems of this kind have ever been asserted with respect to language. Maynard not only notes this discrepancy between estimations of the epistemic capabilities of images and of verbal language on the part of scholars relying on perceptual sciences to explain the epistemic accomplishments of visual representations, but also traces this stance in the analyses of evolutionary palaeoanthropologists who theorise about the origin of the human mind and the capacities of our ancestors to communicate via symbolic means (see Maynard, unpublished paper).

Maynard points out that although scholars of this profession often use cave paintings as evidence of our ancestors' developing capabilities to communicate by using symbols, they more or less immediately deny visual representations their epistemic roles.

Returning to the standard image-to-language reasoning, it is interesting that once we have used image thought-content to infer the advanced linguistic kind in the Ice Age, we put aside and even deny the former. In Philosophy, as in science, the Paleolithic evidence is swiftly awarded the honorific title 'art', allowing us to treat it as something outside the mind [...]

(Maynard, unpublished paper)

Such a negative stance towards images as content providers is expressed for example by David Lewis-Williams, an expert on cave paintings. He writes that:

[t]he best that can be said for pictures is that they trigger memories of information that has been absorbed in different ways, that is, by experience and verbally. So whilst the Upper Palaeolithic images may have sometimes functioned as mnemonics, their capacity to store or convey information was limited. They were not like the hard drives of modern computers.

(Lewis-Williams 2004, 67)

Without doubt, Lewis-Williams suggests an epistemically inferior status for those visual representations produced by our ancestors in comparison with their first verbal utterances. He particularly denies that those images can possess a kind of cognitive content similar to linguistic expressions that allows the storage and transmission of information.

However, there are other scholars working in this field who see less of a discrepancy between the epistemic capabilities of linguistic and visual representations. Mithen's opinion about this topic offers, for example, an interesting contrast to the sceptical attitude diagnosed by Maynard. About the contributions of those early visual representations to the information process he states that:

[i]n summary, although the specific roles that prehistoric artifacts may have played in the management of information about the natural world remain unclear, *there can be little doubt that many of them served to store, transmit and retrieve information*. Major benefits of this will have been enhanced abilities to track long-term change, to monitor seasonal fluctuations and to devise hunting plans. Many of the paintings, carvings and engravings of Modern Humans were tools with which to think about the natural world.

(Mithen 1998, 197, my italics)

I agree with Maynard that the link between those early visual representations and language as tools of communication is apparently better understood along the lines of Mithen's reasoning than of Lewis-Williams's. What reasons are there to deny that those images were meant to transmit information encoded by those early artists?

Taking this as background information for the analysis, it seems reasonable to assume that visual representations can bear a cognitive content and thus play an epistemic role in communication. Moreover, acknowledging this fact also prepares for taking into account the long history of this practice of visual communication. That is, long before mankind developed symbols to record language, a way to transmit information visually via those images had been established. Moreover, not only does visual communication greatly pre-date the use of writing skills, it also has lasted much longer. Our ancestors made use of this practice for around 30,000 years, as Lewis-Williams claims: "[s]till, radiocarbon dating strongly suggests that the Upper Palaeolithic period, that part of the Palaeolithic in which people began to make art, lasted from about 45,000 to 10,000 years ago" (Lewis-Williams 2004, 39), and we still use images in our communicative habits today. Thus, comparing this rather impressive time-span with the few thousand years of our practice of writing, it should come as no surprise that human beings had already developed considerable skill in encoding and decoding visual information. Thus, it is not only the evolutionary development of human vision that allows us to benefit epistemically in processes of deciphering visually presented information, but also this long-term social practice.

4.2.4 Interim results: what can be learnt from theories of perception?

The above discussion of the cognitive content of visual representations in science was motivated by Perini's initial observation that it seems to be common practice in scientific discourse to use images when attempting to transmit essential information about research results, hypotheses and methodology. Hentschel points out that there are 14.8 images per ten pages of text in contemporary scientific journals in biology and 12 images per ten pages of text in physics journals (see Hentschel 2014, 30). It seems, then, that scientists share a particular preference for making use of visual representations to transmit certain information.

Despite this widespread practice, philosophers question the epistemic capabilities of visual representations in scientific discourse, as Perini has made clear. In particular, they express reservations about the suitability of visual means to be regarded as proper components of scientific arguments – whereby 'proper components' refers to the assumed epistemic capacities of images. The above analysis was thus intended to inquire more deeply into the prerequisites of these capacities and, especially, the question about the precise nature of an assumed cognitive content of scientific images had to be answered.

With regard to the epistemic capabilities of visual representations in the context of scientific arguments, proponents of the traditional Fregean point of view seem to regard 'argumentation' as an exclusively verbal activity. Their main objection, then, consists in the fact that images are apparently not propositionally structured. As their line of reasoning suggests that epistemic functions are tightly bound to propositional contents alone, they think one is seemingly entitled to disregard visual representations as epistemic means in this context.

Contrary to this, a different account has been presented here, as it can be assumed that Perini is correct in pointing out that scientists handle those visual elements differently from the way proponents of the Fregean point of view would suggest when affording them a role in scientific discourse, and that it would be wrong to disqualify such practices from an epistemological point of view.

The first suggestion, already argued for in the last section, stipulated starting with a different account of 'argumentation', namely to regard argumentation as a communicative act that can be identified via a certain intention of the speaker and which can be performed by using different representational means. Stripping off the too narrow conception of mere verbal utterances has made it possible to take into account the epistemic achievements of visual representations in the context of scientific discourse more seriously. Apart from Kjørup, whose work we have used as a guideline in this respect, other scholars – in particular those working on argumentation theory – have suggested similar, albeit more broadly conceived methodological approaches to the topic of argumentation.[46]

Yet, although this initial shift in perspectives allows us to take epistemologically seriously the phenomenon of visual representations, the first question that still has to be answered refers to the kind of cognitive content such images might contain and are thus able to transmit in the argumentative context. In this respect, the point has already been stressed that we are apparently able to translate contents from one representational means into another. This holds true also in the case of images and therefore speaks in favour of the thesis that at least some of them contain a cognitive content that can be transmitted in communicative acts.

The question of translatability has triggered another critical point, namely whether scientific images are only auxiliary means in the distribution of content, i.e. whether the choice of a visual presentation of information is only due to the weakness of the human mind to grasp the data otherwise.[47] Philosophers, highlighting this (in principle) reducibility of visually transmitted information, suggest that visual representations do not contribute to the acquiring of *scientific knowledge or other epistemic desiderata*, although, from a *practical point of view*, images may be preferable to other kinds of representations in science. Irving, for example, mentions this sceptical philosophical stance appearing in the discussion about the epistemic value of images in scientific arguments (see Irving 2011, 775).

The suggestion to reduce visually presented content to a presumably epistemically more basic category is critically commented on by Perini. She refers to the reducibility thesis as a putative rationale for why philosophers of science apparently take it for granted that visual representations are not worth analysing as epistemic tools. As Perini states, this neglect reveals "an underlying assumption: The reasoning involved when scientists support hypotheses with figures can be understood without considering visual representations as such" (Perini 2005c, 914). However, neither Perini nor I are convinced by this line of reasoning, as it completely ignores scientists' preferences for visual representations in their reasoning practices.

The question of how to argue for an *independent epistemic status* of scientific images then took us back to the debate about propositionality. If we want to defend the thesis that visual representations can yield knowledge or other epistemic desiderata, we apparently have to discuss first what kind of content – propositional or non-propositional – they can contain and transmit. In the response to this query, we focused on the fact that it is *via perceptual means* that we decipher the content of scientific images. Apart from the sceptics, most contemporary philosophers agree that perception can be regarded as an epistemic source, i.e. as a source of propositional knowledge, thus, pointing to the fact that cognitively accessing the content of images via vision also permits drawing an analogy to everyday perception. That is, if it is admitted that perception in those everyday contexts yields propositional knowledge, then this epistemic mechanism should not be called into question when scientific images are visually investigated. The view has therefore been endorsed that at least some visual representations can transmit a propositional content.

Yet a more detailed look at the topic has made clear that there is more to the epistemic dimension of visual representations than simply propositional knowledge via perception. In particular, Hentschel's example of the first observations of the solar surface has made plain that our visual capabilities are not restricted to already-conceptualised domains. On the contrary, human observers are apparently able to detect vast amounts of information via their visual apparatus, even though they do not possess relevant concepts to name and describe correctly what they have observed. This is not only true with regard to people presumably lacking the relevant individual training in technical vocabulary, it also holds true even if there are no suitable concepts available *per se* – as in the case of the images produced by electron micrographs discussed by Perini.

In this sense, I also partially agree with Grundmann's thesis that our perceptual apparatus allows us to access non-conceptual and thus non-propositional content as well. This seems to be how perception genuinely works. Concepts to classify what we see are often acquired when other people – parents or teachers – explain to us what we have observed. Thus, those concepts are acquired via testimony, that is, by the words of others, that accompany, as explanations, observational situations particularly in our early years. Here, testimony and perception work hand in hand to form the basic knowledge on which our belief system is built. When the relevant concepts are acquired in the way described, it does not seem so far-fetched to assume that we are also able to gain propositional knowledge via perception.[48] The suggestion, then, is that some visual representations can transmit both kinds of contents – propositional and non-propositional alike. In particular, representational images such as photographs or instrumentally produced images such as micrographs, brain scans, or computer graphics, seem to let the observer benefit from her perceptual ability to detect and decipher non-conceptual as well as conceptual contents.

Finally, taking into account that it is via perception that information, transmitted by visual representations, is cognitively accessed, has also made plain why we are obviously particularly skilled in decoding information presented in this way. Empirical evidence supports the thesis that visual perception has developed as a cognitive resource for human beings that has been especially shaped by evolutionary processes. In accordance with this, Hentschel claims that "[a]pproximately 60% of the input into the human brain comes from vision" (Hentschel 2014, 32). Not only do we gain huge amounts of information via our perceptual apparatus, even if we do not possess relevant concepts of what we see, but human vision has also evolved in such a way that we are, for example, particularly good at pattern recognition or detecting colour hues.

Therefore, choosing visual representations to transmit information in scientific discourse also means attempting to benefit from these cognitive advantages provided by those evolutionary processes. Diagrams and graphs, for example, are particularly useful to exploit our ability of pattern recognition. Photographs and other images brought about by causally functioning instruments are especially suitable as evidence, because they can transmit information otherwise only available to the initial observer – the eye-witness, so to speak. Furthermore, acknowledging the capacity of images to transmit even non-conceptual content enables scientists to draw on the further merits of visual representations. On the one hand, scholars can start communicating about phenomena that have not yet been completely conceptualised. That is, visual representations allow scientific collaborative investigations even when no clear concepts of the phenomena are available. They not only allow working out the details, but also the elaboration of the initial concept itself. On the other hand, the fact that images can transmit non-conceptual contents also explains how it might be possible that such visual representations can be used to refute certain hypotheses in science. Their content might reveal to subsequent observers anomalies or new phenomena not expected by the initial investigators. Photographs taken by astronomers can illustrate this point. Ratzka explains that, although those images were initially not intended for this purpose, early photographic recordings are re-evaluated nowadays to check for detections of the movements of asteroids – a phenomenon not known to those early astrophotographers (see Ratzka 2012, 246). Used this way, those pictures refute the initial hypothesis that all the detected entities were stars.

All of these aspects add to the explanation of why many scientists favour visual representations in their communication processes. Moreover, it also suggests an explanation for the reason that certain academic disciplines are more attached to such a visual method of information transmission than others: in particular, empirical sciences apparently profit from the evidential role that visual representations can play in scientific discourse. Moreover, scientific disciplines that produce huge quantities of numerical data – as the results of measurement or simulation processes – benefit from the human cognitive capacity to evaluate visualisations rendered from those data sets.

The final remark in this summary is meant to bring into focus a crucially divergent aspect between everyday and picture perception as highlighted by Maynard. He urges us to keep in mind that it would be wrong to claim a complete identity in this respect. Most of all, we have to be aware of the fact that it is a part of the recipients' background knowledge that visual representations are artefacts, i.e. that they are created and circulated within the community for certain purposes. Scientists will therefore usually take into account the intentions of the communicating party when evaluating visual representations in scientific discourse.

The suggestion, in this regard, was to make use of Scholz's account of image games. This approach explains how a layering of meanings can take place, that is, how initially encoded information can be enriched by the intentions of different people making use of the genuine image in diverse communicative settings. In this way, the apparent tension between regarding scientific images as signs and as entities subjected to a basically perceptual decoding can be relieved. Moreover, as the theory of image games allows us to endorse the artificial character of scientific images despite their perceptual decoding, we can make use of their sign character to help explain why certain visual representations are more easily understandable than others. Some – such as photographs – are more closely related to our capabilities of perceptual decoding than others, for example diagrams whose interpretation requires more training, that is, background knowledge.

Having laid out my hypotheses on the relations between the perceptual abilities of human observers, the cognitive content of visual representations and the functional roles of images in scientific discourse, I will next discuss what effects these findings might have on an epistemological evaluation of scientific images.

4.3 The cognitive value of visualisations

The above considerations about the perceptual deciphering of visually presented information suggest two points about the epistemic status of scientific images. Firstly, the thesis has been defended that at least in those instances where the relevant concepts are at the observer's disposal, propositional knowledge can be gained via picture perception. Moreover, the acquisition of these concepts is often made possible partly by our perceptual abilities. Secondly, the discussion of non-conceptual contents being grasped via perception and the correlated capacity of some visual representations to transmit such a kind of information suggests that, from an epistemological point of view, there is more to the epistemic status of these visual means than being sources of propositional knowledge alone.

In sum, the above analysis shows that claiming a necessary inferiority of visual representations in comparison to other representational means when information transmission is at issue – a claim that proponents of a traditional Fregean point of view might defend – is not justified. On the contrary, in the previous discussion it was pointed out in what sense images might entail a cognitive content which can be used for such knowledge-orientated communicative purposes in science. Acknowledging the fact that 'visual representation' is a broad category of diverse visual phenomena which might also be of mixed qualities when it comes to transmitting their content to a respective audience, it seems reasonable to assume that by and large visual representations have parity with other communicative vehicles expressing thoughts.

In this final section, the focus will be on the question of whether visual representations are particularly suitable to serve specific epistemic functions in science. Can the thesis be defended that, at least in certain instances, images can be regarded as *epistemically superior* to other representational means when it comes to fulfilling particular communicative tasks?[49]

In the following, the question of the precise *cognitive value* of visual representations in scientific discourse will be discussed. Is there a kind of epistemic advantage inherent to visual representation not present in other vehicles of communication? To find an answer to this, the role of images in learning activities in science, i.e. within a paradigmatic epistemic process, will be examined, whereby the concept of learning is meant here in a broad sense and not merely confined to students' education. On the contrary, *learning* is commonly understood as an essentially cognitive activity.[50] Being successful at learning something normally implies two important epistemic desiderata: *knowledge* and *understanding* (see Kosso 2007, 175).

Both of these epistemic desiderata will be discussed separately in the following analysis. The investigation is guided by the questions of how visual representations fit into the epistemic process of learning, what exactly their contributions in this context are, and how they can support the cognitive aims of learning. In this context, some of the insights will be applied that have been gained concerning the cognitive content of visual representations and the perceptual way they are utilised so that their content is understood. That some visual representations can transmit non-propositional content provides a resource to take into account suggestions discussed in epistemology beyond the traditional point of view, which focuses solely on propositional knowledge.

Firstly, with regard to knowledge acquisition, this means that the question to be asked is in what sense visual representations can contribute to (1) *knowledge-how* or to the attainment of (2) *mental images* via phenomenal knowledge or via knowledge by acquaintance. Both ways allow for knowledge representations beyond propositionally structured contents. Moreover, philosophers traditionally assume that a disregard for these kinds of knowledge in epistemological analyses is justified because, among other things, they can neither be transmitted in conversation nor can logical operations be utilised to process them further (see Grundmann 2008, 86). If it can be shown, however, that they are indeed communicable, namely by means of visual representations, this would not only add to the cognitive value of images in scientific discourse, it might also return into focus those other kinds of knowledge in epistemology.

The second dimension of successful learning, namely *understanding*, is related to a discussion in epistemology similarly motivating a critical dispute about the exclusive focus on propositional knowledge. Triggered by the struggles of analytical philosophers to come up with a convincing concept of knowledge in the aftermath of the Gettier-cases, which undermined the traditional analysis of 'knowledge' as justified, true belief, philosophers started discussing what is really to be appreciated as valuable with regard to knowledge by advancing the questions of the epistemic value of knowledge in comparison to mere true belief and whether there are other epistemic desiderata epistemically worthwhile to consider. In this context, the suggestion has been put forward to regard understanding as an epistemic desideratum with an intrinsic epistemic value (see Kvanvig 2003, 186). Some philosophers even propose to replace 'knowledge' with 'understanding' in epistemological analyses. Thus, discussing how visual representations can facilitate scientific understanding, would also mean showing how they can contribute to an epistemic desideratum valuable in its own right, that is, independent of propositional knowledge.

Considering possible contributions of visual representations to both the acquisition of other kinds of knowledge and to the facilitation of understanding in science therefore means investigating the cognitive value of images beyond the classical setting of the distribution of propositional knowledge.

To attain a better understanding of the topic, the analysis will begin with a discussion of approaches from the realm of educational psychology, in which scholars investigate the effects of visual representations on learning processes from an empirical point of view. Choosing this empirical perspective as a starting point for subsequent epistemological discussions has the advantage not only of showing exactly how images can play a part in the cognitive process of learning, but also of highlighting the constraints on their efficiency in this context, constituted for example by learners' characteristics.

4.3.1 Educational psychology

In this section, some theses will be examined concerning the effect of visual representations on learning processes put forward by educational psychologists. A comprehensive overview of this topic is presented by Ioanna Vekiri (see Vekiri 2002). She discusses three different theoretical approaches from the realm of educational psychology to explain the contributions of graphical displays to students' learning processes. Regarding the educational merits of these visualisations, there are, on the one hand, those theories dealing with the positive effects on *remembering information* (see ibid., 262). Accounts on *dual coding* and *conjoint retention* belong to this set of theories. On the other hand, there are approaches, subsumed under the heading of the *visual argument hypothesis*, dealing with the transmission and processing of information offered visually.

Vekiri focuses her analysis exclusively on graphical displays such as diagrams. It can be assumed, however, that other kinds of visual representations show similar effects on learners' cognitive processes. I will point out possibilities of an application to a wider range of images in the discussion of the different theoretical approaches below. Furthermore, we will focus on dual coding and visual arguments accounts alone, as the theory of conjoint retention does not add a new dimension to the topic of enhancing students' cognitive processes. It is based on the dual coding approach and applied to the realm of maps, thus constituting a case rather of application than a completely new theoretical approach, as Vekiri makes plain (see ibid., 292).

Proponents of the theory of *dual coding* suggest that there are two different cognitive subsystems in the human mind: one to process and store verbal information and another to process and store non-verbal, in particular visual, information (see ibid., 266). Vekiri points out that, despite this assumed duality, proponents of this theory nonetheless argue for a linkage between both systems. "Although the two cognitive systems are functionally distinct, they are interconnected. Associative connections can form between the verbal and visual representations, enabling the transformation of each type of information into the other" (ibid., 267). This connection between both systems is, for example, used to explain why people are able to mentally visualise certain events read in a novel. Moreover, Wolfgang Schnotz defends the thesis that the theory of dual coding not only affects graphical displays but also pictorial representations such as photographs (see Schnotz 2002, 107).

A consequence of the theory of dual coding for educational purposes consists in the thesis that it is advantageous to present information both visually and verbally in this context, for example by adding visual illustrations to a text. Two explanations are offered why such a combination of representational means can enhance the student's cognitive process of learning if certain design criteria are met.

The first aspect concerns the retrievability of information processed and stored in such a dual way. Vekiri describes the potential positive effect on learning processes as follows:

> [i]llustrations and other visual materials may contribute to the effectiveness of instructions by enabling students to store the same material in two forms of memory representations, linguistic and visual. When verbal and visual information is presented contiguously in time and space it enables learners to form associations between visual and verbal material during encoding.[51] This may increase the number of paths that learners can take to retrieve information because verbal stimuli may activate both verbal and visual representations.
>
> (Vekiri 2002, 267)

Obviously, this dual method of cognitively storing and accessing information can be valuable not only in the educational setting that Vekiri discusses, but also in the cases of measurement data shown in diagrams and discussed in the related text in a scientific article or textbook, for instance, where the very same effect can obtain, namely that two different memory traces might support the process of learning the relevant results.

A second way in which the dual coding theory can be used to explain an enhancement of students' learning processes, when visual and verbal information is presented in the right manner, concerns a reduction of the cognitive load on the working memory, presumably achieved by offering visual information.

> Dual coding theory claims that visual representations can be accessed as a whole and processed in a simultaneous manner, whereas linguistic representations are hierarchically organized and processed sequentially, one piece of information at a time. It is likely that graphics can improve our memory of verbal material because, owing to working-memory limitations, their mental reconstruction allows faster and more effective processing than does verbal representations.
>
> (Vekiri 2002, 279)

Again, such a positive effect can be assumed to obtain in the scientific setting as well as in educational contexts.

Apparently the theory of dual coding is a very successful approach to explain learning processes. Vekiri points out that recent studies in neuropsychology and cognitive science seem to have proven the theses put forward by proponents of the dual coding theory empirically (see ibid., 267ff.). However, this theoretical approach can account for the relevance of visual representations in the epistemic processes of science to a certain degree only. The following two reasons make especially plain the minor explanatory status of dual coding approaches in this realm.

Firstly, the dual coding theory somehow presupposes a certain redundancy in information presentation, i.e. the described positive effect is achieved best if visual and verbal information overlap significantly. This way of presenting information, however, cannot always be expected to obtain in scientific discourse. Adhering to dual coding by all means would undermine Perini's thesis that scientists use visual representations as proper components of premises and conclusions in scientific arguments. The crux is that Perini's argument presupposes (and I have also tried to show this in the above discussion) that certain kinds of information – for example non-conceptual contents – cannot be transmitted by verbal representations. Thus, not all kinds of information can be as equally well-presented by visual and linguistic representations as is apparently presupposed by proponents of the dual coding theory.

Secondly, the theory of dual coding has an emphasis on remembering and ways of retrieving information from our cognitive system. These are, admittedly, important aspects of learning. However, there are further aspects of visual representations in scientific communication and scientific discourse that draw on cognitive abilities not covered by dual coding. As an example, the possibility of detecting anomalies undermining prevalent scientific hypotheses to explain certain phenomena was pointed out above. Moreover, scientific images often serve as evidence to support certain theses in science and are, as such, subjected to critical investigations by an audience. None of these functions is particularly related to memorising or retrieving information, that is, the subject matter of the dual coding theory. Consequently, the theory of dual coding is only of partial interest when the epistemic status of visual representations in epistemic processes in science is scrutinised.

The more interesting account in the present context is the so-called *visual argument hypothesis*. Here, the label 'visual argument' might be slightly misleading as proponents of this approach are not concerned with arguments in the philosophical sense. Thus, we are not discussing the validity or structure of arguments, namely premises, conclusions and inferential reasoning, rather this psychological approach focuses on the ability of visual representations to transmit information and to enable the recipient to grasp complex relations existing among them.

> Visual argument concentrates on the perceptual and interpretation processes that take place when learners extract meaning from graphical representations. It claims that graphical displays are more effective than text for communicating complex content because processing displays can be less demanding than processing text.
>
> (Vekiri 2002, 262)

Proponents of this account state that visualisations enhance the process of learning at the following levels:

1   Such representations offer information both about their individual elements and their relations (see ibid., 281). Graphical displays make it easier for recipients to learn about those elements and they support inferences about their relations simply by looking at the depictions. This perceptual feature allows for further merits of visual representations in cognitive processes.

2   In particular, it provides for "computational advantages" (ibid.). That is, recipients do not have to search a body of text for the relevant information "and then store it in working memory while searching for the next relevant piece" (ibid., 282) – a process that is "prone to error because working memory has limited capacity and cannot maintain data for a long time without constant attention" (ibid.), as Vekiri points out. Visual representations allow this information to be externalised while keeping it constantly present before the reasoner's eyes. In this way, not only is the cognitive load reduced and capacities thus saved for further reasoning, but also the likelihood of errors is diminished.

3 Relying on their perceptual capabilities in deciphering visually presented information can enable recipients to draw inferences almost automatically about the information offered, instead of involving them in long interpretive activities (see ibid., 282). Just by looking at a bar graph, the student may *simply see* a difference in length expressing a difference in quantity.

4 Visual representations can support the recipients' cognitive processes by providing them with a concrete mental image that can help to work out the solution to a problem in the following way: "[w]hen people reason about a problem using symbolic representations they do not have to mentally carry out all the thinking processes but, instead, they can think of a solution by manipulating parts of visual images. Reasoning often requires consideration and evaluation of alternative possibilities" (ibid.).[52]

5 Finally, and linking the current discussion to what has been previously said about dual coding, presenting information visually "may trigger the recall of relevant knowledge" needed for ongoing reasoning processes (ibid., 283).

This last point, as well as the penultimate one, do not seem to restrict themselves to graphical displays but exhibit an advantage, rather, in the processes of reasoning facilitated by all different kinds of visual representations. I would therefore claim a broader applicability for the visual argument hypothesis than Vekiri does in her article. However, care should be taken not to overemphasise the advantages mentioned, as they are dependent on at least two further conditions.

Firstly, visual representations can be more or less apt for particular communicative purposes. Quite a few scholars discuss how to improve the design of visual representations to have them result in a higher efficiency. Vekiri mentions design guidelines for graphical displays (see ibid., 301ff.). Suggestions with respect to graph design are, for example, put forward by Priti Shah and James Hoeffner (see Shah and Hoeffner 2002, 62f.). Alexander Renkl and Katharina Scheiter discuss proposals of design enhancements on a more general level (see Renkl and Scheiter 2015, online first).[53]

Secondly, learner characteristics play a significant role in creating the epistemic advantages of visual representations in cognitive processes (see Schnotz 2002, 113f.). Although a variety of aspects are discussed in this regard,[54] the most significant factors seem to be related to the following abilities, highlighted by Schnotz:

> [v]isuo-spatial text adjuncts and other forms of visual displays can support communication, thinking, and learning only if they interact appropriately with the individual's cognitive system. Accordingly, the effects of visuo-spatial adjunct aids depend on *prior knowledge, cognitive abilities, and learning skills.*
>
> (Schnotz 2002, 113, my italics)

The topic of prior knowledge will be briefly examined at this point.

Apparently this category contains two different kinds of knowledge, one related to the depictive style of the visual representation presented to the learner and another connected to the informational content of the image at hand. The former aspect seems to be naturally relevant, as beyond naturalistic depictions, ways of visualising information have been increasingly developed. This developmental process has been accelerated significantly by the possibilities offered by IT devices within the last few decades. Thus, it comes as no surprise that people have to learn new depictive styles to correctly decipher information presented with their aid. However, this does not undermine the previous thesis that we can nonetheless draw on our perceptual abilities and, therefore, on evolutionarily manifested advantages of processing visual information when decoding visual representations. For instance, pattern detection can be made to work even though we are not familiar with a particular style of depiction. Thus, what we are facing here seems to be similar to what happens in the course of concept acquisition: we can receive information via our perceptual apparatus even if we do not possess the correct concepts of what we perceive. Yet our abilities as observers can be improved significantly by acquiring the relevant concepts.

The second, content-related aspect of background knowledge seems to be more controversial in the debate. On the one hand, scholars suggest that more background knowledge in the related domain enhances the cognitive efficiency of visual representations used for purposes of information transmission (see Vekiri 2002, 304).[55] On the other hand, it is pointed out that more background knowledge might diminish the cognitive value of visual representations as recipients can, for example, visualise relevant details by reading a text alone (see e.g. Schnotz 2002, 114). Moreover, it is argued that people equipped with a higher degree of background knowledge might also be more tempted to disregard visual representations as relevant sources of new information and simply experience them as an entertaining side-effect. Renkl and Scheiter discuss this learners' bias, that is, the tendency to ignore information presented only visually, as one of the main problems affecting the use of visual representations in educational environments (see Renkl and Scheiter 2015, online first).

It seems reasonable to assume that a higher degree of background knowledge can cause both effects, namely a better understanding and a tendency to neglect information presented in a visual way, as these are no contradictory effects *per se*. Only their contingent combination will, without doubt, undermine the positive effect that images might have in this context. Renkl and Scheiter stress a point important in the context of education. Apparently students have to be instructed to acknowledge the relevance of visually presented information correctly. However, the neglect mentioned here does not extend more generally to the context of scientific discourse. Scientists publishing and reading articles usually acknowledge the relevance of visual representations, as illustrated above, by drawing attention to the invention of the database INSPIRE in the natural sciences that allows the storage and search for visual information separately (see www.projectthepinspire.net, accessed February 16, 2016).

Beyond that, scholars in educational psychology suggest that presenting information visually is particularly helpful to students with low prior domain knowledge. As Schnotz states: "[p]revious research has pointed out that comprehension among learners with low domain knowledge (but sufficient visuo-spatial cognitive skills) is increased when pictures are added to a text" (Schnotz 2002, 114). Again, this seems to be a reasonable claim, as visual representations can guide the learner's attention to notice the relevant details, can highlight relations otherwise overlooked, or present complex information in a significantly simplified fashion. To be concise, students' prior knowledge seems to be a somewhat ambiguous condition influencing the process of learning by using visual means.

Summing up the previous discussion, it can be stated that, after recipients have mastered the initial obstacles to work with visual representations effectively (which is often part of their scientific training), the latter can, according to the visual argument hypothesis, support the cognitive process of learning on at least three different levels: firstly, by showing the relation between individual pieces of information; secondly, by making information directly perceptually accessible; and, thirdly, by enabling a more efficient use of cognitive resources. This last aspect is highlighted by Vekiri. "Also, displays support thinking during problem solving because they reduce the amount of information that must be maintained in working memory" (Vekiri 2002, 288). Moreover, that visual representations indeed bring about these theoretically proclaimed advantages is demonstrated by several empirical studies cited by Vekiri and others.

Provided with this empirical background information concerning the cognitive effectiveness of visual representations, the epistemological analysis can now be continued. What exactly the contributions of images might be to the two components of cognitive processes such as learning, namely (1) to acquire *knowledge* and (2) to achieve *understanding* will be examined in the following two sections.

4.3.2 Visual representations and the varieties of knowledge

Concerning the epistemic functions of visual representations, the discussion hitherto has been about the possibility of gaining *propositional knowledge* via perceptually deciphering the encoded information. By analogical reasoning, my suggestion was that if it is an acceptable thesis that propositional knowledge can be gained via perception, then the same should be allowed for picture perception. This general statement was then limited by the addition of the following constraints.

Firstly, there are kinds of visual representations in science that presuppose more background knowledge than others in order to be interpreted correctly. The ability to decipher information presented in diagrams and graphs, for example, usually presupposes a certain training. In a similar fashion, Fleck has emphasised that observational skills in science relying on the use of instruments have to be learnt, that is, the students have to be trained how to 'see' correctly. In particular, he shows how a lack of training might contribute to misinterpretations of results in microscopy (see Fleck 1986b, 118ff.). In such instances, our perceptual abilities are not sufficient to gather all of the relevant information, as either the phenomenon or the form of its presentation does not belong to what we are evolutionarily familiar with to observe.[56] Here, testimony and perception are equally relevant to gather the information presented by those visual means. Such an interplay between different epistemic sources is not unusual – on the contrary, it seems to be common practice, as Scholz explains (Scholz 2009c).

The second constraint pointed out is closely connected to this co-operative outcome of epistemic sources. It seems that our perceptual apparatus yields propositional knowledge if we have learnt the relevant concepts beforehand. Thus, although our perceptual apparatus often allows us to navigate in the world without difficulty, we often need additional explanations to categorise phenomena correctly – that is, we need the relevant concepts that are often transferred via a combination of showing and telling, via perception and testimony. In the same way, a correct interpretation of certain images might presuppose the prior acquisition of relevant concepts.

Promising as the suggested capability to transmit propositional knowledge already sounds for the possible epistemic prospects to expect from the usage of visual representations in science, I wish nevertheless to examine another epistemic dimension of scientific images in this section. This additional epistemic potential of visual representations draws on two aspects discussed above, namely on the theory of dual coding and on the capacity to transmit non-conceptual content. The first point, put forward and empirically defended in the cognitive sciences, refers to the fact that the human brain can apparently store incoming information both visually and propositionally. Combining this with the thesis that picture perception can also transmit non-conceptual content, the theory of dual coding offers an explanation for why presenting information visually in the scientific discourse can constitute a proper epistemic merit. By using visual means, we can provide others with information, namely non-conceptual, that cannot be transferred otherwise. Moreover, as the theory of dual coding shows, our brain is apparently able to process visual information separately. Thus we are cognitively able to handle this information without translating it into propositionally structured expressions. The implications of this line of reasoning for acquiring the different types of knowledge analysed in epistemology will be discussed in what follows. Although philosophers are mainly concerned with propositional knowledge, i.e. *knowing-that*, there is nevertheless a variety of other epistemic concepts that has caught their attention. Grundmann mentions the following four kinds of knowledge in his introductory work to epistemology (see Grundmann 2008, 86):

1　propositional knowledge (knowing-that)
2　knowledge by acquaintance
3　phenomenal knowledge (knowledge about *qualia*)
4　knowing-how (skills)

Although there is this variety of different kinds of knowledge, Grundmann points out that epistemologists are primarily concerned with knowing-that (see ibid., 71). Eva-Maria Jung explains what reasons are commonly mentioned to justify this prioritisation. She identifies two different arguments put forward as a rejoinder to the claim that epistemologists wrongly focus on propositional knowledge (see Jung 2012, 13). Firstly, philosophers claim a difference in essence between knowing-that and knowing-how. And, as epistemology is exclusively concerned with propositional knowledge, knowing-how simply does not belong in the scope of its analysis. Secondly, knowing-how can be reduced to knowing-that. The former therefore need not be considered as an independent kind of knowledge.

Grundmann analyses different arguments to reduce knowledge of the kinds (2) to (4) of the list above to knowing-that (see Grundmann 2008, 74ff.). He comes to the cautiously formulated conclusion that it might be possible to reduce all of them to propositional knowledge (see ibid., 85). Moreover, he claims that even if such a reduction might not be possible, there are two good reasons[57] supporting the prioritisation of propositional knowledge in epistemology: (a) only propositional knowledge can be communicated and thus shared within a community; moreover, only propositional knowledge can be cognitively processed further via valid inferences, and (b) only propositional knowledge is in line with the aim of truth in epistemology (see ibid., 86).

Now, the above analysis provides argumentative means to broaden the focus of epistemology, as it permits calling into question at least one of the two reasons mentioned by Grundmann. It will be seen in due course that the claim about communicability can be easily rejected. In the same way, the results we hitherto obtained can be used to show that there are no particular difficulties in cognitively processing visual information.

In the following discussion, the focus will be primarily on *knowing-how*. This more detailed analysis will allow a brief comment on *knowledge by acquaintance* and *phenomenal knowledge*. All three concepts are related to learning and the sharing of knowledge more broadly in communicative contexts. However, it can be noted that, whereas knowing-how seems to be of special relevance for educational purposes, in science, and thus in relations between experts and laypeople, phenomenal knowledge might be of relevance to experts *per se*. If it can be shown that at least one of the kinds of knowledge mentioned above can be promoted via visual representations in science, their epistemic relevance – in comparison to linguistic representations – seems then to be proven.

The topic of *knowing-how* as an independent epistemic category, i.e. independent of propositional knowledge, is discussed broadly both in epistemology and in the philosophy of mind. To explain this concept, I will follow Eva-Maria Jung and Albert Newen's suggestion to draw a distinction between *theoretical* and *practical knowledge* (see Jung and Newen 2011, 95). This distinction is explained as follows:

> [t]heoretical knowledge [...] describes a relation between a subject and a proposition thereby being related to a norm of truth. [...] Practical knowledge, instead, describes a relation between a subject and an activity. This knowledge is related to the norm of success: We ascribe some ability to a person if she is able to successfully perform it.
>
> (Jung and Newen 2011, 95)

In addition to this, Jung and Newen point out that practical knowledge also implies a kind of warrant concerning the ability to perform the relevant action. This supplementary condition is necessary to exclude cases of performing some action x to bring about y that happen to be successful by mere chance. It would be counterintuitive to call such instances 'knowledge' (see ibid.).

According to this distinction, knowing-how belongs to the category of practical knowledge. As was highlighted in Grundmann's list of knowledge categories above, knowing-how is commonly regarded as consisting in certain skills. Knowing how to ride a bicycle or how to play the piano are common examples in philosophy. In science, we might think of instances such as knowing how to set up certain experiments, how to use instruments such as microscopes, or how to write a scientific article, etc. Including all kinds of accidentally successful action performances as instances of knowledge-how is inappropriate, as is subsuming all kinds of reflexes under this label. Therefore, Jung suggests as a criterion of demarcation that only actions brought about intentionally belong to the domain of practical knowledge. These intentional acts can be further characterised either by being directed at a particular aim or by certain formal aspects of how to perform the respective action (that is, a conformity to certain rules) (see Jung 2012, 158). Moreover, she claims that these intentional actions can be influenced by processes of learning and modification by the respective subject (see ibid., 159). If the student realises that a certain action does not lead to the intended aim, or only via a variety of unnecessary detours, she can learn to improve her actions (see ibid.).

Now, this characterisation of the object of knowing-how makes clear why the question dominating the respective debate in epistemology is about a possible reduction of knowing-how to knowing-that. If there are rules, say, about how to play the piano correctly, then why not argue that the relevant knowledge simply consists in *knowing that the piano is played in accordance with these rules*? In particular, philosophers convinced that propositional knowledge is the only category relevant to epistemological discussion suggest different strategies, for example, like the one drawing on rule-following just mentioned, to reduce knowing-how to knowing-that.[58] As Jung points out (see ibid., 13), these attempts enable them to maintain their project to exclusively analyse propositional knowledge without being forced to ignore the epistemic phenomenon of knowing-how.

Tempting as this strategy might seem, Jung also shows that there are at least two major difficulties that proponents of such reductive approaches have to face. On the one hand, they have to explain that, although an epistemic subject might know all the relevant rules about how to perform action x correctly, she is still not able to do so (the so-called 'knowledge-action-gap'). On the other hand, they have to account for the fact that there are actions which we perform to reach a goal that are not guided by rules (see ibid., 48). Jung takes these findings as important hints that although "practical knowledge might involve the knowledge of regulative propositions concerning the action", a complete reduction of knowing-how to propositional knowledge is not possible (Jung and Newen 2010, 124).

Contrary to such reductive approaches, she emphasises the relevance of knowing-how as an object of inquiry in epistemology in addition to propositional knowledge (see Jung 2012, ch. 1.5). She supports this initial conviction with further arguments bolstering the dichotomy between propositional and practical knowledge. Both with regard to contents transmitted and to the aims pursued by their means, the two kinds of knowledge differ essentially (see ibid., ch. 3.3). The following list summarises her theses on the topic.

- *Content*: the object of knowing-how are intentional actions. This kind of knowledge is always related to certain contexts and epistemic subjects. In this sense, the contents of knowing-how cannot be objectified completely, contrary to propositional knowledge, whose content is thus expressible by propositional means.
- *Aim*: the aim of knowing-how consists in successful action performances. Consequently, it can be assumed that the acquisition and deployment of knowing-how to perform such actions are guided by a particular norm, namely that of successful action performance (contrary to the norm of truth in the case of propositional knowledge).

I agree with Jung that knowing-how and knowing-that should be regarded as distinct categories in epistemology. Despite my general sympathy with her approach, two critical remarks have to be added about her criteria to characterise practical knowledge. Firstly, her thesis that the content of knowing-how cannot be objectified, i.e. that it necessarily contains subjective and private elements, does not seem to be convincing. In particular, there arises a certain tension to another claim of hers, namely that knowing-how can be taught, particularly by showing how to practice certain actions, and can thus be acquired in educational processes (see ibid., 72). If, however, teaching is possible in such a visual format – this issue will be returned to in due course – there have to be at least some paradigmatic instances of the action in question that can be demonstrated in order for it to be copied by the student. A partial objectification thus seems to be possible.

Secondly, although it seems to make sense to distinguish between propositional and practical knowledge along the lines of different norms, there remain some doubts about whether she has chosen the correct ones for her contrastive project. Especially regarding propositional knowledge, the debates in epistemology show that there is no consensus amongst philosophers concerning the status of truth. An example: Alvin I. Goldman, who proposes a "veritistic approach" in epistemology, that is, who emphasises the relevance of true belief as the predominant aim in our knowledge-seeking enterprises (see Goldman 1999, ch. 3), has been constantly criticised by others who object that he is wrong to put such a stress on truth alone. A similar objection might be raised to Jung's account if she claims truth to be the decisive criterion on the part of propositional knowledge in order to enable the relevant distinction between both knowledge categories.

Despite these critical remarks, however, Jung's argumentation seems to be quite convincing. Especially, the fact that her work is located at the interface between epistemology and the philosophy of mind allows her to elaborate another interesting thesis. In an earlier article, Jung and Newen had already pointed out that the constant misunderstandings of Gilbert Ryle's concept of knowing-how consist partly in neglecting the fact that his approach not only aims at a semantic analysis of the term, but also poses the question of "whether all mental cognitive processes can be analyzed in terms of propositional knowledge" (Jung and Newen 2011, 84f.). They argue that Ryle's project, which is embedded in the philosophy of mind, is also meant to tackle the topic of how knowledge is represented in the mind so that the latter can process it. Jung and Newen elaborate on this analysis of representational modes of knowledge in their own hypotheses about the distinction between propositional and practical knowledge.

They suggest three different modes of representation, namely "(i) propositional representations, (ii) sensorimotor representations and (iii) image-like representations" (ibid., 96). They characterise the first category as "language-like" (ibid.). Jung specifies this later as implying the ability to bear truth values and as being conceptual in kind (see Jung 2012, 164).

Contrary to this, the second category is considered as being non-conceptual in kind. Moreover, as Jung and Newen point out, the latter is also closely connected with certain qualities of our environment that we perceive and which trigger certain actions (see Jung and Newen 2011, 97). These representational means are deeply intertwined with our perceptual abilities. Jung explains that the human body is the object of sensorimotor representations. What is represented in the mind about a particular action are its expected duration, the kind and quantity of bodily forces to perform this action and certain motoric rules (see Jung 2012, 171). Although practical knowledge can entail rules that are expressible by linguistic means, and although the respective skills might at least partly be acquired by such propositional knowledge, Jung defends the claim that practical knowledge is usually represented in the sensorimotor way. She presents two aspects to support this claim: on the one hand, if the performance of certain skills is interrupted or disabled, for example because of certain diseases, we will note that we will not have recourse to propositional knowledge for guidance. On the contrary, attempts to consciously focus on the action in question will usually be experienced as disruptive rather than as helpful. On the other hand, Jung claims that small children and animals who do not possess a language are able to perform reasonable actions nonetheless and can thus be said to possess practical knowledge of certain kinds (see ibid., 177). Of course, assuming that knowing-how and knowing-that are represented differently in the human mind also lends further support to Jung's thesis that knowing-how and knowing-that are two independent epistemic categories.

Finally, Jung and Newen suggest a third category of mental representations, which they call "image-like". Contrary to sensorimotor representations, they are independent of concrete situations, as they can also be triggered by our imagination. Furthermore, they are "systematically connected with perceptual images and sensorimotor representations" and "with other image-like patterns" (Jung and Newen 2011, 98). The interesting thing about this third category that Jung highlights is the possibility of making use of image-like representations in educational contexts. She points out that, whereas sensorimotor representations cannot be consciously accessed, image-like ones can, and are thus employable as a medium to transfer the relevant knowledge for educational purposes (see Jung 2012, 180). To begin again with a critical comment on this: it does not quite convince that we cannot consciously access the sensorimotor representations of certain actions. From my point of view, it is exactly this that happens if people are asked to show or demonstrate the performance of certain skills. Admittedly, they will not consciously process all the details of this action, but they can call to mind the way they usually perform it in order to demonstrate it. This seems to suggest that they are at least aware of what is essential to this action to be demonstrated so that their students can copy them. Again, this line of reasoning is already implied in Jung's own work when she discusses Edward Craig's pragmatistic conception of knowledge. In this context, she highlights the fact that Craig's concept of the good informant, which Craig takes to lie at the heart of our ordinary concept of knowledge, not only entails classical testifiers but also people who can answer the question they were asked not by telling, but only by showing how to do something (see Jung 2012, 72).

Despite this critical remark, Jung makes a good point in arguing that image-like representations are particularly useful for educational purposes. Different components, already discussed above, play a role in this and now require piecing together.

A start can be made with what seems to be most beneficial when using visual representations for educational purposes: it is not only that we are able to make use of images to transfer practical knowledge at all – which is not possible with the aid of linguistic representations. It is apparently also the case that, from a methodological point of view, such an educational practice permits a considerable expansion in the size of the audience. Whereas a teacher can address only a limited number of students by a direct demonstration of a particular action, showing how to do x via images not only allows a broader audience to be reached on a synchronic level – for example, by live-streaming a lecture – but also diachronically, for example by means of images showing how to set up a particular experiment in a textbook.

After practising with the aid of such images, the students can make use of them for mental training, as Jung explains (see ibid., 181). 'Mental training' means that people are able to imagine certain actions and perform them mentally so that their actual performance of these actions will be later enhanced.

From an epistemological point of view, these features constitute proper epistemic merits. As explained above, knowing-how cannot be translated into linguistic expressions, that is, traditional methods of teaching are not possible. Showing people directly how to do x might be a way out of this dilemma, but it considerably restricts the number of people that can be addressed by this demonstration. A methodology of utilising images in this way not only permits the transfer of practical knowledge to be made public, but also enables teachers to professionalise it. As images, moving or non-moving, can be recorded and stored, these educational means are not only reusable in a variety of instances, they can also be subjected to performance ratings which might suggest modifications to the initial images in order to better meet the learners' requirements.

Of course, the issue discussed here has already acquired the utmost importance for people concerned with electronic learning and virtual reality. Max Hoffmann and his colleagues, for example, propose utilising virtual-reality tools in engineering studies (see Hoffmann et al. 2015). They claim that in this way more students can be given practical training, even though their universities might not be able to offer them the relevant training in a real laboratory because they lack the financial resources to do so. Thus, Hoffmann et al. point out another dimension in which to broaden the scope of practical training, namely via virtual-reality tools.

Finally, the whole process can also be reversed, that is, visual representations can be used to let experts learn about the implicit knowledge of practitioners. That such a kind of knowledge is often present and can play significant roles in cognitive processes has been pointed out by Eugene S. Ferguson who explains, for example, how non-conceptual thinking has guided the design and development of machines by craftsmen and designers (see Ferguson 1977). Now, as Hoffmann indicates,[59] by experiencing virtual-reality simulations, craftsmen or technicians might then be able to indicate why they think that certain newly developed machines do not fit their requirements, despite their lacking the conceptual ability to communicate this linguistically to engineers. The reasons that these epistemic merits are possible need recapitulating. Two aspects seem to be relevant: firstly, visual representations are cognitively accessed via perception. This enables the transmission of non-conceptual content. Secondly, Jung's discussion of different ways to represent knowledge in the human mind also makes plain why we can process information presented visually quite easily and without the necessity of translation. It simply matches the way we think about certain aspects of the world. Here we can link our considerations to some results from the cognitive sciences, namely to the theory of dual coding – a connection that is also suggested by Jung (see Jung 2012, 178). Proponents of this theory suggest that information is encoded in two different ways in the human mind, namely propositionally and visually. This hypothesis seems to be supported by empirical studies on certain brain lesions that disable one of the two possible ways to store information. An example are patients suffering from aphasia, that is, the loss of their ability to communicate linguistically,[60] who might nonetheless be able to communicate by visual means (see Sacks 2010, 45f.).

To summarise the results so far, visual representations can play an essential, even indispensable part in educational processes focusing on knowing-how. Consequently, we are entitled to maintain that some scientific images possess an epistemic status that is independent of other representational means and thus not epistemically reducible to them. Having discussed visual representations in the context of propositional and practical knowledge, another of their epistemic merits will now be examined that is connected with categories two and three of Grundmann's list above: *knowledge by acquaintance* and *phenomenal knowledge*.

What I am about to suggest is best grasped by starting with a negative point. It was explained above that the dual coding theory suggests two different ways of mentally processing and storing information: one dealing with propositional, the other one with visually presented information. Apparently the student's mind works best if both ways are activated during the process of learning (see Eitel and Scheiter 2015, 153). As Alexander Eitel and Katharina Scheiter argue from an educational-psychology point of view, this effect – also known as the "multimedia effect" – obtains not only when text and image are presented simultaneously to the student, but also when presented sequentially (see ibid., 154). However, what happens if one of these ways is completely omitted? More specifically, what happens if the student in question only receives propositional information about subject x and no visual clues, or vice versa, only visual ones but no propositional information? Does she nonetheless acquire a kind of knowledge in either of these cases? Regarding this question, it seems relevant to consider two aspects, namely the background knowledge of the particular student and her preferred style of learning. The latter refers to the fact that some people are more apt to process visual information than others, who prefer linguistic explanations (see e.g. Kirby, Moore and Schofield 1988).

To start with the presentation of mere visual information, it can be assumed that people with less background knowledge will have greater difficulty cognitively processing the information presented, and thus understanding what they are supposed to learn. A reference can be made here to what has already been said about image games: the student might be able to decipher what is shown in the image, but she might not at all understand what her teacher, by presenting this information, intends to tell her. As Kjørup has pointed out, most images are in need of linguistic anchoring in order to be useful in a communicative situation – such as in educational processes. The second case, that is the presentation of mere linguistic information, is illustrated by Wartenberg's example of guidebooks for bird-watchers. Suppose, for the sake of argument, that such guidebooks do not contain images but only linguistic descriptions and explanations. What kind of information will be lost in this scenario? What might be learnt from images, but not from the corresponding text? Suppose that you opened this somewhat informationally impoverished book on the page describing wrens. Suppose further that you have never seen a wren – neither in the wild or depicted in any way – before reading this entry. Now consider the following two questions: do you think that after reading the entry you would be able to recognise a wren if you saw it in the wild? Could you imagine its appearance? It can be assumed that you would not be able to do either of these. Of course, your background knowledge about other birds might help to exclude some completely odd mental images of a wren. For example, the entry tells you that it is smaller than the common house sparrow. Thus, it would be rather unlikely that you would imagine a bird the size of a dove after reading it. Nonetheless, the text alone will not provide you with the kind of information necessary to create a mental image of this bird. Images, on the other hand, can easily provide us with this kind of information. That is, *they can acquaint us with physical entities* of which we have no genuine experiences such as those gained by watching them in the wild or in a zoo.

This example thereby brings together aspects from both categories of knowledge, that is, knowledge by acquaintance and phenomenal knowledge, without completely agreeing with either so that we could adopt the respective label for the current case. Let me explain.

On the one hand, the above example apparently has a lot in common with Frank Jackson's famous *Mary argument* in the philosophy of mind. It reads as follows:

> Mary is confined to a black-and-white room, is educated through black-and-white books and through lectures relayed on black-and-white television. In this way she learns everything there is to know about the physical nature of the world. She knows all the physical facts about us and our environment, in a wide sense of 'physical' [...]. If physicalism is true, she knows all there is to know. For to suppose otherwise is to suppose that there is more to know than every physical fact, and that is just what physicalism denies. [...] It seems, however, that Mary does not know all there

is to know. For when she is let out of the black-and-white room or given a color television, she will learn what it is like to see something red, say. This is rightly described as *learning* [...]. Hence, physicalism is false.

(Jackson 1986, 291, his italics)

There are similarities as well as dissimilarities to the bird-guide example. What is similar is the way Mary and the prospective bird-watcher learn about their subject matter first. Both of them get cognitive access in a somewhat limited way only. Mary has to learn about colours without ever experiencing one and, likewise, the prospective bird-watcher has to learn about wrens. Both of them lack certain qualitative information in what they learn about their subject matter. The difference between both examples is, of course, the object of learning. Jackson's argument is about *qualia* – i.e. the question *what it is like* to experience something, say, red, whereas the bird-watcher example is about physical objects in general.

I do not want to dwell on Jackson's argument, as my point is not about *qualia* and physicalism in the philosophy of mind.[61] Despite the difference in direction, Jackson's argument nevertheless makes it very plausible that, in some instances, linguistic descriptions alone will not suffice to provide us with the necessary information to construe a correct mental representation of the entity in question – neither of a certain colour nor of the appearance of a particular bird.[62] Images, however, often allow us to construct a mental representation without great effort. Such mental images can then be used, for instance, to recognise examples of the same species in the wild by comparing the mental image and the visual appearance of the bird in question. Yet because of the directional difference between Jackson's argument and the bird-watcher example, I am reluctant to call what has been learnt by the student about the visual appearance of the wren phenomenal knowledge.

Similar difficulties arise with regard to the second category, namely *knowledge by acquaintance*. Nonetheless, it can be assumed that it describes very well what happens cognitively when dealing with instances like the two examples just mentioned. Bertrand Russell, who introduced this concept in epistemology and compares it to what he calls "knowledge by description", defines it in the following way: "I say that I am acquainted with an object when I have a direct cognitive relation to that object, i.e. when I am directly aware of the object itself" (Russell 1910–1911, 108). Obviously, such a direct cognitive relation to the object in question is what is missing in the case of Mary and the prospective bird-watcher above. Neither of them has direct access to their object of interest, namely colours or wrens. This relational character of 'acquaintance', that is, the relation between the epistemic subject and the object in question, is particularly stressed by Russell as one of its constitutive characteristics (see ibid., 109).

Moreover, as Ali Hasan and Richard Fumerton point out, knowledge by acquaintance "is knowledge of something and logically independent of knowledge that something is so-and-so" (Hasan and Fumerton 2014, sect.

1). Summarising Russell's account, they add that "for Russell acquaintance is nonjudgemental or nonpropositional" (ibid.). These are the aspects of knowledge by acquaintance that shall be emphasised here. Apparently this conceptual framework fits well with what has hitherto been stated about the acquisition of non-conceptual content via perception – either in a direct way or mediated via visual representations – and its processing and storage in a separate cognitive subsystem, as indicated by proponents of the theory of dual coding.

Unfortunately, just as I cannot adopt the concept of phenomenal knowledge for my purposes, neither can I wholeheartedly subscribe to the theory of knowledge by acquaintance. The dispute is about the objects of these knowledge-relations. Russell suggests that "sense-data" should be regarded as the paradigmatic instance of objects of acquaintance (see Russell 1910–1911, 109). He explains: "[w]e shall say that we have acquaintance with anything of which we are directly aware, without the intermediary of any process of inference or any knowledge of truths" (Russell 1912, ch. 5).

Yet I am reluctant to commit my approach to the sense-datum theory because this account entails many well-known problems.[63] According to this theory, for example, sense-data are solely bound to private subjective experiences, which leaves unresolved the question of whether we can ever know that we are talking about the same objects if they are given to us in this private and subjective way (see Grundmann 2008, 475), despite the fact that people have no particular difficulty in agreeing about what they perceive, for example how many chairs and tables they see in a particular room. Thus, there has to be an object of reference that guarantees this inter-subjective agreement on the facts. Hence, an account of knowledge by acquaintance that focuses on sense-data as an appropriate epistemology cannot be adopted here.

At first sight this might be a rather unsatisfying result, as neither of the two approaches can be accepted as the proper epistemic framework for the purpose of explaining what kinds of knowledge are transmitted via images. However, neither the theory of phenomenal knowledge nor the account of knowledge by acquaintance was designed for this purpose, and consequently criticising them for not accounting for this phenomenon is inappropriate. However, this discussion was not intended as a critique. On the contrary, the above has shown in what sense both accounts might, despite those difficulties, contribute to a better theoretical understanding of how it is possible to acquire information relevant to constructing a mental image of the entity in question. Images can transmit a non-propositional content. In some instances, this content entails that all relevant information constructs a corresponding mental image. Both theoretical approaches, dealing with questions about either phenomenal knowledge and knowledge by acquaintance, support the thesis that there is more to human cognition than mere propositional knowledge. They suggest models for how to comprehend the acquisition of non-propositional knowledge. Consequently, the proposal is to take those approaches as a starting point to develop a proper epistemological account concerning information distribution and acquisition by means of images. To be concise, visual representations are relevant in scientific discourse as they can be used to convey a mental presentation of the appearance of the entity depicted. They can, so to speak, foster an inner picture of it by transmitting non-conceptual information that is processed in a separate cognitive subsystem. This information will, however, be lost if scientific images are replaced by mere verbal descriptions.

In order to elucidate the role of imagination in science, Tamar Szabó Gendler's discussion of the intimate connection between imagination and counterfactual reasoning should be taken into account (see Gendler 2013, ch. 4.4). "It has been argued that imagination plays a central role in figuring out what would happen – or what would have happened – had things been different from how they in fact are or were" (ibid.). If we can assume that there is such a connection, the role of imagination in science becomes obvious at once. *Counterfactual reasoning* is a main feature of the scientific enterprise itself. It is needed to invent and perform experiments, to invent theories and ways of testing them. Furthermore, it is a basic requirement that natural laws permit counterfactual reasoning within their scope, that is, they have to remain valid under altered circumstances. Predictions can be given in the form of counterfactual statements, for example. In all of these contexts and scientific tasks, imagination plays an important role, and so does the knowledge conveyed by visual representations concerning their object of depiction.

Thus, the capacity to transmit non-conceptual content and to make it directly accessible via perception allows us to explain in what sense visual representations can contribute to the acquisition of practical knowledge and how they can acquaint us with entities so that we acquire mental images of them. Beyond that, the fact that we cognitively access visual information by perceptual means explains in what sense images can contribute to another epistemic desideratum, namely to *scientific understanding*. This is what we will discuss in the next section.

4.3.3 Visual representations and scientific understanding

Apart from the acquisition of (propositional) knowledge, learning is normally associated with the aim of understanding. Peter Kosso points out that solely memorising propositions is not what is expected of our students – especially not in science (see Kosso 2007, 175); they are not usually required to parrot hypotheses and statements during an examination, for example. On the contrary, scientific training ideally means having students partake in the community of researchers, that is, enabling them to apply acquired knowledge to new questions, to reflect critically on this information and, if necessary, to correct some of its components. The aspect of understanding now acquires relevance here. Are there any particular contributions, then, that visual representations can make with respect to scientific understanding? Possible answers to this question will be presented following a brief discourse of this incipient point on the epistemic relevance of understanding itself.

Scientific understanding is commonly regarded as an ability to coherently fit new items into one's knowledge system and to apply the newly acquired information to solve further tasks and puzzles. Wesley C. Salmon phrases this in the following way: "[...] we have scientific understanding of phenomena when we can fit them into the general scheme of things, that is, into the scientific world-picture" (Salmon 1993, 12f.). But how exactly should this fitting-relation be conceived? What does Salmon suggest when he claims that 'to understand something' means 'being able to fit it into "the general scheme of things"'? An answer to this question is offered by Jonathan L. Kvanvig who emphasises that this fitting-relation is the crucial difference between *knowledge* and *understanding*. He states:

> [...] that understanding requires, and knowledge does not, an internal grasping or appreciation of how the various elements in a body of information are related to each other in terms of *explanatory, logical, probabilistic, and other kinds of relations that coherentists have thought constitutive of justification.*
>
> (Kvanvig 2003, 192f., my italics)

In a similar fashion, Kosso suggests that understanding in science goes beyond a mere additive compilation of evidence on a certain matter (see Kosso 2007, 179). That there is something important lacking in cases when evidential facts are collected in science without the achievement of understanding is highlighted by several examples. For instance, he draws his readers' attention to how the phenomenon of contagion was discovered (see ibid., 182) and points out that, although Thucydides in ancient Greece reported in detail how people became infected with the plague through nursing patients already sick, he did not infer the mechanism of contagion lurking in the background. As Kosso puts it:

> [h]is knowledge, however, stops with the isolated fact of the disease somehow being transmitted from one person to another. [...] He is not credited with the first proposal of the germ theory, since he did not understand the process of infection.
>
> (Kosso 2007, 183)

It is in this sense that Kosso claims that scientific observation alone might yield factual knowledge but not understanding (see ibid., 184). Taking the above discussion into account, this statement seems surprising – but only at first sight. Two clarifying remarks should be added at this juncture.

Firstly, Kosso's thesis that observation – respectively (visual and pictorial) perception – can yield propositional knowledge if concepts are available is acceptable, and yet (secondly) his statement is still puzzling in that the way he phrases his ideas about scientific observation seems to suggest that the latter is more or less independent of theoretical considerations which, of course, it is not. Elsewhere, he has discussed in detail the intertwining of observation and scientific theories (see e.g. Kosso 1988; 1993). Thus, to understand his claim correctly, we might interpret him in the following way: observation is theory-related in different ways. Yet it might be the case that the theories used are not adequately embedded within the broader network of scientific theories that he calls attention to. If this happens – either by deliberately screening off other theoretical assumptions that might lead to a rejection of the theory at hand or unwittingly as a consequence of the observer's own scientific training, a proper understanding of the phenomenon at hand will be blocked as in the Thucydides example above.

However, even though I agree with Kosso's suggestion, another constraint to his thesis should be added, namely that understanding is usually a gradual matter. This assumption seems to be perfectly in line with Kosso's ideas. If less background knowledge is available that the scientist can use to interpret her observations, her understanding of the phenomenon at hand will also be affected in a negative way. It can be stated, then, that Thucydides did understand that the disease observed was transmitted from person to person, but he did not understand how this happened. Consequently, his understanding of disease transmission was only partial, and not, as Kosso suggests, completely lacking.

The topic of understanding has rather recently started to attract the attention of epistemologists. This new focus of research is an aftermath of the challenge posed by the Gettier-cases concerning the analysis of knowledge. One consequence of this discussion has been to bring into focus the epistemic aims and values a subject might pursue when seeking knowledge (see Jung 2012, ch. 2.4.3). It is here that the debate about understanding as an epistemic desideratum valuable in its own right begins (see Pritchard and Turri 2014, sect. 5). Moreover, it is also here that we get a clearer grasp of what motivates Kosso's critique on regarding the striving for evidence as the correct aim of science. Philosophers concerned with the topic of understanding usually point out that *truth* is but one aim valued in epistemic projects. Thus, amassing true beliefs might be a laudable facet of science, but does not reveal its whole epistemic enterprise.

In particular, this debate has been fuelled by Kvanvig's work (see Kvanvig 2003). He discusses the question of whether understanding is a species of knowledge. Kvanvig starts his analysis by pointing out that it is commonly assumed that understanding and knowledge are closely connected (see ibid., 188). It is usually said that if a student understands that x, she also knows that x. In instances like this, the student will possess true beliefs that are propositionally structured if she understands the respective information. In this sense, propositions can also play a role. This assumed intimate connection between knowledge and understanding is now challenged by epistemologists, some of whom even call for a replacement of 'knowledge' by 'understanding' in epistemology as a consequence of this critical discussion.[64]

What is at stake here is the *factivity* of understanding. What does this mean? Obviously, we can have knowledge without understanding (see ibid., 191), as the above example of the student learning by rote shows. But does this claim also hold the other way around, i.e. do we need knowledge as a basis for understanding? Usually it is admitted that there are factive and non-factive usages of the term 'understanding'. Factive understanding implies truth in the same way that knowledge does, as Kvanvig points out (see ibid., 190). Non-factive understanding, then, is either due to "misspeaking or to the expression of propositions that do not involve the concepts of knowledge or understanding central to epistemological inquiry" (ibid.). Statements such as, 'I understand that he was not able to attend the conference', illustrate this latter case.

However, non-factive understanding is often more or less immediately ruled out as a candidate for epistemological investigations, which is Kvanvig's line of argument (see ibid., 190f.). He explicitly draws our attention to factive understanding and identifies two kinds, namely:

> propositional understanding and objectual understanding. The propositional sort occurs when we attribute understanding in the form of a propositional operator, as in understanding that something is the case, and the objectual sort occurs when understanding grammatically is followed by an object [...]
>
> (Kvanvig 2003, 191)

Although the second sort is not straightforwardly propositional in kind, he also thinks that it is factive, as we have to have true beliefs about the object in question in order to be attributed with an understanding of that very object (see ibid.). Moreover, Kvanvig argues that other varieties of understanding – such as "understanding why, when, where, and what are explicable in terms of understanding that something is the case" (ibid., 189). He does not consider understanding-how as a relevant concept in epistemology – similar to what was noted above about the concept of knowing-how – as it is patently more closely related to practical concerns than to theoretical ones (see ibid., 190).

The feature of factivity seems to suggest that understanding and knowledge are somehow on a par, that is, from an epistemological point of view, understanding does not seem to contribute anything new. If I know that Pluto has five satellites, I do understand that Pluto has five satellites. Both – knowing that p and understanding that p – presuppose that p is true, as Kvanvig makes clear. Despite this initial similarity, however, he defends the claim that there is also an important distinction, namely:

> [...] once we move past its factivity [of understanding, N.M.], the grasping of relations between items of information is central to the nature of

understanding. By contrast, when we move past the factivity of knowledge, the central features involve nonaccidential connections between mind and world.

(Kvanvig 2003, 197)

Thus, from Kvanvig's point of view, understanding is independently epistemologically valuable because it involves the grasping of coherence relations amongst different true beliefs, and this grasping also contributes to the systematising and organising of our belief system (see ibid., 202).

Other philosophers defend conceptions of understanding that diverge even more radically from the concept of knowledge than Kvanvig's claims suggest. In particular, many scholars are less convinced of the factive status of understanding. Catherine Z. Elgin, for example, discusses understanding, amongst other things, in the context of scientific endeavour, and points out that it would not make much sense to demand factivity in this setting.

The growth of understanding often involves a trajectory from beliefs that, although strictly false, are in the right general neighborhood to beliefs that are closer to the truth. The sequence may terminate in true beliefs. But even the earlier steps in the sequence should fall within the ambit of epistemology. For they are, to an extent – often to a considerable extent – cognitively valuable.

(Elgin 2007, 37)

Her finding can be read both diachronically and synchronically. Diachronically, this affects the phenomenon of progress in science in general. Scientific realists point out that our current scientific theories are at least approximately true, and hence that we can think of the history of science as a developmental process towards truth. Of course, scientists might have deviated somewhat in the past (the phlogiston theory, for example), but nonetheless our theories and thus our knowledge of the phenomena, and also our understanding of them, has more or less constantly increased. Insisting on the factivity of understanding, that is, maintaining that understanding that p implies that p is true, would then expel such approximations to truth from the scope of 'understanding' completely, which would be a rather counterintuitive consequence.

A similar case can be made for the synchronic level in science. What is critically discussed here by philosophers is the usage of *idealisations* in cognitive processes in science (see e.g. Elgin 2007; Mizrahi 2012). Idealisations are, for example, brought about by laboratory conditions relevant to most experiments yielding observational data. Moreover, they also obtain due to making use of models or *ceteris paribus* laws in processes of reasoning. Therefore, idealisations are often the starting point for cognitive processes in science, though they are, strictly speaking, not true. They are simplifications of actual phenomena or processes. Particular aspects are intentionally omitted in these cases so that the amount of information is reduced. Can we nonetheless understand the (actual) phenomenon or subject matter in question? Our intuitions suggest that at least a partial understanding is possible, but a theory demanding the factivity of this cognitive achievement would deny this.

A corollary of what has been stated so far about the concept of understanding and its potential relation to factivity is that either we have to admit that images can yield propositional knowledge to be understood in an epistemologically relevant sense or that they simply do not belong to the scope of 'understanding'. Again, this seems to be a rather counterintuitive consequence of such an approach to 'understanding'. In this context, Elgin is correct in reminding us that "[w]e also understand pictures, words, equations, and diagrams. Ordinarily these are not isolated accomplishments; they coalesce into an understanding of a subject, discipline, or field of study" (Elgin 1993, 14). Although content with Elgin's statement that we use the term 'understanding' also with respect to those different vehicles of information, I suggest some clarifying remarks at this point. Earlier in this analysis, Scholz's suggestion was introduced that there is a variety of levels involved when speaking about 'understanding a picture' (see Scholz 1993). Bearing this in mind, we should at least make a distinction between the two levels of understanding implied in Elgin's quotation. On the one hand, we understand the respective vehicle of information (which might imply a variety of sub-levels of understanding, as pointed out by Scholz) and on the other, we understand the contribution that the content of this informational vehicle makes concerning the development of a theory or even, as Elgin states, "a subject, discipline, or field of study".

Anyway, both dimensions of understanding share the same difficulty if we focus on visual representations as the respective informational vehicles and nonetheless strive to maintain the claim about the factivity of understanding. As Elgin points out, images are not propositionally structured, and thus lack the capacity to bear truth values in the traditional Fregean sense (see Elgin 1993, 27). This provides an interesting twist to the starting point of processes of understanding. In the case of idealisations in science, the initial step consists in propositional statements that cannot be called 'true' in the strict sense, since they entail simplifications, etc. Visual representations, at least from the traditional point of view, are neither true nor false – nonetheless, as Elgin pointed out above, they can constitute the starting point for a process of understanding.

The thesis is, then, that by contributing to our understanding of scientific phenomena, visual representations fulfil an important epistemic task. Focusing on 'understanding' instead of 'knowledge' regarding the epistemic capacities of images in science allows us to support the claim that visual representations can play crucial roles in cognitive processes in this context by another argument. This line of reasoning runs as follows.

Visual representations are correctly taken as *heuristic tools* in this context, namely in the sense of supporting the cognitive process of learning. Yet, not only do they enable students to acquire propositional knowledge, but they also allow them to achieve an understanding of the information presented. Acknowledging the fact that understanding is an epistemic desideratum in its own right now enables a particular twist in the argumentation: whereas pointing to the heuristic function of visualisations usually implies the devaluation of their epistemic status, we can defend the opposite point of view. *If visual representations can facilitate understanding, and understanding is independently epistemically worthwhile, then it can be stated that images can make a substantial epistemic contribution.*

However, this line of reasoning presupposes (1) that understanding is epistemically worthwhile in its own right. So far, we have only pointed out that it is a distinct epistemic desideratum – but why should we strive for understanding? (2) We also have to show how visual representations can support scientific understanding. That they apparently do play an important role here is empirically supported by different studies carried out by educational psychologists (see e.g. Müller et al. 2012; Schnotz 2002; Vekiri 2002). Yet the question remains how it works. How exactly do images support scientific understanding? The remaining part of this section is devoted to answering these two questions.

The above examples of learning processes demonstrate that understanding adds an important epistemic dimension to our knowledge-seeking enterprises, and thus that epistemologists are well-advised to consider more seriously understanding as an epistemic desideratum in its own right. However, I do not agree with the broader thesis that the concept of knowledge should be replaced by 'understanding' in epistemology, because, although usually intimately linked, they are nonetheless quite independent of each other. As a corollary of this relationship, both concepts are relevant to consider when theorising about epistemic achievements, projects and practices. A start can thus be made by explaining what exactly understanding can add to the epistemic project in science.

Philosophers who consider understanding as relevant to epistemology commonly stress its additionally epistemic value which is spelled out in the grasping of certain connections in a body of information. Kosso expresses this benefit in the following way: "[u]nderstanding reveals the larger landscape and includes the ability to apply one idea to other situations without being given detailed instructions" (Kosso 2007, 176).

Here, Kosso points out two particular achievements of understanding: firstly, what he calls *revealing the larger landscape*, that is, understanding how bits of information in one's area of research fit together. For example, students in the philosophy of science might learn about Karl Popper's account on falsification first, and afterwards about Kuhn's theory of scientific revolutions. Learning in addition that these are successive approaches to explain (amongst other things) what demarcates science from pseudoscience will allow students to understand the sequential development of these theories much better, namely as a consequence of a shift in focus on the topic amongst philosophers of science. The second achievement of understanding that Kosso points out is that it enables scientists to apply their knowledge to answer new questions in their field of expertise. In this sense, understanding is an important goal of scientific education because, in the long run, it will enable students to do their own research.

From Kosso's perspective, this second epistemic accomplishment of understanding in science is a direct consequence of the first, that is, the understanding of connections (see ibid., 182). Earlier in this section, we mentioned Kvanvig's theses concerning the different kinds of relations possible here. Yet what *relata* do they connect? Here, Kvanvig only speaks vaguely about "various elements in a body of information" (Kvanvig 2003, 192). Does Kosso offer a more precise account in the context of science? He suggests that "[t]he achievement of understanding is in apprehending the connections between theories and the global coherence among concepts" (Kosso 2007, 179). Thus, the *relata* that he points out are, on the one hand, theories and, on the other, concepts.

Admittedly, understanding can be addressed as a particular phenomenon in both instances. However, clarifying the exact nature of the *relata* in question also depends on the degree of understanding that the recipient is supposed to acquire. It apparently makes a difference, for example, in the context of education whether students are required to understand a particular law and its applications or to grasp developments in the theoretical descriptions of a phenomenon. In particular, it is not always a set of theories that is linked in cases where understanding is attributed. Nevertheless, it can be taken that understanding is particularly valuable as it is based on realising the connections between, for instance, theoretical statements. One of the epistemic benefits that such a kind of understanding implies, without going into detail, is that it allows scientists to explain new phenomena by using metaphors[65] – which is also pointed out by Fleck. He explains how the invention of metaphors is made possible by the scientist's simultaneous membership of different thought collectives. More precisely, Fleck defends the claim that such a metaphoric reasoning is possible by making use of background knowledge taken from popular science (see Fleck 1979, 112).[66] The scientist is enabled to make this explanatory transition just by noticing the connection between two concepts or two theoretical approaches. Moreover, by introducing metaphors in scientific discourse we not only get epistemic access to a phenomenon that previously was not conceptualised at all, but it might also become the seed of a completely new theory. Here, the epistemic relevance of understanding becomes more than obvious. Now, in what sense can we state that visual representations support such understanding in science?

To answer this question it is appropriate to refer back to the results of educational psychology which explicitly deal with the contributions of visual representations in the context of learning. To summarise the results of the above discussion, those approaches suggested the following three ways in which visual representations influence the process of learning and how they can thus enhance understanding:

1 *Levels of information*: visual representations are not only able to transmit information about particular items, but also about relations among them.
2 *Visual deciphering*: visual information is predominantly grasped by making use of our perceptual apparatus. For example, comparative tasks amongst visually presented items can be performed in this way without the need to engage in long interpretations. Just by looking at a bar graph, the student may *simply see* the difference in length expressing a difference in quantity, etc.
3 *Cognitive processing*: visual information can not only reduce the cognitive load on the working memory, but also allows for a more economic hand-ling of our cognitive resources in general.

In the following, we have to analyse in what sense these epistemic virtues can not only support learning in general, but also understanding in science in particular.

The first point seems to be obvious. It was said that scientific understanding is about grasping connections – especially relations between concepts, theories and the like. As Kosso puts it: "[u]nderstanding [...] is entirely a matter of fitting into a pattern. Understanding depends on coherence" (Kosso 2007, 181). The patterns that Kosso mentions here can thereby appear on different levels. They can link individual concepts or statements, or individual concepts and theories, or theories to theories. Such patterns can support inductive and deductive reasoning, as they might, for instance, reveal relations of entailment or hierarchy. Moreover, those patterns can also link unknown items to already known ones – which seems to be the case that Kosso has in mind when talking about scientific understanding.

Now, visual representations can serve exactly this purpose. As Matthew T. McCrudden and David N. Rapp point out with respect to image design for educational purposes: "[a]n effective visual display as designed for educational purposes has two main functions: (1) to communicate important information and (2) to communicate relations about information via spatial arrangements" (McCrudden and Rapp 2015, online first). Thus, the obvious part to be played by visual representations is to show the connections mentioned by Kosso, i.e. to *literally visualise* them. Tree diagrams are a striking example in this context. Students are not only expected to learn something about particular items, but also about their relations. Visualisations can highlight such relations in an immediate fashion, and thus support the cognitive process of understanding.

McCrudden and Rapp discuss this contribution of visual representations to the epistemic processes of students' education under the labels of "organ-ization" and "integration". The organisation of information in a given image can be accomplished at different levels, and thus allows the learner to draw relevant inferences. The authors mention three kinds in particular, namely "temporal inference", "hierarchical inference" and "relational inference". The sequential depictions in Figure 3.3, for instance, will allow the learner to draw temporal inferences about the life cycle of a frog. The tree diagram mentioned in the previous example can allow for hierarchical inferences between concepts. And the diagrams of the detection of gravitational waves, discussed in the introduction to this book (see Figure 1.2), can illustrate the last kind of inference. Those diagrams allow the recipient to compare the two recorded signals to each other and to the curve theoretically predicted for such events.

In addition to those inferences, mentioned by McCrudden and Rapp, there are a variety of spatial relations among objects that a recipient can infer by regarding images. The *Pioneer plaque* (see Figure 2.10) can illustrate this. The depiction at the bottom shows the place of the spacecraft's origin, and thus how it is related to our solar system.

Integration is closely related to organisation. McCrudden and Rapp point out that what is implied here is the relation between newly presented items and the students' background knowledge. This seems to be exactly what Kosso suggests for the case of understanding, namely to connect theories, i.e. newly learnt ones to formerly acquired ones. Now, how can visual representations facilitate this epistemic achievement? McCrudden and Rapp explain that there are two ways in which integration can take place: an active and a passive one. In particular, the second one offers an explanation of how images can serve the purpose at hand. They can entail clues that will (or, in the case of education, are supposed to) activate the students' prior knowledge. Active integration, on the other hand, is guided by the student's expectation. That is, the learner assumes that the visually presented information is somehow related to a field of prior acquired knowledge.

Although McCrudden and Rapp have a point in highlighting the fact that visual representations can contribute both to the organisation and the integration of information in the processes of learning, it seems that the latter, i.e. supporting integration, is not a special achievement of visual representations in particular, but can be gained by other representational means as well.

It might be objected that the epistemic achievements just explained are only of a secondary quality, as they are intentionally brought about by the teacher who utilises the visual representation and thus presents the information depicted in a way that the above-mentioned advantages of organisation and integration can be exploited. Visual representations contribute to understanding, but only in the context of previously known facts. Hence, they cannot make contributions to constituting a genuinely new kind of knowledge and understanding, but are only vehicles that pass on known information.

We have to admit that images can be used in this way, which holds, incidentally, for all other kinds of representational tools as well. However, we do not have to agree with the thesis that this is the only way that images can make contributions to epistemic processes and to understanding in particular. The rationale for this claim is closely connected to the second aspect concerning how images can facilitate learning, namely *visual deciphering*. This perceptual mode of access also enables recipients to make use of correlated skills that have developed in the course of evolution. In accordance with this, Zachary C. Irving argues for a fundamental role of visual representations concerning scientific understanding (see Irving 2011). He discusses the difference between visual and numerical representations and highlights the fact that, because of the limitations of human cognitive capacities, the former are particularly useful for the understanding of large data sets. His primary example concerns scatter plots which, according to Irving, are especially useful for detecting patterns among the data (see ibid., 780f.). As an example let us take a look at the Hertzsprung-Russell diagram (see Figure 4.5), showing the correlation between the temperature and magnitude of stars. Obviously, by simply looking at the plot, we can literally see how magnitudes, temperatures and luminosities of stars are related and which of these relations are the most common.

I agree with Irving that some visual representations are especially valuable as they enable pattern detection among data. However, this merit should be related to our abilities as visual observers and our resulting cognitive setup, but I am reluctant to discuss this as a kind of cognitive limitation. That

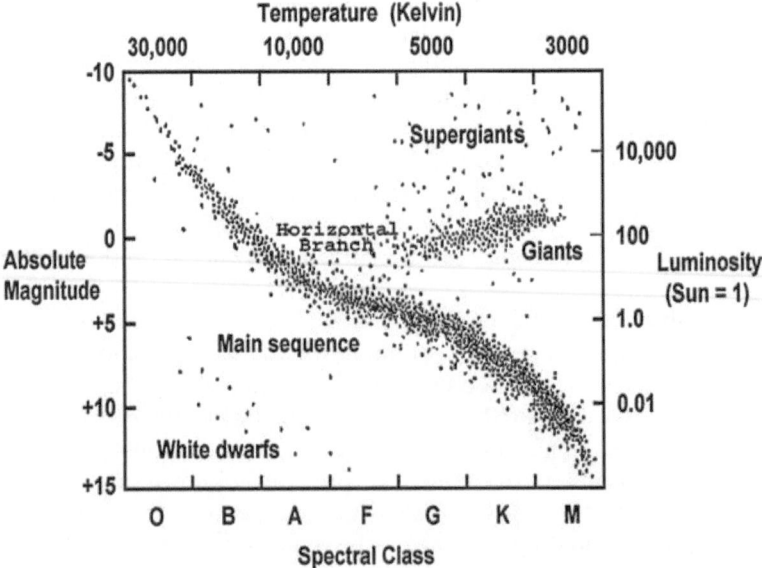

Figure 4.5 Hertzsprung-Russell (H-R) diagram.
Source: NASA/CXC/SAO.

human observers are particularly apt to perform this task is, for example, suggested by projects such as *Galaxy Zoo*. Such projects demonstrate, contrary to Irving's example, that human observers can detect patterns whose identification is simply not possible at the level of numerical data. That human beings are particularly skilled in the task of pattern detection is undoubtedly a consequence of evolutionary processes. Thus, making information available in a way that also activates these skills can enhance our understanding by connecting the cognitive processing of information to these abilities.

Consequently, the grasping of connections is not necessarily a result of a previously intended design. Contrary to such a sceptical approach, it can be assumed that not all relations detected with the aid of visual representations are previously known and this is because of the excellence of human observers in pattern detection. The capacity also to transmit non-conceptual information via images can contribute to such visual discoveries or detections of anomalies.

The final point concerning the ability of visual representations to enhance understanding is related to the aspect of *cognitive processing*. As McCrudden and Rapp explain, "[l]earners have limited processing resources. Of particular relevance to visual displays are the resources associated with attention and working memory" (see McCrudden and Rapp 2015, online first). What is important here is a more economic handling of those cognitive resources necessary for processing incoming information. Educational psychologists suggest that visualisations can constitute a kind of relief for our cognitive system which is achieved as follows.

On the one hand, visually presented information can guide our attention, as McCrudden and Rapp point out. In particular, by making use of 'signalling' techniques, important information can be highlighted in an image and the learner's attention can be directed towards it. The diagram of the yearly average number of sunspots is a good example (see Figure 4.6). An arrow has

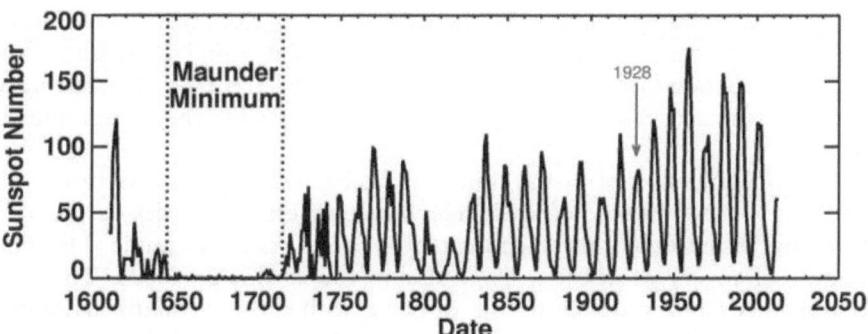

Figure 4.6 Diagram: sunspots.

Source: NASA. For a colour version of this image see www.nasa.gov/images/content/352130main_ssn_yearly_lg.jpg

been added to indicate the peak of 1928 that researchers considered being similar to the emerging cycle of sunspots in 2009. This signal directs the viewer's attention directly to the relevant information.

On the other hand, and more importantly, visual representations can keep information and relations among the data available while we think about problem solutions. We do not have to store all the information in our working memory during this process.[67] In this sense, visualisations might provide the necessary cognitive resources to work out the relevant connections in order to fully understand a particular topic. This method of supporting the cognitive processing of information by visual representations is, for example, highlighted by McCrudden and Rapp. They explain that images can, for instance, reduce the effort needed to select important information, as they include signals which make this information more salient. "Similarly, a display improves processing efficiency, when it helps a learner organize important information more quickly with the display than in its absence or if the display is not designed well (e.g., related ideas are not near one another)" (McCrudden and Rapp 2015, online first). All of these aspects of information transmission with the aid of visual representations can enhance scientific understanding – also without a prior translation into propositional statements. The discussion above has made clear *how images can facilitate scientific understanding*. Our analysis has made plain that images often provide more efficient ways to achieve understanding than other representational means, for example, by visually organising the relevant information and by literally keeping it before the recipients' eyes. However, although our previous results do support the thesis that visual representations can play positive epistemic roles by facilitating scientific understanding, they leave open the question of whether there are instances that inevitably call for the utilisation of images to allow students to understand what has been presented. Or to put it differently, do we have to admit that there are always other ways to achieve scientific understanding that work equally well?

There are different answers to this question. Firstly, as has been pointed out by educational psychologists, there are different types of students. Some prefer verbal descriptions, whereas others rely on visual representations. Thus, a moderate thesis would be to claim that the latter type of learners are somehow in need of images to get a clear grasp of what they are supposed to learn.

The phenomenon of visual thinking has been discussed in section 3.1.1 and the suggestion made that images can play essential roles as tools of thinking in the exploratory context of science. As an example, Dr John Snow's discovery of cholera transmission routes was analysed. This discovery was made possible by his working with different maps arranging various facts that could offer a possible explanation for the spreading of cholera among the local population. Here it became clear that it was by working with those maps (see Figure 3.1) that Snow finally understood how the disease was transmitted. Barbara Tversky highlights this capacity of visual representations as tools of thinking. She discusses Snow's case and points out that it is often the case that "[i]n science [...] the underlying phenomena generating the data are not always known" (Tversky 2015, 111). This was patently the case when Snow worked out his hypothesis by using maps.

On a more general level, Tversky claims that it is often due to the ambiguity of visual representations that they are such an effective means to develop problem solutions. She states that "[m]essy diagrams, then, can be crucial for thinking through problems, arousing and considering multiple possibilities. [...] Because ambiguity allows reconfiguration and reinterpretation, ambiguous sketches promote discoveries and inferences" (ibid.). Thus, what is commonly regarded as a shortcoming of visual representations, namely their apparent ambiguity, turns out to be particularly advantageous in the context of visual thinking.[68]

Cases of visual thinking, then, suggest clear instances where images are inevitable tools to achieve understanding. Yet this doesn't imply that images are indispensable in processes of achieving understanding more generally. It seems to be a mere contingent fact about the make-up of the reasoners' minds that they rely on images in their cognitive processes.

Secondly, there are cases where images are indeed indispensable to scientific understanding because they transmit information that cannot be expressed otherwise. At least some images are capable of transmitting non-propositional content, such as micrographs or depictions in guidebooks in biology. If this information is essential to understand the phenomenon at hand correctly, for example how to discern between different birds of prey during their flight, it seems that images providing this information can be called indispensable to the respective process of understanding. The example afforded by Perini is pertinent here. She points out that electron micrographs can "represent very complicated structural properties, even when there are no linguistic terms for the same features" (Perini 2005c, 921). That is, if the structures visible in such a micrograph are to be communicated, the utilisation of the image clearly becomes necessary since there are no proper terms available to translate those features. Moreover, if this communicative exchange finally leads to an understanding of the reasons for the functions of those structures, it becomes clear in what sense certain images can indeed be indispensable to the process of achieving scientific understanding.

As a concluding remark, we will discuss an objection to this claim put forward by Henk W. de Regt (see de Regt 2014). He agrees with the general idea that visual representations can facilitate scientific understanding. However, de Regt also claims that "visualization is a very effective way to achieve scientific understanding but it is *not* indispensable – there are other ways to reach the same goal" (ibid., 378, my italics). From his point of view, visualisability constitutes a theoretical quality that can unfold only in combination with particular skills of the working scientists (see ibid., 380). Moreover, the acquisition of these skills is part of educational processes within a particular community (see ibid., 393f.). In this sense, it is not a merely subjective characteristic of understanding but shared by members of a group. However, as a consequence of the necessity to learn depictive conventions beforehand, some scientists will not be able to profit from certain visualisations when trying to understand certain phenomena. They simply do not possess the relevant background knowledge to interpret them correctly, whereas others equipped with this information will benefit from the visual representation at hand. To illustrate his thesis, de Regt discusses the case of Feynman diagrams in physics (see ibid., 389ff.).

The educational process that de Regt mentions in this context coincides with Fleck's (and also Kuhn's, as de Regt mentions (see ibid., 394)) ideas about scientific education. His approach is similar to theirs in that he also discusses the necessity to work with experienced scientists or practitioners to learn to see correctly, i.e. to make correct observations in their scientific fields (see Fleck 1979, 54, 104; Fleck 1986b, 118).

De Regt is clearly right in pointing out that not all kinds of visual representations are equally intelligible to the human mind. The theory of image games allows us to acknowledge this fact without great difficulty. As there are varied people and varied intentions involved in such games, a layering of meanings can occur when visual representations are used in communicative contexts. To understand those image games correctly, we then have to have the relevant background knowledge about the contexts in which those images are used.

Moreover, it can be agreed that, as tools of communication, visual representations have been developed in the course of their application, and thus have been adapted to particular requirements of information transmission – just as it is also the case with all other kinds of tools of communication, for instance the usage of false colour photographs in astronomy. Here, scientists have to learn that those colours can, for example, indicate different altitudes (see Figure 2.12) or different kinds of radiation (see Figure 2.2). Thus, de Regt has a point in highlighting the fact that scientists have to acquire certain skills and background knowledge to interpret such images correctly.

However, and contrary to de Regt, I think that, as image perception is just a special case of perception in general, it allows us to utilise the cognitive advantages provided by the evolutionary development of our visual apparatus. We are able to grasp non-conceptual content via images and we are particularly skilled in pattern detection. The example of the discovery of gravitational waves, discussed in the introduction to this book, is relevant at this point. Although the observer might not be familiar with the particular style of depiction of the diagram presented as evidence of this discovery (see Figure 1.2) and the theoretical assumptions supplying the background to this depiction, she might nonetheless be able to see the similarity between both curves. Thus, a kind of basic information seems to be transferable by visual means that does not presuppose prior education in interpretation.

This way of arguing, however, implies that people somehow deprived of the possibility of visually accessing the world, that is, in particular blind people, will not be able to acquire certain qualitative dimensions of non-conceptual information connected to the visual mode of accessing them. Information about colour and brightness might be paradigmatic examples here, whereas spatial information might also – at least partially – be accessed by tactile or aural means.

4.3.4 Interim results: scientific images as a source of knowledge and understanding

Whereas in the previous two sections our considerations have mainly concerned the possibility of transmitting and acquiring propositional knowledge via using visual representations in scientific discourse, this last section is meant to broaden the focus.

From an epistemological point of view, especially when recalling the prevailing sceptical attitude of traditional analytical philosophers towards this topic, that scientific images can indeed yield propositional knowledge under certain circumstances is a result of some interest. Nonetheless, such a conclusion still does not seem to be an entirely satisfactory answer to Perini's question of why scientists use these images in the presentations of their results. If transmitting propositional knowledge were the only undertaking, linguistic expressions could clearly also be employed. Why therefore be concerned with visual representations? This laid out the framework in this last section to further analyse the cognitive value of scientific images. The question of whether additional merit is inherent in the practice of utilising visual representations in communicative contexts that cannot be dealt with by other kinds of representation or not is investigated at this juncture.

The approach to this topic is based on the analysis of the *process of learning* as a paradigmatic cognitive process. A successful accomplishment of this cognitive undertaking yields two separate epistemic desiderata: knowledge and understanding. Both are not only relevant in educational contexts, but also of major importance in scientific epistemic practices in general. The questions asked are whether and how visual representations can contribute to these epistemic achievements. Table 4.1 summarises the main results of the above analysis on this topic.

The focus on the cognitive process of learning chosen for this analysis permits results obtained in empirical case studies in educational psychology and adjacent disciplines such as cognitive science and neuroscience to be taken into account. Amongst other things, it is made clear that certain constraints on the epistemic efficiency of images are a consequence of the recipients' cognitive set-up. That is, the characteristics of students (for example, background knowledge, preferred form of information presentation, etc.) are relevant considerations, as are the challenges posed by the interplay between different representational means deployed to transmit information – the problem of

Table 4.1 The cognitive value of visual representations in processes of learning

Cognitive achievements	Potential cognitive achievements via visual representations	Capacities of visual representations enabling these achievements
Knowledge	Propositional knowledge (knowing-that)	Perceptual mode of information access (presupposition: possession of relevant concepts)
	Practical knowledge (knowing-how)	Perceptual mode of information transmission allows communication of visual demonstrations
	Phenomenal knowledge / knowledge by acquaintance	Transmission of non-conceptual content supports mental images
Understanding	Grasping of connections between concepts and theories	Perceptual mode of information transmission can trigger retrieval of relevant background knowledge (dual coding theory)
		Perceptual mode of information transmission makes the display of relations and computational advantages possible (visual argument hypothesis)

integration (see e.g. Renkl and Scheiter 2015, online first). Thus, the results obtained in educational psychology made particularly plain that it is not exclusively due to the kind of representational means deployed in the context of learning, but to other contextual matters as well, that the latter can be effective in the way intended.

Moreover, the above discussion of two theoretical approaches, namely the dual coding theory and the visual argument hypothesis presented in detail by Vekiri (see Vekiri 2002), not only illustrates that visual representations can indeed have a positive impact on learning processes, but also reveals insights into how exactly they influence the recipients' cognitive apparatus to enable this positive outcome. The main aspects are summarised in Table 4.1.

At first sight, it might seem puzzling that the results obtained in educational psychology are only attributed to the facilitation of understanding, but that this first impression might somehow be misleading is due to the following reasons.

Firstly and admittedly, the capacity to offer information about individual elements and their relations, illustrated by proponents of the visual argument hypothesis, also affects the acquisition of knowledge, in particular knowing-that. In this sense, their results are also of relevance to this epistemic achievement of learning. As these kinds of contribution of visual representations to knowledge acquisition are, however, implied by the recipients' perceptual way of accessing the encoded information, they are not mentioned separately in the table. A similar point has to be made about the possibility of storing visual information in the human brain, a thesis entailed in the dual coding theory. This clearly belongs to the capacity of images to transmit non-conceptual content. This is why it is not listed as a separate point. Furthermore, the remaining aspects of the processing of visual information and its cognitive merits seem to belong rather to the facilitation of understanding than to the mere amassing of units of information.

Highlighting the fact that images often contribute to the achievement of scientific understanding does not imply a devaluation of the epistemic capacities of visual representations. This has been one of the main arguments presented above. On the contrary, the thesis was defended that this is one of the epistemic merits that make visual representations epistemically worthwhile when used in scientific discourse. Philosophers such as Elgin, Kvanvig and Kosso are right in pointing out that understanding bears an intrinsic epistemic value. Examples show convincingly that something important is lacking if knowledge is acquired without understanding. Thus, giving substance to the claim that visual representations can bring about this valuable epistemic desideratum, as the theses of educational psychologists based on and proven by empirical studies do, is of great importance to the project of correctly explaining the epistemic status of scientific images.

The second epistemic merit resulting from the above analysis is related to the epistemic desideratum of knowledge. In addition to the capacity to convey propositional knowledge under certain circumstances, visual representations can also transmit practical knowledge, i.e. a kind of knowledge usually regarded as not communicable and, therefore, as epistemologically inferior to propositional knowledge (see Grundmann 2008, 86). Moreover, showing that visual representations can be used to transmit this kind of knowledge suggests that some standard of correctness can be applied to it. In this sense, the second objection mentioned by Grundmann, namely that knowing-how is not related to truth in any epistemologically relevant sense, can be called into question. Our rejoinder to this objection has been to show, by again drawing on results from educational psychology, that there are patently at least some paradigmatic instances of a correct and successful application of practical knowledge that can be transmitted via visual representations. One purpose of empirical studies in educational psychology also consists in an approach to these optimal ways of visually presenting certain information. Images are constantly subjected to redesigning processes to improve them for educational purposes. Thus, the standards of correctness suggested for practical knowledge also bear on image design and its subsequent development in educational contexts.

Beyond that, it has been argued that using visual representations to transfer practical knowledge also considerably increases the number of recipients. Direct demonstrations of knowing-how usually only reach a small group of people. Making use of images, on the other hand, not only allows the transmission of knowing-how on a synchronic level to a broader audience, but also on a diachronic level, as they can e.g. be published in textbooks or distributed in electronic learning courses and employed in students' education for several years.

Finally, an additional epistemic merit that visual representations can occasion is to provide recipients with concrete mental images. These images can be mentally manipulated to search for solutions to problems, as proponents of the visual argument hypothesis claim. Closely related to this point is also the fact that mental images play a significant role in that they enable counterfactual reasoning in science. Weighing and deciding between alternatives presupposes a clear understanding of the facts at hand – and, as has been shown above, sometimes this understanding cannot be achieved properly without visual representations that might also provide the scientists with a clear visual conception, i.e. a mental image, of the entity in question. Last but not least, Jung makes us aware that these images can also be used as tools for mental training to improve the deployment of skills obtained later (see Jung 2012, 181).

All of these virtues are based on the fact that we make use of our visual perceptual apparatus to cognitively access information presented in scientific images. This way of deciphering the content allows us to draw on special skills related to vision, such as pattern detection, and we can gain access to non-conceptual content transmitted by visual means.

None of the three epistemic achievements discussed above, that is, the facilitation of scientific understanding, the transmission of knowing-how and the conveyance of mental images, can be properly attained if we make use of other representational means in scientific discourse. These are then the three merits that answer Perini's question about why it is that scientists use visual representations in the explanatory context of science.

4.4 Summary

In this chapter, the discussion has been on the question of the exact epistemic status of visual representations in the explanatory context of science. The investigation has focused on a comparative task: how the epistemic status of scientific images in comparison with other representational means – chiefly with linguistic expressions – is to be conceived.

The starting point of the analysis was a detailed discussion of Perini's theses on the topic. Her analytical approach to the subject matter at hand paved a way to revealing the diverse epistemic threads connected to the usage of visual representations in scientific discourse. One of the first aspects discussed in detail was the possible role of images in scientific arguments. Here, Perini insisted on their independent epistemic relevance. She made us aware of the fact that, although scientists treat visual representations as proper parts of scientific arguments, philosophers of science tend to neglect their epistemic value.

The next step was to get a better understanding of the nature of the philosophical problem lurking in the background and usually put forward as a rationale for this neglect in the philosophy of science. It became clear that at least two closely related aspects play a role in this. On the one hand, Steinbrenner points out that analytical philosophy is largely focused on language. On the other hand, this focus is still being passed on in the development of this philosophical school, in particular by maintaining Frege's theses on thoughts – later called 'propositions' – as the truth-bearing contents of sentences that can be transmitted and be subjected to logical operations such as negations. Having initially set too narrow a focus on language might have tempted some philosophers to regard propositions themselves as language-like entities. However, even if we are not trapped within this narrow-minded framework, Frege's argument that only propositions are proper candidates to bear truth values makes it rather difficult to maintain the claim that visual representations can make serious epistemic contributions in the context of scientific arguments.

Perini's solution to this problem is based on Goodman's theory of depiction. His account constitutes the framework that she employs to clarify whether images are indeed bearers of truth values and therefore permissible components of scientific arguments. Earlier in this book, however, Goodman's account has been rejected as a suitable approach to theorising about scientific images. Consequently, my solution diverges from the one suggested by Perini, although her analytical approach to the topic motivated my own reasoning.

In the following, we analysed (a) whether visual representations possess a cognitive content of some kind, i.e. a content that can be transmitted via images in communicative contexts, and (b) we investigated the precise nature of such a presumed content. Maynard's ideas on this topic were utilised as a guideline for this investigation. He emphasises that it is counterintuitive to point out, on the one hand, the intimate relation between the development of images and language in human prehistory and to argue, on the other, for an inferior status of images in comparison to language today. Maynard urges us to reconsider the question why it is that we willingly admit that linguistic expressions can transmit thoughts, but are reluctant to allow visual representations to fulfil the same epistemic task.

Taking his challenge seriously, the suggestion was made to pay closer attention to the fact that our predominant way of accessing visually presented information is via perception. The question of what kind of information we are able to process cognitively via perception then seems to be related naturally to queries about the kinds of cognitive content transferable via scientific images. With this respect, two findings from epistemology were highlighted. On the one hand, contemporary philosophers usually regard perception as an epistemic source, yielding propositional knowledge. On the other hand, that we are able to perceive entities for which we do not have concepts, particularly manifested by Hentschel's example of the first observations of the granulation of the solar surface, makes it plausible to claim that the content of perception can be non-conceptual in kind. Contrary to Grundmann, who took this to support the thesis that it is only non-conceptual – and, thus, non-propositional – content that can be transmitted via perception (see Grundmann 2008, 492f.), I argued for a more moderate claim, namely that the content of visual perception can be both propositional and non-propositional in kind. This seems to be possible because we acquire concepts by a mixture of testimony and perception. Initially we receive non-conceptual content via perception. This process is enriched in the course of our life by the concepts we acquire, which thus make the acquisition of propositional knowledge via perception possible, and constitutes a long-term development of our perceptual skills.

The analyses presented in sections one and two of this chapter made plain that scientific images can indeed yield propositional knowledge. Yet inquiring into the perceptual mode of accessing information transmitted by visual means also revealed further merits that can be brought about when using visual representations in scientific discourse. By taking into account that our perceptual apparatus has been shaped by evolutionary processes, it can be called our predominant way of cognitively accessing the world. Here, not only were our abilities of pattern detection and colour differentiation pointed out, but also that we can acquire non-conceptual content by perceiving the world.

The epistemic advantages of pattern detection for scientific purposes are immediately clear. Moreover, the human ability to grasp non-conceptual content via perception also suggests another way in which visual representations can contribute to scientific arguments. Non-conceptual content can be particularly relevant when images are regarded as evidence of observational results. Here, another epistemic merit of scientific images became salient: they can support the detection of anomalies to prevalent scientific hypotheses, because they can entail content that is not already conceptualised in the vocabulary of the theory at hand.

The epistemic merits employed by our perceptual apparatus to cognitively access visually presented information also laid out the framework for the last section of this chapter. Whereas the previous discussions suggested that some visual representations can indeed have parity with linguistic expressions when it comes to the transmission and acquisition of propositional knowledge, we finally inquired into the possibility that sometimes scientific images can exhibit epistemic merits that cannot be achieved by using other representational means. The discussion centred around the process of learning as a paradigmatic cognitive process, and that successfully accomplishing the process of learning yields both knowledge and understanding.

The contributions that visual representations can make to bring about these two epistemic desiderata were examined. On the one hand, because of the capacity to transmit non-conceptual content and the perceptual mode to access visually presented information, scientific images can supply recipients not only with propositional knowledge, but also with knowing-how and mental images. Thus, not only can some images transmit kinds of knowledge usually regarded as incommunicable (see Grundmann 2008, 86), but they also allow us to bring the information thus gained to a cognitive effectiveness relevant in scientific reasoning, such as the manipulation of mental images to work on solutions to problems. On the other hand, by taking into account empirical results from educational psychology and adjacent academic disciplines, it became comprehensible in what sense images can contribute to scientific understanding, defined as the grasping of connections between concepts and theories as propounded by Kosso and others (see Kosso 2007, 179). Moreover, by acknowledging that understanding is intrinsically epistemically worthwhile, another important epistemic merit of visual representations, namely to facilitate understanding by being integrated into scientific discourse, was thereby revealed.

To be concise, this analysis not only demonstrated how scientific images can contribute to scientific arguments and yield propositional knowledge, it also made clear in what sense at least some visual representations in certain contexts can facilitate epistemic achievements that are not attainable via other representational means. Of course, only a claim about possibilities shall be defended here. Visual representations and also other vehicles of communication are subjected to certain constraints when realising their epistemic effectiveness. As we learnt from educational psychology, learner characteristics play an important role in this respect, as does the design of visual representations. Not all ways of visually presenting information are likewise appropriate to any epistemic context. Admittedly, as visual representations have become a tool of expressing human thoughts – 'thoughts' here in a broad sense – they have also become subject to developmental processes of refinement and modification, as linguistic symbols have also. Visual skills have to be trained to decipher correctly the many different ways of visually presenting information.

Having pointed out that visual representations can indeed be epistemically effective, and acknowledging the fact that they play an increasing role in science and society, some new responsibilities for scientists that appear in this constellation will be discussed briefly in the next chapter.

Notes

1 Taking ethics of science into account, we should add that it is not only epistemic practices that are subjected to such evaluations and improvements, but also practices that give rise to moral questions.
2 It has to be added, however, that the considerations on justificatory reasons and practices, presented in the previous chapter, already exceeded a merely descriptive approach. We not only investigated how scientists *do* justify their epistemic practices, but also discussed the question of how they *should* justify their practices. Thus, the normative dimension was already inherent to our previous line of reasoning.

3 The model of information transmission used in the following discussion is based on related discussions in social epistemology, in particular in the debate about knowledge by testimony. For an introduction to the "transmission model of testimony" see Gelfert (2014, ch. 7).

4 This focus on knowing-that in epistemology, however, has not stayed uncontested. Eva-Maria Jung, for example, tries to make a case for including also "practical forms of knowledge", as she calls it, referring thereby to knowing-how (see Jung 2012, ch. 1.5).

5 This argument is presented in a similar fashion by Maynard (2009, sect. 2).

6 Relevant discussions of the topic are especially connected to attempts at establishing a picture language, as mentioned in section 2.2 of this book.

7 This is a translation of his paper *Der Gedanke. Eine logische Untersuchung* published in German in 1918/1919 (see Frege 1993, 30ff.).

8 It has to be added that not all philosophers of science would agree that *truth* is the (ultimate) goal of science. Usually it is scientific realists who defend a claim along these lines, although most of them admit that other epistemic desiderata and also practical concerns might play a role. Thus, Ilkka Niiniluoto points out that there is no reason to assume that the cognitive aim of science has to be "one-dimensional", that is, exclusively directed towards truth (see Niiniluoto 2015, ch. 2.4). Moreover, scientific anti-realists such as Bas C. van Fraassen explicitly deny that truth is the goal of science. Van Fraassen claims that scientists' sole aim in their epistemic endeavours is "empirical adequacy" (see van Fraassen 1980, 12).

9 He adds however that the captions of pictures can indeed be true or false, and discusses some interesting examples in this respect (see Gombrich 2004, 59ff.).

10 Contrary to my suggestion, Richter accepts Goodman's theory as the guiding model to make sense of the concept of scientific visualisations and tries to modify Perini's work within this framework (see Richter 2014, ch. 4.2.2ff.). Having made my reasons sufficiently clear why I think that Goodman's theory of depiction is not a suitable means to make use of in this context, Richter's approach will be put aside without further discussion.

11 In logics, *arguments* are usually considered to be purely linguistic in kind (see Salmon 1983).

12 Meanwhile there is an extensive branch of studies flourishing on what might be called 'diagrammatic reasoning', some of them with an even more wide-ranging interest in diagrams and their role in epistemic reasoning than merely considering those mentioned above (see e.g. Giardino and Greenberg 2015; Shin, Lemon and Mumma 2014; Shin 2015; Stenning 2002; Tversky 2011 and references therein). Moreover, there are investigations into the possibilities of establishing completely visual logics (see e.g. Bagusche 2012).

13 A fascinating example is Oliver Byrne's book dedicated to the attempt to offer complete visual demonstrations and proofs of Euclid's theorems – also meant as an attempt to introduce this complex material without using linguistic expression, and therefore supposed to be more easily comprehensible than a linguistic tractate on the topic might be (see Byrne 2013). Again, a lively debate also started about the relevance of visualisations in mathematics (see e.g. Brown 1996; Mancosu, Jørgensen and Pedersen 2005; Sazdanovic forthcoming and references therein).

14 Richter mentions this as option (iii) of how images can play a role in scientific argumentation (see Richter 2014, 159).

15 IUPAC is an abbreviation of 'International Union of Pure and Applied Chemistry'.

16 Unfortunately, Scholz does not explain why he makes a distinction between contexts of usage and cultural contexts. It might be objected that cultural contexts are simply descriptions of contexts of usage. Ethnologists might offer explanations as to why we make use of certain images, and these explanations could then be called 'cultural contexts'.

17 Some of the ideas that are presented in this section are also discussed in Mößner (2013b).

18 Ana Laura Nettel and Georges Roque call this clear-cut distinction between persuasion and argumentation, between *pathos* and *logos*, into question. In particular, they doubt that the role of visual representations can exclusively be characterised as a kind of visual rhetoric. Contrary to this, they point out that although visualisations are often regarded as emotionally influencing, these representations can nevertheless play an argumentative part (see Nettel and Roque 2012, 67).

19 This is also the way Vögtli and Ernst term the epistemic role of visual representations in argumentative contexts, in particular the role of visualisations produced by scientific instruments (see Vögtli and Ernst 2007, 36).

20 A similar debate takes place in the philosophy of film (see Wartenberg 2011b, 12f.). Can what has been visually presented in a film be translated or paraphrased? One might argue that, as films are audiovisual media, the verbal argument can simply be peeled out of the mix. Yet this is not what the philosophy of film is about. Wartenberg makes this clear when stating that "[t]he crux of the debate is whether films within the standard genres of filmmaking – from fiction films to documentaries and even avant-garde films – can actually do more than raise a philosophical question or record a philosophical argument, whether some films should really be counted as doing philosophy on their own" (Wartenberg 2011a, 551). Proponents of film philosophy try to show that some films are capable of philosophising by filmic means only. That is, films are said to make significant contributions to philosophical debates. These contributions are transmitted in the audiovisual style of the medium itself and cannot be translated into other representational means. Stephen Mulhall, for example, defends this ambitious thesis (see Mulhall 2008).

21 An overview is offered by McGrath (2014).

22 Laurence BonJour (2013) offers an overview on different epistemological approaches to this topic in BonJour.

23 For more information about the space mission see www.nasa.gov/mission_pages/newhorizons/main/index.html, accessed May 24, 2016.

24 The topic of understanding as an epistemic desideratum in science will be dealt with in more detail in section 4.3.3 of this book.

25 Abstract paintings, for example, do not belong to this set of visual representations in Sober's sense. Neither does he extend his theory to maps, diagrams, or graphs that mix different representational modes. Blair's account is thus much broader in scope.

26 This point is also highlighted by Dieter Mersch who argues that, although images can show entities (for example, people and their actions), they cannot show a negative state of affairs (see Heßler and Mersch 2009, 19ff.).

27 At least not without the aid of further symbolic auxiliary means, such as crossing out the content of a particular image to express that what it shows is not the case.

28 A discussion of recent approaches to this topic is provided by Gendler and Hawthorne (2009); Nanay (2014) and Schantz (2009).

29 Scholz (2004) offers an insightful discussion of the common metaphor of *epistemic sources*.

30 Philosophical sceptics have to be mentioned as an exception here, as some of them do not regard perception as a reliable source of information about the external world. However, scepticism seems to be a philosophical problem in its own right that needs not be restricted to particular epistemic sources, it being about the *possibility of knowledge* in general. More information about how scepticism is related to perception is offered by Grundmann (2008, ch. 7.1).

31 Eric Margolis and Stephen Laurence (2014) discuss different approaches to the concept of concepts.

32 Further reasons are discussed by José Bermúdez and Arnon Cahen (2015, sect. 4.1).

33 It has to be added that Peacocke does not speak about 'representation' in this context, as he takes this to involve figurative meaning. That is, the image of an aurochs in the Altamira cave can also be meant to represent success in hunting, not a particular animal. Such instances of a more wide-ranging meaning attribution via representation, Peacocke explicitly wants to exclude from his theory of depiction (see Peacocke 1987, 383). Thus, when speaking about 'representational qualities' in the following discussion of his ideas, I will use the concept in accordance with those constraints that result from his theoretical approach.

34 Admittedly, this visual experience has to be repeated several times to learn the relevant characteristics.

35 Artists, particularly fresco painters, are confronted with the problem of how to trick the human eye in a way that their paintings, situated on ceilings etc. and therefore often regarded from unfavourable angles, acquire their appropriate proportions nonetheless. Here it is the reverse, that is, spatial properties of the frescoes are willingly distorted so that their shapes appear without distortion in the recipient's perceptual experience of them. Another example of tricking the eye in the domain of art is *anamorphoses*.

36 With regard to this difficulty see in particular his explanations in Fleck (1986d).

37 The topic of anomalies in observations is most famously discussed by Kuhn, who suggests that anomalies constitute bones of contention that will finally necessitate a shift in the prevalent paradigm (see Kuhn 1996, ch. 6).

38 I have discussed this in more detail with regard to the training of journalists and their abilities to check their sources of information for accuracy.

39 The following website offers access to more than 40 projects of this kind, see www.zooniverse.org, accessed January 11, 2016.

40 What decisions have to be made by the user in her classificatory task is visualised in the decision tree produced by Coleman Krawczyk (University of Portsmouth), see https://data.galaxyzoo.org/gz_trees/gz_trees.html, accessed January 12, 2016.

41 Marcel Boumans (2016) also discusses the relevance of human vision or, as he puts it, of the eye as "a reliable tool for judgment" in science.

42 As an interesting spin-off, it can be added that this interrelationship between the indicative colouring of potential prey and the predators' ability to perceive those colours triggered another evolutionary process which somehow aims at undermining the initial effect. Harmless potential prey have developed visual

characteristics of dangerous or poisonous species to lead their predators astray. Hoverflies, for example, look like wasps.

43 Schawelka also discusses these phenomena, subsumed under the label of 'reflectance'. He thereby refers to the ability of objects to reflect, absorb or transmit photons. Which of these abilities are realised depends on two aspects, namely (1) the structure or texture of the respective object and (2) the energy of the respective photon, i.e. the wavelength of the light (see Schawelka 2007, 45).

44 Of course, it is not only in regard to colour perception that scientists have tried to overcome the limitations of the human eye, as the astrophysicist Thorsten Ratzka, for example, has pointed out in his retrospective description of human observations of celestial objects (see Ratzka 2012).

45 His main examples are drawings, but he also applies his analytical results to other kinds of images, in particular to photographs and diagrams.

46 An overview of different accounts in the context of argumentation theory is presented by Jens E. Kjeldsen (2015, ch. 2).

47 Downes mentions this line of reasoning in his epistemological analysis of scientific images and attributes it to Pierre Duhem by stating that it was "[...] Pierre Duhem's position that it was only human weakness that resulted in visual aids being required in the service of science" (Downes 2012, 117). Obviously, Downes hereby refers to Duhem's theses on the two kinds of the human mind which the latter presents when discussing the relevance of models in science (see Duhem 1998, ch. 4). From his point of view, it is due to a missing capacity of reasoning in abstract terms that some people need models or visual means of another kind to imagine the relevant relations in order to understand physical laws and theories correctly.

48 I thank Johann Marek for helpful discussions of this topic.

49 Some of the ideas presented in the following section have been published in Mößner (2015).

50 Of course, there may also be practical elements involved, for instance the practising of certain skills to acquire the relevant expertise for deploying them later on.

51 A multiplicity of ways to process and encode visual information mentally is also suggested by some of the case studies presented by the neuropsychologist Oliver Sacks. In *"The Mind's Eye"* (Sacks 2010, 202ff.), for example, he discusses the medical histories of different people who went blind in the course of their lives. Interestingly, some of them kept their ability to construct mental images, that is to visualise objects, in their minds, whereas others totally lost this ability. These latter patients were nevertheless able to learn empirical facts about their environments with the aid of their other senses. These medical case studies speak in favour of the thesis that there are at least two different cognitive subsystems for processing and encoding information, which can also be used separately if one of the systems is damaged or takes on new tasks from other parts of the brain.

52 The advantage of visual representations to trigger visual imagings that can be further manipulated is also discussed by Max J. Kobert who offers some helpful examples (see Kobert 2010).

53 Of course, there are innumerable design guides offering advice in this respect. In our previous discussion we have already learnt, for instance, about Frankel and DePace's suggestions (2012).

54 Michael P. Verdi and Raymond W. Kulhavy, for instance, mention *gender* as a relevant factor to consider when theorising about the educational efficiency of maps

(see Verdi and Kulhavy 2002, 33f.). Schnotz points out that the recipient's *age* can play a role (see Schnotz 2002, 113).

55　Regarding map comprehension, see e.g.Verdi and Kulhavy (2002, 33).

56　Hausken's and Roskies's analyses of fMR images in medical practices show nicely what happens in interpretive processes if people (experts or laypeople) wrongly assume that they are familiar with the results of new imaging technologies (see Hausken 2015; 2017; Roskies 2007; 2008).

57　Actually, he mentions three reasons. The possibility of cognitively processing knowledge can be regarded as a reason on its own.

58　Jung discusses a variety of these strategies in detail (see Jung 2012, ch. 1.4.3.1 and 1.4.3.2).

59　Personal conversation with Max Hoffmann.

60　There are different variants of this disease. Some can be so severe that the patient even loses her ability to understand linguistic expressions.

61　A critical discussion of Jackson's arguments, objections and rejoinders is offered by Martine Nida-Rümelin (2015).

62　Again it has to be added that, in particular, the student's background knowledge plays a major role with this respect. It might fill the gaps left open by the description at hand so that a proper mental image can nonetheless be construed by the learner. A similar case can be made for images. The information transmitted by them might be as incomplete as the information transmitted by a linguistic description. Consequently, images serving educational purposes have to be carefully chosen.

63　The most famous ones are discussed by Michael Huemer (2011, sect. 3).

64　For a critical discussion of such approaches see e.g. Koppelberg (1993).

65　A compilation of important works on the topic of metaphors is presented in Ortony (1998).

66　I discussed this point in more detail in Mößner (2013a).

67　In this sense, images might be regarded as a kind of extended memory system, so to speak, though I do not want to relate this to the debate about the extended mind here. An introduction to this debate is offered by Holger Lyre (2010).

68　Further examples of visual thinking in scientific and technological practices are discussed by Hentschel (2014, ch. 10).

5 Outlook

New responsibilities?

At first glance it might seem puzzling to finish a book dealing with epistemological problems of visual representations in science by pointing to responsibilities, which is usually thought of as an ethical topic. Nonetheless there are indeed good reasons to choose this way of concluding my study on the epistemology of scientific images.

A start can be made by briefly summarising the main results of the above analysis. This recapitulation will enable us to comprehend why highlighting the responsibilities of scientists in this context not only seems to be a reasonable conclusion, but also even a mandatory consequence of the results of this investigation. However, only some motivating reasons will be presented here that guided the decision to stress this point about responsibilities in science, but they will not be spelled out in detail. The latter task is left open to further research on this topic.

In a nutshell, this study was meant to shed some light both on the concept and on the epistemology of visual representations in science. The analysis took three steps, one with regard to conceptual and two with respect to epistemological questions.

The suggestion about the conceptual issue (1) was to think of scientific images as *signs* suitable for storing and transmitting information. But how can the encoding and decoding of visual information be explained best? Here, the analysis of types of theories of depiction allowed us to single out Newall's mixed theory of depiction as a suitable explanatory account. Newall proposes to take into consideration on the one hand non-veridical seeing as the particular way of decoding information presented via images, and on the other, he points out that there are different standards of correctness due to the diverse ways of visually encoding information considered relevant for achieving the resulting interpretation of the image in question. The fact that visual representations in science are produced in a variety of ways – with and without the aid of technical devices and instruments – has made Newall's mixed approach particularly useful as a template to theorise further about the concept of scientific images. Its allowance for different standards of correctness enables us to fill in the diverse ways of visually encoding information in this context.

The decoding of visual information has played a crucial role at different stages of my investigation at hand. Its basic set-up, however, remains the same throughout. The ability to cognitively work with images seems to be an offspring of our capacities for visual perception. It has been suggested to take resemblance relations as informing our cognitive practices of decoding visual information in science at the fundamental level. Resemblance relations make plain the connection between ordinary perception and image perception. We understand visual representations more easily than any other kind of representational means because of this initial connection to perception. It is via resemblance relations that we can exploit their perceptual basis to make images a vehicle of information. However, as images are broadly used in scientific communication, increasingly more conventions have been developed – hand in hand with the different technologies to produce images – telling us how to encode and decode those images correctly. Here, the relevant point was to make clear that in the same way that we modify and refine the stock of other symbolic means, such as natural language, so do we also develop visual representations due to the demands appearing in their contexts of usage.

Although this first part was predominantly about clarifying some conceptual aspects, the discussion already highlighted certain features of those visual phenomena that are also of relevant consideration in the epistemological setting. These points were connected to the question in what sense visual representations can transmit information about their object of depiction – a characteristic most relevant to their usage in scientific epistemic processes. This question has been considered more thoroughly in the second part of my analysis.

Here (2), the different functional roles were examined that can be fulfilled by visual representations in scientific processes. As diverse as scientific images are with respect to their appearances, they also serve a variety of different functional purposes – some epistemic, some non-epistemic. The first suggestion, then, was to make a distinction between two different main contexts of usage, namely the exploratory and the explanatory contexts, to deal with this diversity. This differentiation not only allowed more precision in determining what particular functions visual representations fulfil in science, it also made clear that many reservations held by philosophers about the epistemic capacities of scientific images are a result of their confusion of those contexts, thereby allowing the transference of problems from one domain to the other.

The functional roles of visual representations that we identified in the exploratory context were the following: images can serve (a) as the object of research, (b) as a surrogate for the object of research and (c) as tools of thinking. In the explanatory context, the concrete functional roles depend on the purposes of communication and the communicating agents involved. Transmitting information to the interested public or to funding officers puts the focus on rather non-epistemic functions of images, such as attracting attention, than on epistemic ones.

Moreover, contrasting functional roles of visualisations in the exploratory context with their roles in the explanatory setting not only helped to put a clearer focus on epistemological problems arising in these contexts, but also made it possible to suggest more precise answers to the question of what might justify the epistemic practices of scientists making use of visual representations. In the explanatory setting, we can make use of justificatory models developed in the context of the debate about knowledge by testimony. In the exploratory context, background knowledge played a significant role to justify knowledge claims based on visual representations. What was relevant to know in this context was, for example, how reliably the instrument producing those images works; that is, knowledge concerning the causal connection between instrument and specimen is a relevant consideration. Furthermore, the informativeness (for example, resemblance) of the respective image defined by a function mapping the results has to be taken into account. Finally, coherence considerations about the fitting of the results obtained into the broader theoretical network at hand can be of significance.

Analysing the functional roles of visual representations in accordance with the two contexts had already illustrated that social processes have a crucial influence on epistemic roles played by images in science. Thus, we took into account Fleck's theoretical considerations on communication processes in science to explain the diversity of appearances and the functions of images in this context.

Finally (3), the epistemic status of visual representations was discussed, particularly in comparison with other representational means such as linguistic expressions. The question that has been considered at this point was whether images can make independent contributions to epistemic processes. This investigation was particularly motivated by Perini's theses on the topic. She made plain that scientists regard visual representations as proper parts of scientific arguments. This epistemic practice raises the question of whether images can transmit propositional knowledge although their content is apparently not propositionally structured. Perini also points out that philosophers often regard this as such an unsurmountable problem that they simply deny that images can play any relevant epistemic role – a claim that clashes sharply with scientific practices. These conflicting assessments of the epistemic capabilities of scientific images were the starting point of the analysis in the third part of this book.

What became particularly important to the discussion of images as components of scientific arguments was a clarification of the exact nature of the cognitive content encoded and transmitted by visual representations. Taking as one of the main characteristics of propositions the idea that they are expressible in linguistic statements, but are not linguistic statements themselves, the question about the nature of the cognitive content of visual representations in science came down to the issue of whether the epistemic subject has the relevant concepts at her disposal to formulate such propositional statements. The suggestion in this regard was to reason by analogy with how we acquire knowledge via perception, as this is the basic way we decipher images – contrary to the thesis that we always have to *read* them, which presupposes the prior acquisition of related skills (see Scholz 2009b, 41ff.).

This line of reasoning made clear that both ordinary vision and image perception, in particular initially, yield a non-conceptual content which can be transformed into a conceptual one if the relevant concepts are later acquired by the epistemic subject. Here it became obvious that an analysis of the capacities and characteristics of epistemic sources should pay due attention to their mutual interactions (see Scholz 2009c). That is, the acquisition of concepts, for example, usually involves both perception and testimony. It is in this sense that I suggested that the cognitive content of visual representations in science can be regarded as both conceptual and non-conceptual in kind. Acknowledging that we cognitively access the content transmitted by images via vision allows us to benefit from certain capacities of our visual apparatus that have been specially shaped by processes of evolution. Two such aspects were of particular importance in the context of scientific images, namely the ability to detect patterns and to transmit non-conceptual content via images. Both features help explain why visual representations are particularly useful as providers of information in scientific arguments. Finally, these characteristics also offer a rationale for why scientific images can sometimes create a particular epistemic merit not achievable by other representational means. As our analysis made clear, those merits are especially related to the transmission of knowing-how, the supply of mental images and to the facilitation of scientific understanding.

This short summary of the main results of the above analysis reveals the rationale in my thinking that it is relevant to conclude by highlighting responsibilities of the agents involved. Whereas the second part of this investigation demonstrated the ubiquity of images in diverse scientific practices, the third part not only showed that visual representations can be epistemically efficacious, but also highlighted their particular epistemic capacities in the explanatory context of science. However, the question then arises why this should constitute the basis for requesting responsibilities.

Let me begin with a clarifying remark. Usually, as Grundmann points out, when we speak about epistemic responsibilities, the agent addressed is the one who forms a belief, or is about to alter one already held, on the basis of new incoming information (see Grundmann 2008, 238f.). Consequently, this topic is discussed with respect to epistemic justification and is broadly known as the "deontological model" of the latter (see ibid., 239). However, the point that is to be defended here is not primarily about belief-formation, but is only indirectly related to this topic. In the current context, the agent who is urged to act in an epistemically responsible way is *the one providing the information* in question, the testifier, epistemologically speaking.

Elsewhere (see Mößner 2010b, ch. 2) I have pointed out that philosophers discuss the topic of epistemic responsibility particularly with respect to testimony in the sense just mentioned. In this context, I defended the idea that the speaker – the testifier – acts as a source of knowledge. Thus, if this is the aim of the speaker in question, if she *wants* her audience to learn the information intended to be transmitted by her utterance, she is advised to act accordingly – for example, to utter her statement honestly – and, thus, to fulfil her epistemic responsibilities (see ibid., ch. 2.4.1). In some settings, such responsibilities coincide with what people have learnt during their training and apprenticeship. For instance, journalists usually acquire relevant epistemic tools, for instance background knowledge on how to check their sources for reliability before distributing information gained by these sources, during their practical training (see ibid., ch. 5.5.3). Similarly this seems also to apply to scientists, whose position as information providers equals that of journalists in importance. Let me explain.

There are several points involved here that can be grasped best when recalling the different agents whose significance was highlighted in the discussion of Fleck's analysis of scientific communication processes. He introduces the distinction between inter- and intracollective communication processes, the former including members of different communities in the process of communication, the latter referring to an exchange of ideas between members of the same collective. Moreover, intercollective thought exchange can be subdivided further into, on the one hand, communicative acts between experts and, on the other, between experts and laymen. Finally, Fleck's investigation also taught that different feedback loops can influence the whole process of communication. How those processes occur and what they imply for the process of communication is important to be borne in mind while examining in the following Fleck's theses on the exchange between experts and laymen.

Popularising scientific ideas and results is the aim intended when experts communicate with laymen in an epistemically relevant sense.[1] The claim that epistemological responsibility should be required of experts in this context is based on two aspects. Firstly, as the previous discussion has shown, images possess a particularly persuasive power, which is why producers and distributors of such images should take into account the limited abilities of their lay audience to assess visually presented information with regard to epistemic qualities. Ottino highlighted this difficulty in his case study about depictions of nano-robots (see Ottino 2003). He suggests that scientists producing such images should take into consideration that people might believe in the existence of such machines as a result of seeing their depictions, although it is simply impossible to build them in the way that they appear in those images. Thus, what becomes clear is that visual representations are a powerful means to influence public opinion. The second reason why scientists are asked to act in an epistemologically responsible way when transmitting information to laymen via visual representations is related to science and scientific knowledge being highly esteemed in Western societies (see e.g. Chalmers 1999, xix). Poser explains this by calling attention to there being different aspects involved in the assumption that science broadly influences a diversity of domains in human life (see Poser 2001, 11f.), two of which are particularly relevant in our current setting.

1   Scientists play important roles as expert witnesses in legal affairs and in policy counselling. Kitcher explicitly draws attention to the role of science and scientists in the political setting of democratic societies (see Kitcher 2011). His detailed discussion of the difficulties arising because of the biased debate about global warming in the US media (see ibid., sect. 31, esp. pp. 186f.) makes it clear that scientific knowledge is commonly regarded as a crucial basis for considerations on the designing of future living conditions (for example, regarding economic and ecological commitments).
2   The world view not only of scientists, but also that of ordinary people, depends significantly on scientific results. As Poser states, our conceptions of the world have changed radically in the past due to scientific discoveries (see Poser 2001, 12). This is illustrated best by taking a look at what happened during the shift from the Ptolemaic to the Copernican system. It can be assumed that such *scientific revolutions*, as Kuhn calls them (see Kuhn 1996), will also occur in the future and affect human life to no less an extent than those events in the past.

This particular status of scientific expertise (and experts) in human society and the effects it has on public opinion constitutes the rationale for my claiming the particular epistemic responsibility of scientists. They are regarded as an important source of information – of knowledge. Consequently, they should – ideally – care about the people who rely on their words. Moreover, as images are especially appealing in particular to ordinary recipients, scientists should handle those representational means thoughtfully. That is, they should contemplate the reasons for using them in certain contexts and what they want to express with their aid. They should avoid the distribution of misleading images. Of course, we have to concede that errors might occur, as scientific results are fallible. However, taking into account the particular role that scientists play in our societies, it seems fair to ask them at least to avoid careless image practices that might provoke misinterpretations of scientific results, evidence, or theories.

What can happen otherwise is shown in Kitcher's analysis of the global warming dispute. Here, political and economic interests have played a significant role in fuelling the debate and creating the public impression that no consensus has yet been reached amongst experts on how serious the developments are, whether global warming is anthropogenically brought about, and thus how urgent it is to decide on political measures to regulate the emission of greenhouse gases. Kitcher expresses it concisely by stating that:

> [p]ublic decision making about urgent issues and questions involving scientific complexities is often stalled because the 'consensus of experts'

is questioned: climate scientists who have spent decades trying to warn of the risks to our descendants are frustrated by the constant need to reiterate the explanations.

(Kitcher 2011, 25)

In this context, some self-proclaimed opponents of the thesis about global warming demonstrate nicely the persuasive power of images. In 1998, Michael Mann and his colleagues published a diagram that became famous as the so-called "hockey stick diagram" (see Mann, Bradley and Hughes 1998). In this diagram, data was plotted showing developments of Northern Hemisphere average temperature between 1400 and 1995. The appearance of this diagram resembles the shape of a hockey stick at rest. That is, whereas the temperature remained more or less stable and thus developed horizontally between 1400 and 1850, there was a significant vertical rise in the temperature curve starting at 1850. Approximately at the same time, namely in 1850, people started making use of oil, gas and other fossil resources to gain energy in a steadily increasing way. Consequently, the diagram was used in the scientific debate as evidence supporting the thesis that global warming is significantly influenced by human behaviour (see Blasberg and Kohlenberg 2012). However, Mann's thesis did not stay uncontested. Opponents criticised the statistical methods used to develop this diagram (see ibid.), and they also pointed out some mistakes present in the data. But even though experts from the National Academy of Sciences subsequently not only corrected the data, but also confirmed the correctness of the general claim about the significance of the anthropogenic influence on global warming, "denialists continue to claim fraud with respect to the 1998 paper, as if nothing has happened since", as Elizabeth Anderson points out (Anderson 2011, 152). Moreover, those denialists continue to make frequent use of Mann's diagram to accuse him of scientific fraud (see Blasberg and Kohlenberg 2012) and even produced a corresponding *YouTube* clip to ridicule Mann's scientific work.[2]

However, it is not only the rhetorical power of visual representations that calls for their responsible handling in science, it is also that, in our digital age, scientists not only have a vast repertoire of diverse images at their disposal, for example when searching the internet, but also powerful software tools to manipulate them in a positive or negative sense. As Hentschel expresses this regarding the case of astrophotography:

[d]igitized images in combination with state-of-the-art image-processing software allows a 'cleaning up' of the data (by substracting the bias-frame and dividing the flat field image as well as covering up cosmetic defects), which operations are equivalent to the former habit of retouching conventional photographs. Moreover, retrospective image sharpening, contrast enhancing, contour smoothening, and so on became possible.

(Hentschel 2000, 31)

It might be here that questions of quality, that is, what Hentschel calls a "cleaning up of the data" – by, for example, cropping a photograph – can come into conflict with the recipients' actual *visual literacy*. That is, the audience might not be able to read the representation correctly; they are not able to understand the relevance of the changes just mentioned due to a lack of training.

This is an important aspect that scientists, producing and distributing visual representations, should keep in mind, namely that, in the context of popular science, their products will be regarded by people who might not know, for instance, how to discern between important and non-important parts of the image. This also implies that they are not able to assess whether the tailoring of parts of an image affects its meaning in significant ways. At least, this lack of background knowledge calls for an appropriate embedding into a larger communicative context, for instance by adding legends, arrows, captions and clear links in the corresponding text. It is here that the investigations and experimental results of educational psychology and related disciplines can be indicated and, if possible, the experience of experts in these disciplines transferred to other domains. It is also here that suggestions to enhance the visual literacy of the recipients, as demanded for example by Scholz (see Scholz 2009b, 163ff.) and Sachs-Hombach (see Sachs-Hombach 2003), are needed.

The above considerations invoke the responsibilities of scientists to laymen addressed in an epistemologically relevant sense. There are at least two different groups of laymen: namely the interested public and students who are taught to become future experts in the relevant domains themselves. Here it becomes clear that the more traditional distinction between two clearly separated contexts of responsibilities, namely responsibilities towards the scientific community and responsibilities towards society at large, mentioned by Thomas Reydon (see Reydon 2013, 68), is somewhat muddled by actual scientific practice.

What adds to this already complex situation is another fact that showed up in Fleck's considerations on the topic, namely the presence of certain feedback loops triggered by the inter- and intracollective *wandering of ideas* and the respective *circulation of images*. That is, images which have been produced to teach laymen about certain research results and which were adapted to this purpose, for example by means of simplifications and omissions, might return to the esoteric circle of the scientific community in later stages of the exchange when people no longer think of them in terms of educational means (see Fleck 1979, 117). Images can cross the borders of expertise more easily, and hence also affect experts, although initially they were intended to transmit information to a totally different audience.

Vögtli and Ernst explain how to conceive of this circulation of images when discussing the *process of copying* visual representations in science (see Vögtli and Ernst 2007, ch. 2.2). They highlight the fact that often one and the same image is reused in different contexts. Several reasons might be responsible for this practice. Pragmatic considerations might play a role, but in many instances it will simply be the impossibility of producing new images, as the relevant data and the necessary technical skills to do so are not available (see ibid., 99). Moreover, Vögtli and Ernst highlight two crucial consequences of this practice of copying: firstly, this kind of utilisation of visual representations is, more often than not, related to a kind of de-contextualisation of the image in question (see ibid., 101). That is, the visual representation is copied, but the corresponding legends or captions are not. This can contribute to misleading interpretations of the respective image. Furthermore, it might also support the impression that new research results are transmitted by the image in question, although it does not present the current state of research.

Secondly, the reiterated use of particular visual representations creates what Vögtli and Ernst call *canonical images* or *canonical icons* (see ibid., ch. 2.1). These are images that are usually interpreted in a particular way and no longer allow for divergent ways of regarding the phenomenon at hand (see ibid., 78). As Vögtli and Ernst point out, those images do not transmit their content in a neutral and objective way. On the contrary, by transferring fixed points of view they set constraints on the abilities of scientists to think critically. The dynamics on which such a process is based can be grasped when reconsidering Weiss's discussion of the developmental history of microscopy and how the limitations of representations produced by technological devices restricted both scientists' observations of and their respective conceptual thinking about their research object, in this case the human cell (see Weiss 2012). In particular, Weiss's contemplation on the scientific fallacy concerning the cytoskeleton in human cells shows the crucial influence that images can have on research processes – although the image in question does not even attain canonical level.

It is those social dynamics and processes, apparent in such practices as image copying and the pictorially induced guidance of research, that constitute the reasons for the claim that scientists' epistemological responsibilities with regard to their image practices not only concern laypeople, but also have to take into account their scientific peers. Some of those images are able to influence research programmes on a much broader level than their producers, however well-intentioned, might have considered possible.

Notes

1 Of course, there is a variety of other ways to communicate with laymen: attempts to raise research funds might furnish another reason, as might the wish simply to appear important. However, neither of these ways is directed at transmitting information in an epistemically relevant sense, which is why they are absent from the above discussion.
2 See e.g. "Hide the decline – satire on global warming alarmists", www.youtube.com/watch?v=WMqc7PCJ-nc, accessed March 4, 2016.

Bibliography

Abbaneo, D., G. Abbiendi, M. Abbrescia, S. Abdullin, A. Abdulsalam, B. S. Acharya, D. Acosta et al. 2012. "A New Boson with a Mass of 125 GeV Observed with the CMS Experiment at the Large Hadron Collider." *Science* 338 (6114): 1569–75. doi:10.1126/science.1230816.

Abbott, B. P., R. Abbott, T. D. Abbott, M. R. Abernathy, F. Acernese, K. Ackley, C. Adams et al. 2016. "Observation of Gravitational Waves from a Binary Black Hole Merger." *Physical Review Letters* 116 (6): 061102. doi:10.1103/physrevlett.116.061102.

Achenbach, Joel. 2015. "Scandals Prompt Return to Peer Review and Reproducible Experiments." *The Guardian,* February 7. www.theguardian.com/science/2015/feb/07/scientific-research-peer-review-reproducing-data.

Adelmann, Ralf. 2009a. "Implizite Bild- und Medientheorien. Reflexionen beim Verfertigen naturwissenschaftlicher Visualisierungen." In *Datenbilder: zur digitalen Bildpraxis in den Naturwissenschaften,* edited by Ralf Adelmann, Jan Frercks, Martina Heßler and Jochen Hennig, 162–78. Bielefeld: transcript-Verlag.

Adelmann, Ralf. 2009b. "Orbits. Visuelle Modellierungen der Marsoberfläche am Deutschen Zentrum für Luft- und Raumfahrt." In *Datenbilder: zur digitalen Bildpraxis in den Naturwissenschaften,* edited by Ralf Adelmann, Jan Frercks, Martina Heßler and Jochen Hennig, 23–64. Bielefeld: transcript-Verlag.

Adelmann, Ralf, Jan Frercks, Martina Heßler and Jochen Hennig, eds. 2009. *Datenbilder: zur digitalen Bildpraxis in den Naturwissenschaften.* Bielefeld: transcript-Verlag.

Anderson, Elizabeth. 2011. "Democracy, Public Policy, and Lay Assessments of Scientific Testimony." *Episteme* 8 (02): 144–64. doi: 10.3366/epi.2011.0013.

Andreas, Holger. 2013. "Theoretical Terms in Science." In *The Stanford Encyclopedia of Philosophy,* Summer 2013 Edition, edited by Edward N. Zalta. The Metaphysics Research Lab, Center for the Study of Language; Information, Stanford University. http://plato.stanford.edu/archives/sum2013/entries/theoretical-terms-science/.

Arnheim, Rudolf. 1997. *Visual Thinking.* Repr. Berkeley et al.: University of California Press.

Assmann, Jan. 2009. "Altägyptische Bildpraxen und ihre impliziten Theorien." In *Bildtheorien. Anthropologische und kulturelle Grundlagen des Visualistic Turn,* 1st ed., edited by Klaus Sachs-Hombach, 74–103. Suhrkamp-Taschenbuch Wissenschaft 1888. Frankfurt/Main: Suhrkamp.

Atkin, Albert. 2013. "Peirce's Theory of Signs." In *The Stanford Encyclopedia of Philosophy,* Summer 2013 Edition, edited by Edward N. Zalta. The Metaphysics

Research Lab, Center for the Study of Language; Information, Stanford University. http://plato.stanford.edu/archives/sum2013/entries/peirce-semiotics/.

Audi, Robert. 2002. "The Sources of Knowledge." In *The Oxford Handbook of Epistemology*, edited by Paul K. Moser, 71–94. Oxford et al.: Oxford University Press.

Austin, John L. 1998. *Zur Theorie der Sprechakte*. 2nd ed., edited by Eike von Savigny. Stuttgart: Reclam.

Bagusche, Stefan. 2012. "Nicht-sprachliche Logiken." In *Visualisierung und Erkenntnis: Bildverstehen und Bildverwenden in Natur- und Geisteswissenschaften*, edited by Dimitri Liebsch and Nicola Mößner, 113–47. Cologne: Herbert von Halem.

Baigrie, Brian Scott. 1996a. "Descartes's Scientific Illustrations and 'la grande mecanique de la nature'." In *Picturing Knowledge: Historical and Philosophical Problems Concerning the Use of Art in Science*, edited by Brian Scott Baigrie, 86–134. Toronto Studies in Philosophy. Toronto et al.: University of Toronto Press.

Baigrie, Brian Scott, ed. 1996b. *Picturing Knowledge: Historical and Philosophical Problems Concerning the Use of Art in Science*. Toronto Studies in Philosophy. Toronto et al.: University of Toronto Press.

Bailer-Jones, Daniela M. 2009. *Scientific Models in Philosophy of Science*. Pittsburgh, PA: University of Pittsburgh Press.

Bantinaki, Katerina. 2009. "Depiction." In *A Companion to Aesthetics*, 2nd ed., edited by Stephen Davies, 238–41. Blackwell Companions to Philosophy 3. Malden, Mass. et al.: Wiley-Blackwell.

Bantinaki, Katerina. 2014. "What is a Picture? Depiction, Realism, Abstraction, by Michael Newall." *Mind* 123 (491): 944–7. doi:10.1093/mind/fzu114.

Barceló Aspeitia, Axel Arturo. 2012. "Words and Images in Argumentation." *Argumentation* 26: 355–68.

Bartelborth, Thomas. 1996. *Begründungsstrategien: ein Weg durch die analytische Erkenntnistheorie*. Berlin: Akademie-Verlag.

Bermúdez, José and Arnon Cahen. 2015. "Nonconceptual Mental Content." In *The Stanford Encyclopedia of Philosophy*, Fall 2015 Edition, edited by Edward N. Zalta. The Metaphysics Research Lab, Center for the Study of Language; Information, Stanford University. http://plato.stanford.edu/archives/fall2015/entries/content-nonconceptual/.

Bigg, Charlotte, ed. 2009. *Atombilder: Ikonografie des Atoms in Wissenschaft und Öffentlichkeit des 20. Jahrhunderts*. Abhandlungen und Berichte / Deutsches Museum, N.F., 25. Göttingen: Wallstein-Verlag.

Bishop, Michael A. 1992. "Theory-Ladenness of Perception Arguments." *PSA: Proceedings of the Biennial Meeting of the Philosophy of Science Association* 1: 287–99.

Blair, J. Anthony. 1996. "The Possibility and Actuality of Visual Arguments." *Argumentation and Advocacy* 33: 23–39.

Blair, J. Anthony. 2004. "The Rhetoric of Visual Arguments." In *Defining Visual Rhetorics*, edited by Charles A. Hill and Marguerite Helmers, 41–61. Mahwah, New Jersey: Lawrence Erlbaum Associates.

Blasberg, Anita and Kerstin Kohlenberg. 2012. "Die Klimakrieger. Wie von der Industrie bezahlte PR-Manager der Welt seit Jahren einreden, die Erderwärmung finde nicht statt. Chronologie einer organisierten Lüge." *Die Zeit*, November 22. www.zeit.de/2012/48/Klimawandel-Marc-Morano-Lobby-Klimaskeptiker.

Boehm, Gottfried. 2008. *Wie Bilder Sinn erzeugen: die Macht des Zeigens*. 2nd ed. Berlin: Berlin University Press.

Bogen, James. 2002. "Experiment and Observation." In *The Blackwell Guide to the Philosophy of Science*, edited by Peter K. Machamer and Michael Silberstein, 128–48. Blackwell Philosophy Guides 7. Malden, Mass. et al.: Blackwell.

BonJour, Laurence. 2013. "Epistemological Problems of Perception." In *The Stanford Encyclopedia of Philosophy*, edited by Edward N. Zalta, Spring 2013 Edition. The Metaphysics Research Lab, Center for the Study of Language; Information, Stanford University. http://plato.stanford.edu/archives/spr2013/entries/perception-episprob/.

Bosinski, Gerhard. 2009. "Das Bild in der Altsteinzeit." In *Bildtheorien. Anthropologische und kulturelle Grundlagen des Visualistic Turn*, edited by Klaus Sachs-Hombach, 1st ed., 31–73. Suhrkamp-Taschenbuch Wissenschaft 1888. Frankfurt/Main: Suhrkamp.

Boumans, Marcel. 2016. "Graph-Based Inductive Reasoning." *Studies in History and Philosophy of Science Part A* 59: 1–10. doi:10.1016/j.shpsa.2016.05.001.

Boyd, Richard. 2010. "Scientific Realism." In *The Stanford Encyclopedia of Philosophy*, edited by Edward N. Zalta, Summer 2010 Edition. The Metaphysics Research Lab, Center for the Study of Language; Information, Stanford University. http://plato.stanford.edu/archives/sum2010/entries/scientific-realism/.

Bredekamp, Horst. 2003. "A Neglected Tradition? Art History as Bildwissenschaft." *Critical Inquiry* 29 (3): 418–28. doi:10.1086/376303.

Bredekamp, Horst. 2009. *Galilei der Künstler: der Mond, die Sonne, die Hand*. 2nd ed. Berlin: Akademie-Verlag.

Bredekamp, Horst, Birgit Schneider and Vera Dünkel, eds. 2008. *Das Technische Bild: Kompendium zu einer Stilgeschichte wissenschaftlicher Bilder*. Berlin: Akademie-Verlag.

Brown, James Robert. 1996. "Illustration and Inference." In *Picturing Knowledge: Historical and Philosophical Problems Concerning the Use of Art in Science*, edited by Brian Scott Baigrie, 250–68. Toronto Studies in Philosophy. Toronto et al.: University of Toronto Press.

Brown, James Robert. 1997. "Proofs and Pictures." *The British Journal for the Philosophy of Science* 48 (2): 161–80.

Brown, James Robert. 2004. "Peeking into Plato's Heaven." *Philosophy of Science* 71 (5): 1126–38. doi:10.1086/425940.

Brown, James Robert and Yiftach Fehige. 2011. "Thought Experiments." In *The Stanford Encyclopedia of Philosophy*, edited by Edward N. Zalta, Fall 2011 Edition. The Metaphysics Research Lab, Center for the Study of Language; Information, Stanford University. http://plato.stanford.edu/archives/fall2011/entries/thought-experiment/.

Bunge, Mario. 2010. "Reading Measuring Instruments." *Spontaneous Generations: A Journal for the History and Philosophy of Science* 4 (1): 85–93. doi:10.4245/sponge.v4i1.11725.

Burri, Regula Valérie. 2008. *Doing Images: Zur Praxis Medizinischer Bilder*. Bielefeld: transcript.

Byrne, Oliver. 2013. *The First Six Books of the Elements of Euclid: In Which Coloured Diagrams and Symbols Are Used Instead of Letters for the Greater Ease of Learners*. Edited by Werner Oechslin. Reprint of London, Pickering, 1847. Cologne: Taschen.

Campbell, John. 2009. "Manipulating Colour: Pounding an Almond." In *Perceptual Experience*, edited by Tamar Szabó Gendler and John Hawthorne, Repr., 31–48. Oxford et al.: Clarendon Press.

Carey, Benedict. 2014. *How We Learn: Throw Out the Rule Book and Unlock Your Brain's Potential*. London: Pan Books.

Carrier, Martin. 1994. *The Completeness of Scientific Theories: On the Derivation of Empirical Indicators Within a Theoretical Framework: The Case of Physical Geometry*. Dordrecht et al.: Kluwer Academic Publishers.

Carter, Matt. 2013. *Designing Science Presentations: A Visual Guide to Figures, Papers, Slides, Posters, and More*. 1st ed. Amsterdam: Academic Press.

Cartwright, Lisa. 1997. *Screening the Body: Tracing Medicine's Visual Culture*. 2nd ed. Minneapolis, Minn. et al.: University of Minnesota Press.

Carusi, Annamaria, Aud Sissel Hoel, Timothy Webmoor and Steve Woolgar, eds. 2014. *Visualization in the Age of Computerization*. Routledge Studies in Science, Technology and Society. Hoboken: Routledge Chapman & Hall.

Chakravartty, Anjan. 2007. *A Metaphysics for Scientific Realism: Knowing the Unobservable*. Cambridge et al.: Cambridge University Press.

Chakravartty, Anjan. 2010. "Informational Versus Functional Theories of Scientific Representation." *Synthese* 172 (2): 197–213. doi:10.1007/s11229-009-9502-3.

Chakravartty, Anjan. 2013. "Scientific Realism." In *The Stanford Encyclopedia of Philosophy*, edited by Edward N. Zalta, Summer 2013 Edition. The Metaphysics Research Lab, Center for the Study of Language; Information, Stanford University. http://plato.stanford.edu/archives/sum2013/entries/scientific-realism/.

Chalmers, Alan Francis. 1999. *What Is This Thing Called Science?* 3rd ed. Buckingham: Open University Press.

Chang, Hasok. 2007. *Inventing Temperature: Measurement and Scientific Progress*. 1st issued as paperback. Oxford et al.: Oxford University Press.

Coady, C. A. J. 1992. *Testimony: A Philosophical Study*. Oxford et al.: Clarendon Press.

Crane, Tim. 2009. "Is Perception a Propositional Attitude?" *The Philosophical Quarterly* 59 (236): 452–69. doi:10.1111/j.1467-9213.2008.608.x.

Curd, Martin, J. A. Cover and Christopher Pincock, eds. 2013. *Philosophy of Science: The Central Issues*. 2nd ed. New York et al.: W. W. Norton & Company.

Daston, Lorraine and Peter Galison. 1992. "The Image of Objectivity". *Representations* 40: 81–128.

Daston, Lorraine and Peter Galison. 2007. *Objektivität*. Edited by Christa Krüger. 1st ed. Frankfurt/Main: Suhrkamp.

Daston, Lorraine and Elizabeth Lunbeck, eds. 2011. *Histories of Scientific Observation*. Chicago et al.: University of Chicago Press.

Dauthendey, Max. 1981. "Mein Vater wird Photograph." In *Die Wahrheit der Photographie: klassische Bekenntnisse zu einer neuen Kunst*, edited by Wilfried Wiegand, 23–40. Frankfurt/Main: Fischer.

de Regt, Henk W. 2014. "Visualization as a Tool for Understanding." *Perspectives on Science* 22 (3): 377–96. doi:10.1162/posc_a_00139.

Descartes, René. 1994. *Meditationen über die Grundlagen der Philosophie: mit den sämtlichen Einwänden und Erwiderungen*. Edited by Artur Buchenau. Philosophische Bibliothek 27. Hamburg: Meiner.

Dommann, Monika. 2004. "Vom Bild zum Wissen: eine Bestandsaufnahme wissenschaftshistorischer Bildforschung." *Gesnerus* 61: 77–89.

Downes, Stephen M. 2012. "How Much Work Do Scientific Images Do?" *Spontaneous Generations: A Journal for the History and Philosophy of Science* 6 (1): 115–30.

Dretske, Fred I. 1999. *Knowledge and the Flow of Information*. Stanford, Calif.: CSLI Publications.

Duhem, Pierre. 1998. *Ziel und Struktur der physikalischen Theorien*. Edited by Friedrich Adler, Ernst Mach and Lothar Schäfer. Hamburg: Meiner.

Dünkel, Vera. 2008. "Röntgenblick und Schattenbild: Zur Spezifik der frühen Röntgenbilder und ihren Deutungen um 1900." In *Das Technische Bild: Kompendium zu einer Stilgeschichte wissenschaftlicher Bilder*, edited by Horst Bredekamp, Birgit Schneider and Vera Dünkel, 136–47. Berlin: Akademie-Verlag.

Eaton, Marcia. 1980. "Truth in Pictures." *The Journal of Aesthetics and Art Criticism* 39 (1): 15–26. doi:10.2307/429915.

Eco, Umberto. 1994. *Die Suche nach der vollkommenen Sprache*. Edited by Burkhart Kroeber. München: Beck.

Eitel, Alexander and Katharina Scheiter. 2015. "Picture or Text First? Explaining Sequence Effects When Learning with Pictures and Text." *Educational Psychology Review* 27 (1): 153–80. doi:10.1007/s10648-014-9264-4.

Elert, Glenn. 2014. The Physics Hypertextbook. http://physics.info/.

Elgin, Catherine Z. 1993. "Understanding Art and Science." *Synthese* 95: 13–28.

Elgin, Catherine Z. 2007. "Understanding and the Facts." *Philosophical Studies* 132 (1): 33–42. doi:10.1007/s11098-006-9054-z.

Faulkner, Paul. 2011. *Knowledge on Trust*. Oxford et al.: Oxford University Press.

Feagin, Susan L. 1998. "Presentation and Representation." *The Journal of Aesthetics and Art Criticism* 56 (3): 234–40.

Ferguson, Eugene S. 1977. "The Mind's Eye: Nonverbal Thought in Technology." *Science* 197 (4306): 827–36. doi:10.1126/science.197.4306.827.

Fleck, Ludwik. 1979. *Genesis and Development of a Scientific Fact*. Edited by Thaddeus J. Trenn, Robert K. Merton and Fred Bradley. Chicago et al.: University of Chicago Press.

Fleck, Ludwik. 1986a. *Cognition and Fact: Materials on Ludwik Fleck*. Edited by Robert S. Cohen. Boston Studies in the Philosophy of Science 87. Dordrecht et al.: Reidel.

Fleck, Ludwik. 1986b. "Problems of the Science of Science [1946]." In *Cognition and Fact: Materials on Ludwik Fleck*, by Ludwik Fleck, edited by Robert S. Cohen, 113–27. Boston Studies in the Philosophy of Science 87. Dordrecht et al.: Reidel.

Fleck, Ludwik. 1986c. "The Problem of Epistemology [1936]." In *Cognition and Fact: Materials on Ludwik Fleck*, by Ludwik Fleck, edited by Robert S. Cohen, 79–112. Boston Studies in the Philosophy of Science 87. Dordrecht et al.: Reidel.

Fleck, Ludwik. 1986d. "To Look, To See, To Know [1947]." In *Cognition and Fact: Materials on Ludwik Fleck*, by Ludwik Fleck, edited by Robert S. Cohen, 129–51. Boston Studies in the Philosophy of Science 87. Dordrecht et al.: Reidel.

Fleck, Ludwik. 2011. *Denkstile und Tatsachen: Gesammelte Schriften und Zeugnisse*. Edited by Sylwia Werner and Claus Zittel. 1st ed. Suhrkamp-Taschenbuch Wissenschaft 1953. Berlin: Suhrkamp.

Frank, Gustav and Barbara Lange. 2010. *Einführung in die Bildwissenschaft: Bilder in der visuellen Kultur*. Einführung. Darmstadt: Wissenschaftliche Buchgesellschaft.

Frankel, Felice and Angela H. DePace. 2012. *Visual Strategies: A Practical Guide to Graphics for Scientists and Engineers*. New Haven; London: Yale University Press.

Franklin, Allan. 1999. *Can That Be Right? Essays on Experiment, Evidence, and Science*. Dordrecht et al.: Kluwer.

Freeman, Michael. 2009. *Alles über Digitalfotografie: Nahaufnahme, Stillleben, schwarzweiss, Belichtung, Licht & Beleuchtung, Porträt, Druck, Kamera & Zubehör, Farbe, Landschaft*. Cologne: Evergreen GmbH.

Frege, Gottlob. 1956. "The Thought. A Logical Inquiry." *Mind* 65 (259): 289–311.

Frege, Gottlob. 1993. *Logische Untersuchungen*. Edited by Günther Patzig. 4th ed. Göttingen: Vandenhoeck und Ruprecht.

Fricker, Elizabeth. 1994. "Against Gullibility." In *Knowing from Words: Western and Indian Philosophical Analysis of Understanding and Testimony*, edited by Bimal Krishna Matilal and Arindam Chakrabarti, 125–61. Dordrecht et al.: Kluwer Academic.

Fricker, Elizabeth. 1995. "Telling and Trusting: Reductionism and Anti-Reductionism in the Epistemology of Testimony." *Mind* 104: 393–411.

Fricker, Elizabeth. 2002. "Trusting Others in the Sciences: A Priori or Empirical Warrant?" *Studies in History and Philosophy of Science* 33: 373–83.

Frigg, Roman and James Nguyen. 2016. "Scientific Representation." In *The Stanford Encyclopedia of Philosophy*, edited by Edward N. Zalta, Winter 2016 Edition. The Metaphysics Research Lab, Center for the Study of Language; Information, Stanford University. http://plato.stanford.edu/archives/win2016/entries/scientific-representation/.

Gaiger, Jason. 2014. "The Idea of a Universal Bildwissenschaft." *Estetika: The Central European Journal of Aesthetics* 2: 208–29.

Gelfert, Axel. 2014. *A Critical Introduction to Testimony*. London: Bloomsbury Publishing.

Gendler, Tamar Szabó. 2013. "Imagination." In *The Stanford Encyclopedia of Philosophy*, edited by Edward N. Zalta, Fall 2013 Edition. The Metaphysics Research Lab, Center for the Study of Language; Information, Stanford University. http://plato.stanford.edu/archives/fall2013/entries/imagination/.

Gendler, Tamar Szabó and John Hawthorne, eds. 2009. *Perceptual Experience*. Repr. Oxford et al.: Clarendon Press.

Giardino, Valeria and Gabriel Greenberg. 2015. "Introduction: Varieties of Iconicity." *Review of Philosophy and Psychology* 6 (1): 1–25. doi:10.1007/s13164-014-0210-7.

Giovannelli, Alessandro. 2010. "Goodman's Aesthetics." In *The Stanford Encyclopedia of Philosophy*, edited by Edward N. Zalta, Summer 2010 Edition. The Metaphysics Research Lab, Center for the Study of Language; Information, Stanford University. http://plato.stanford.edu/archives/sum2010/entries/goodman-aesthetics/.

Golan, Tal. 2002. "Sichtbarkeit und Macht: Maschinen als Augenzeugen." In *Ordnungen der Sichtbarkeit*, edited by Peter Geimer, 1st ed., 171–210. Suhrkamp-Taschenbuch Wissenschaft 1538. Frankfurt/Main: Suhrkamp.

Goldberg, Sanford C. 2010. *Relying on Others: An Essay in Epistemology*. Oxford et al: Oxford University Press.

Goldman, Alvin I. 1999. *Knowledge in a Social World*. Oxford et al.: Clarendon Press.

Goldman, Alvin I. and Dennis Whitcomb, eds. 2011. *Social Epistemology: Essential Readings*. Oxford et al.: Oxford University Press.

Gombrich, Ernst H. 1969. "The Evidence of Images." In *Interpretation: Theory and Practice*, edited by Charles S. Singleton, 35–68. Baltimore: Hopkins Press.

Gombrich, Ernst H. 1972. "The Visual Image." *Scientific American* 227 (3): 82–96. doi:10.1038/scientificamerican0972-82.

Gombrich, Ernst H. 2004. *Kunst und Illusion: zur Psychologie der bildlichen Darstellung*. Edited by Lisbeth Gombrich. 6th ed. Berlin: Phaidon.

Gooding, David C. 2010. "Visualizing Scientific Inference." *Topics in Cognitive Science* 2 (1): 15–35. doi:10.1111/j.1756-8765.2009.01048.x.

Goodman, Nelson. 1976. *Languages of Art: An Approach to a Theory of Symbols*. 2nd ed., Indianapolis: Hackett.

Goodwin, William. 2009. "Visual Representations in Science." *Philosophy of Science* 76 (3): 372–90.

Gordin, Michael D. 2012. *The Pseudoscience Wars: Immanuel Velikovsky and the Birth of the Modern Fringe*. Chicago: University of Chicago Press.

Graham, Peter J. 2006a. "Liberal Fundamentalism and Its Rivals." In *The Epistemology of Testimony*, edited by Jennifer Lackey and Ernest Sosa, 93–115. Oxford: Clarendon Press.

Graham, Peter J. 2006b. "Testimonial Justification: Inferential or Non-Inferential?" *The Philosophical Quarterly* 56 (222): 84–95.

Groarke, Leo. 2015. "Going Multimodal: What Is a Mode of Arguing and Why Does It Matter?" *Argumentation* 29 (2): 133–55. doi:10.1007/s10503-014-9336-0.

Groß, Dominik and Tobias Heinrich Duncker, eds. 2006. *Farbe – Erkenntnis – Wissenschaft: zur epistemischen Bedeutung von Farbe in der Medizin*. Berlin et al.: Lit.

Groß, Dominik and Sabine Müller. 2006. "Mit bunten Bildern zur Erkenntnis? Neuroimaging und Wissenspopularisierung am Beispiel des Magazins 'Gehirn & Geist'." In *Farbe – Erkenntnis – Wissenschaft: zur epistemischen Bedeutung von Farbe in der Medizin*, edited by Dominik Groß and Tobias Heinrich Duncker, 77–92. Berlin et al.: Lit.

Grundmann, Thomas. 2008. *Analytische Einführung in die Erkenntnistheorie*. Berlin et al.: de Gruyter.

Haddock, Adrian, Alan Millar and Duncan Pritchard, eds. 2010. *Social Epistemology*. Oxford et al.: Oxford University Press.

Hardwig, John. 1985. "Epistemic Dependence." *The Journal of Philosophy* 82 (7): 335–49.

Harré, Rom. 2010. "Equipment for an Experiment." *Spontaneous Generations: A Journal for the History and Philosophy of Science* 4 (1): 30–8. doi:10.4245/sponge.v4i1.11334.

Harris, Paul L. 2012. *Trusting What You're Told: How Children Learn from Others*. Cambridge, Massachusetts; London, England: Belknap Press.

Harth, Manfred and Jakob Steinbrenner, eds. 2013. *Bilder als Gründe*. Cologne: Herbert von Halem.

Hasan, Ali and Richard Fumerton. 2014. "Knowledge by Acquaintance Vs. Description." In *The Stanford Encyclopedia of Philosophy*, edited by Edward N. Zalta. The Metaphysics Research Lab, Center for the Study of Language; Information, Stanford University. http://plato.stanford.edu/archives/spr2014/entries/knowledge-acquaindescrip/.

Hausken, Liv. 2015. "The Visual Culture of Brain Imaging." *Leonardo* 48 (1): 68–9. doi:10.1162/leon_a_00899.

Hausken, Liv. 2017. "The Media Aesthetics of Brain Imaging in Popular Science." In *Reasoning in Measurement*, edited by Nicola Mößner and Alfred Nordmann, 57–72. London and New York: Routledge.

Hennig, Jochen. 2009. "Epistemologie des Schattens." In *Datenbilder: zur digitalen Bildpraxis in den Naturwissenschaften*, edited by Ralf Adelmann, Jan Frercks, Martina Heßler and Jochen Hennig, 195–207. Bielefeld: transcript-Verlag.

Hennig, Jochen. 2011. *Bildpraxis: visuelle Strategien in der frühen Nanotechnologie*. Bielefeld: transcript-Verlag.

Hentschel, Klaus. 2000. "Drawing, Engraving, Photographing, Plotting, Printing: Historical Studies of Visual Representations, Esp. in Astronomy." In *The Role of Visual Representations in Astronomy: History and Research Practice*, edited by Klaus Hentschel and Axel D. Wittmann, 11–43. *Acta Historica Astronomiae, Vol. 9*. Thun et al.: Deutsch.

Hentschel, Klaus. 2014. *Visual Cultures in Science and Technology: A Comparative History*. Oxford et al.: Oxford University Press.

Hentschel, Klaus and Axel D. Wittmann, eds. 2000. *The Role of Visual Representations in Astronomy: History and Research Practice*. 1st ed. *Acta Historica Astronomiae, Vol. 9*. Thun et al.: Deutsch.

Hesse, Mary B. 1970. *Models and Analogies in Science*. 2nd ed. Notre Dame, Ind.: University of Notre Dame Press.

Heßler, Martina. 2006. "Von der doppelten Unsichtbarkeit digitaler Bilder." *zeitblicke* 5 (3). http://nbn-resolving.de/urn:nbn:de:0009-9-6474.

Heßler, Martina and Dieter Mersch. 2009. "Einleitung: Bildlogik oder Was heißt visuelles Denken?" In *Logik des Bildlichen: zur Kritik der ikonischen Vernunft*, edited by Martina Heßler and Dieter Mersch, 8–62. Bielefeld: transcript-Verlag.

Hoffmann, Max, Lana Plumanns, Laura Lenz, Katharina Schuster, Tobias Meisen and Sabina Jeschke. 2015. "Enhancing the Learning Success of Engineering Students by Virtual Experiments." *Learning and Collaboration Technologies* 9192: 394–405. doi:10.1007/978-3-319-20609-7_37.

Hopkins, Robert. 2003. "Pictures, Phenomenology and Cognitive Science." *Monist* 86 (4): 653–75. doi:10.5840/monist200386434.

Hopkins, Robert. 2009. *Picture, Image and Experience: A Philosophical Inquiry*. Cambridge et al.: Cambridge University Press.

Huemer, Michael. 2011. "Sense-Data." In *The Stanford Encyclopedia of Philosophy*, edited by Edward N. Zalta, Spring 2011 Edition. The Metaphysics Research Lab, Center for the Study of Language; Information, Stanford University. http://plato. stanford.edu/archives/spr2011/entries/sense-data/.

Hyman, John. 2006. *The Objective Eye: Color, Form, and Reality in the Theory of Art*. Chicago et al.: University of Chicago Press.

Irving, Zachary C. 2011. "Style, but Substance: An Epistemology of Visual Versus Numerical Representation in Scientific Practice." *Philosophy of Science* 78 (5): 774–87.

Jackson, Frank. 1986. "What Mary Didn't Know." *The Journal of Philosophy* 83 (5): 291–5.

Jarnicki, Paweł. 2016. "On the Shoulders of Ludwik Fleck? On the Bilingual Philosophical Legacy of Ludwik Fleck and Its Polish, German and English Translations." *The Translator* 22 (3): 271–86. doi:10.1080/13556509.2015.1126881.

Joyce, Kelly Ann. 2008. *Magnetic Appeal: MRI and the Myth of Transparency*. Cornell Paperbacks. Ithaca: Cornell University Press.

Jung, Eva-Maria. 2012. *Gewusst wie? eine Analyse praktischen Wissens*. Berlin et al.: de Gruyter.

Jung, Eva-Maria and Albert Newen. 2010. "Knowledge and Abilities: The Need for a New Understanding of Knowing-How." *Phenomenology and the Cognitive Sciences* 9 (1): 113–31. doi:10.1007/s11097-009-9129-3.

Jung, Eva-Maria and Albert Newen. 2011. "Understanding Knowledge in a New Framework: Against Intellectualism as a Semantic Analysis and an Analysis of the Mind." In *Knowledge and Representation*, edited by Albert Newen, Andreas Bartels and Eva-Maria Jung, 79–105. Stanford, Calif.: CSLI Publications.

Kapust, Antje. 2009. "Phänomenologische Bildpositionen." In *Bildtheorien. Anthropologische und kulturelle Grundlagen des Visualistic Turn*, edited by Klaus

350 Bibliography

Sachs-Hombach, 255–83. Suhrkamp-Taschenbuch Wissenschaft 1888. Frankfurt/ Main: Suhrkamp.

Kemp, Martin. 2003. *Bilderwissen: die Anschaulichkeit naturwissenschaftlicher Phänomene.* Edited by Jürgen Blasius. Cologne: DuMont.

Kemp, Martin. 2012. "'The Testimony of My Own Eyes': The Strange Case of the Mammal with a Beak." *Spontaneous Generations: A Journal for the History and Philosophy of Science* 6 (1): 43–9. doi:10.4245/sponge.v6i1.17157.

Kirby, John R., Phillip J. Moore and Neville J. Schofield. 1988. "Verbal and Visual Learning Styles." *Contemporary Educational Psychology* 13: 169–84.

Kitcher, Philip. 1993. *The Advancement of Science: Science Without Legend, Objectivity Without Illusions.* New York et al.: Oxford University Press.

Kitcher, Philip. 2011. *Science in a Democratic Society.* Amherst, NY: Prometheus Books.

Kitcher, Philip and Achille Varzi. 2000. "Some Pictures are Worth 2 ℵ 0 Sentences". Philosophy 75 (4): 377–81. doi:10.1017/S0031819100000450.

Kjeldsen, Jens E. 2015. "The Study of Visual and Multimodal Argumentation." *Argumentation* 29 (2): 115–32. doi:10.1007/s10503-015-9348-4.

Kjørup, Søren. 1978. "Pictorial Speech Acts." *Erkenntnis* 12 (1): 55–71. doi:10.1007/BF00209915.

Kjørup, Søren. 1989. "Die sprachliche Verankerung des Bildes." *Zeitschrift für Semiotik* 11 (4): 305–17.

Kjørup, Søren. 2009. *Semiotik.* UTB. Munich et al.: W. Fink.

Kjørup, Søren. 2013. "Schnüffeln nach Spionen: Verrat, Vertrauen und fotografische Beweise." In *Bilder als Gründe*, edited by Manfred Harth and Jakob Steinbrenner, 11–34. Cologne: Herbert von Halem.

Kobert, Max J. 2010. "Zur Bedeutung anschaulichen Denkens." In *In Bildern denken? kognitive Potentiale von Visualisierung in Kunst und Wissenschaft*, edited by Ulrich Nortmann and Christoph Wagner, 129–38. Paderborn et al.: Fink.

Koppelberg, Dirk. 1993. "Should We Replace Knowledge by Understanding? A Comment on Elgin and Goodman's Reconception of Epistemology." *Synthese* 95 (1): 119–28. doi:10.1007/bf01064671.

Kosso, Peter. 1988. "Dimensions of Observability." *The British Journal for the Philosophy of Science* 39 (4): 449–67. doi: 10.1093/bjps/39.4.449.

Kosso, Peter. 1993. *Reading the Book of Nature: An Introduction to the Philosophy of Science.* Reprinted. Cambridge et al.: Cambridge University Press.

Kosso, Peter. 2006. "Detecting Extrasolar Planets." *Studies in History and Philosophy of Science Part A* 37 (2): 224–36. doi:10.1016/j.shpsa.2005.05.001.

Kosso, Peter. 2007. "Scientific Understanding." *Foundations of Science* 12 (2): 173–88. doi:10.1007/s10699-006-0002-3.

Kosso, Peter. 2010. "And Yet It Moves: The Observability of the Rotation of the Earth." *Foundations of Science* 15 (3): 213–25. doi:10.1007/s10699-010-9175-x.

Kuhn, Thomas S. 1996. *The Structure of Scientific Revolutions.* 3rd ed. Chicago, Ill. et al.: University of Chicago Press.

Kulvicki, John V. 2006. *On Images: Their Structure and Content.* Oxford: Clarendon Press.

Kulvicki, John V. 2010a. "Knowing with Images: Medium and Message." *Philosophy of Science* 77 (2): 295–313. doi:10.1086/651321.

Kulvicki, John V. 2010b. "Pictorial Diversity." In *Philosophical Perspectives on Depiction*, edited by Catharine Abell and Katerina Bantinaki, 25–51. Oxford: Oxford University Press.

Kvanvig, Jonathan L. 2003. *The Value of Knowledge and the Pursuit of Understanding*. Cambridge et al.: Cambridge University Press.

Lackey, Jennifer. 2008. *Learning from Words: Testimony as a Source of Knowledge*. Oxford et al.: Oxford University Press.

Lackey, Jennifer and Ernest Sosa, eds. 2006. *The Epistemology of Testimony*. Oxford: Clarendon Press.

Ladyman, James. 2003. *Understanding Philosophy of Science*. London et al.: Routledge.

Lake, Randall A. and Barbara A. Pickering. 1998. "Argumentation, the Visual, and the Possibility of Refutation: An Exploration." *Argumentation* 12: 79–93. doi: 10.1023/A:1007703425353.

Larkin, Jill H. and Herberta A. Simon. 1987. "Why a Diagram Is (Sometimes) Worth Ten Thousand Words." *Cognitive Science* 11: 65–99. doi: 10.1111/j.1551-6708.1987. tb00863.x.

Lehnert, M. D., N. P. H. Nesvadba, J.-G. Cuby, A. M. Swinbank, S. Morris, B. Clément, C. J. Evans, M. N. Bremer and S. Basa. 2010. "Spectroscopic Confirmation of a Galaxy at Redshift Z = 8.6." *Nature* 467 (7318): 940–2. doi:10.1038/nature09462.

Lehrer, Keith. 2004. "Representation in Painting and Consciousness." *Philosophical Studies* 117 (1/2): 1–14. doi:10.1023/b:phil.0000014522.51852.fb.

Levinson, Jerrold. 1998. "Wollheim on Pictorial Representation." *The Journal of Aesthetics and Art Criticism* 56 (3): 227–33. doi: 10.2307/432362.

Lewis-Williams, David. 2004. *The Mind in the Cave: Consciousness and the Origins of Art*. London: Thames & Hudson.

Liebsch, Dimitri and Nicola Mößner, eds. 2012. *Visualisierung und Erkenntnis: Bildverstehen und Bildverwenden in Natur- und Geisteswissenschaften*. Cologne: Herbert von Halem.

Lopes, Dominic McIver. 2006. *Understanding Pictures*. Repr. Oxford: Oxford University Press.

Löwy, Ilana. 2008. "Ways of Seeing: Ludwik Fleck and Polish Debates on the Perception of Reality, 1890–1947." *Studies in History and Philosophy of Science Part A* 39 (3): 375–83. doi:10.1016/j.shpsa.2008.06.009.

Lyre, Holger. 2010. "Erweiterte Kognition und mentaler Externalismus." *Zeitschrift für philosophische Forschung* 64 (2): 190–215. doi:10.3196/004433010791655570.

Mancosu, Paolo, Klaus Frovin Jørgensen and Stig Andur Pedersen, eds. 2005. *Visualization, Explanation and Reasoning Styles in Mathematics: Meeting "Mathematics as Rational Activity"*. Dordrecht: Springer.

Mann, Michael E., Raymond S. Bradley and Malcolm K. Hughes. 1998. "Global-Scale Temperature Patterns and Climate Forcing over the Past Six Centuries." *Nature* 392 (6678): 779–87. doi:10.1038/33859.

Margolis, Eric and Stephen Laurence. 2014. "Concepts." In *The Stanford Encyclopedia of Philosophy*, edited by Edward N. Zalta, Spring 2014 Edition. The Metaphysics Research Lab, Center for the Study of Language; Information, Stanford University. http://plato.stanford.edu/archives/spr2014/entries/concepts/.

Matilal, Bimal Krishna and Arindam Chakrabarti, eds. 1994. *Knowing from Words: Western and Indian Philosophical Analysis of Understanding and Testimony*. Dordrecht et al.: Kluwer Academic.

Mayer-Schönberger, Viktor and Kenneth Cukier. 2013. *Big Data: A Revolution That Will Transform How We Live, Work, and Think*. Boston et al.: Houghton Mifflin Harcourt.

Maynard, Patrick. 1989. "Talbot's Technologies: Photographic Depiction, Detection, and Reproduction." *The Journal of Aesthetics and Art Criticism* 47 (3): 263–76. doi: 10.2307/431006.

Maynard, Patrick. 2000. *The Engine of Visualization: Thinking Through Photography.* Ithaca, NY et al.: Cornell University Press.

Maynard, Patrick. 2005. *Drawing Distinctions: The Varieties of Graphic Expression.* Ithaca, NY: Cornell University Press.

Maynard, Patrick. 2009. "Darwin's Diagrams." http://diagrambasedreasoning. wordpress.com/2009/03/08/second-invited-speaker-patrick-maynard/.

Maynard, Patrick. 2010. "Working Light." *Philosophy of Photography* 1 (1): 29–34. doi:10.1386/pop.1.1.29/7.

Maynard, Patrick. 2011. "What Drawing Draws on: The Relevance of Current Vision Research." *Rivista Di Estetica* 47 (2): 9–29.

Maynard, Patrick. 2017. "Photo Mensura." In *Reasoning in Measurement*, edited by Nicola Mößner and Alfred Nordmann, 41–56. London and New York: Routledge.

Maynard, Patrick. unpublished paper. "Our Graphic Minds: Graphic Cognitive Artifacts and the 'New Science of Mind'."

McCrudden, Matthew T. and David N. Rapp. 2015, online first. "How Visual Displays Affect Cognitive Processing." *Educational Psychology Review.* doi:10.1007/ s10648-015-9342-2.

McGrath, Matthew. 2014. "Propositions." In *The Stanford Encyclopedia of Philosophy*, edited by Edward N. Zalta, Spring 2014 Edition. The Metaphysics Research Lab, Center for the Study of Language; Information, Stanford University. http://plato. stanford.edu/archives/spr2014/entries/propositions/.

McIntosh, Gavin. 2003. "Depiction Unexplained: Peacocke and Hopkins on Pictorial Representation." *British Journal of Aesthetics* 43 (3): 279–88. doi: 10.1093/ bjaesthetics/43.3.279.

Meskin, Aaron and Jonathan Cohen. 2008. "Photographs as Evidence." In *Photography and Philosophy: Essays on the Pencil of Nature*, edited by Scott Walden, 70–90. Malden, MA: Blackwell.

Meynell, Letitia. 2008. "Why Feynman Diagrams Represent." *International Studies in the Philosophy of Science* 22 (1): 39–59. doi: 10.1080/02698590802280902.

Mitchell, William J. T. 1994. *The Reconfigured Eye: Visual Truth in the Post-Photographic Era.* 1st paperback ed. Cambridge, Mass. et al.: MIT Press.

Mitchell, William J. T. 2008. *Bildtheorie.* Edited by Gustav Frank and Heinz Jatho. Frankfurt/Main: Suhrkamp.

Mithen, Steven J. 1998. *The Prehistory of the Mind: A Search for the Origins of Art, Religion and Science.* London: Phoenix.

Mithen, Steven J. 2007. *The Singing Neanderthals: The Origins of Music, Language, Mind, and Body.* Cambridge, Massachusetts: Harvard University Press.

Mizrahi, Moti. 2012. "Idealizations and Scientific Understanding." *Philosophical Studies* 160 (2): 237–52. doi:10.1007/s11098-011-9716-3.

Morgan, Mary S. and Margaret Morrison, eds. 1999. *Models as Mediators: Perspectives on Natural and Social Science.* Cambridge et al.: Cambridge University Press.

Mößner, Nicola. 2010a. "Testimoniale Akte neu definiert – ein zentrales Problem des Zeugnisses anderer." *Grazer Philosophische Studien* 80: 151–78. doi: 10.1163/18756735-90000875.

Mößner, Nicola. 2010b. *Wissen aus dem Zeugnis anderer: der Sonderfall medialer Berichterstattung.* Paderborn: Mentis.

Mößner, Nicola. 2011. "Thought Styles and Paradigms—a Comparative Study of Ludwik Fleck and Thomas S. Kuhn." *Studies in History and Philosophy of Science* 42: 362–71. doi:10.1016/j.shpsa.2010.12.002.

Mößner, Nicola. 2013a. "Das Beste aus zwei Welten? Ludwik Fleck über den sozialen Ursprung wissenschaftlicher Kreativität." In *Simultaneität: Modelle der Gleichzeitigkeit in den Wissenschaften und Künsten*, edited by Philipp Hubmann and Till Julian Huss, 111–31. Bielefeld: transcript.

Mößner, Nicola. 2013b. "Können Bilder Argumente sein?" In *Bilder als Gründe*, edited by Manfred Harth and Jakob Steinbrenner 35–57. Cologne: Herbert von Halem.

Mößner, Nicola. 2013c. "Photographic Evidence and the Problem of Theory-Ladenness." *Journal for General Philosophy of Science* 44 (1): 111–25. doi:10.1007/s10838-013-9219-3.

Mößner, Nicola. 2013d. "Review of: Michael Newall: What Is a Picture? Depiction, Realism, Abstraction." *Zeitschrift Für Philosophische Forschung* 67 (3): 498–501.

Mößner, Nicola. 2015. "Visual Information and Scientific Understanding." *Axiomathes* 25: 167–79. doi:10.1007/s10516-014-9246-7.

Mößner, Nicola. 2016. "Scientific Images as Circulating Ideas: An Application of Ludwik Fleck's Theory of Thought Styles." *Journal for General Philosophy of Science* 47 (2): 307–29. doi:10.1007/s10838-016-9327-y.

Mößner, Nicola. 2017. "Visual Data: Reasons to Be Relied on?" In *Reasoning in Measurement*, edited by Nicola Mößner and Alfred Nordmann, 99–110. London and New York: Routledge.

Mößner, Nicola and Alfred Nordmann, eds. 2017. *Reasoning in Measurement*. London and New York: Routledge.

Moylan, Elizabeth. 2015. "Inappropriate Manipulation of Peer Review." BioMed Central Blog, March. http://blogs.biomedcentral.com/bmcblog/2015/03/26/manipulation-peer-review/.

Mulhall, Stephen. 2008. *On Film*. 2nd ed. London et al.: Routledge.

Müller, Andreas, Jochen Kuhn, Alwine Lenzner and Wolfgang Schnotz. 2012. "Schöne Bilder in den Naturwissenschaften: motivierend, anregend oder doch nur schmückendes Beiwerk?" In *Visualisierung und Erkenntnis: Bildverstehen und Bildverwenden in Natur- und Geisteswissenschaften*, edited by Dimitri Liebsch and Nicola Mößner, 207–36. Cologne: Herbert von Halem.

Müller, Sabine. 2007. "Visualisierung in der astronomischen Digitalfotografie mit Hilfe von Falschfarben." In *Vom Bild zur Erkenntnis? Visualisierungskonzepte in den Wissenschaften*, edited by Dominik Groß and Stefanie Westermann, 93–110. Kassel: Kassel University Press.

Müller, Sabine and Dominik Groß. 2006. "Farben als Werkzeuge der Erkenntnis. Falschfarbendarstellungen in der Gehirnforschung und in der Astronomie." In *Farbe – Erkenntnis – Wissenschaft: zur epistemischen Bedeutung von Farbe in der Medizin*, edited by Dominik Groß and Tobias Heinrich Duncker, 93–116. Berlin et al.: Lit.

Munzner, Tamara. 2008. "Process and Pitfalls in Writing Information Visualization Research Papers." In *Information Visualization*, edited by A. Kerren, J. Stasko, J.-D. Fekete and C. North, 134–53. Berlin, Heidelberg: Springer Science + Business Media. doi:10.1007/978-3-540-70956-5_6.

Nanay, Bence. 2011. "Perceiving Pictures." *Phenomenology and the Cognitive Sciences* 10 (4): 461–80. doi:10.1007/s11097-011-9219-x.

Nanay, Bence. 2011. ed. 2014. *Perceiving the World*. Oxford: Oxford University Press. doi:10.1093/acprof:oso/9780195386196.001.0001.

Nemeth, Elisabeth. 2011. "Scientific Attitude and Picture Language. Otto Neurath on Visualisation in Social Sciences." In *Image and Imaging in Philosophy, Science, and the Arts: Proceedings of the 33rd International Wittgenstein-Symposium in Kirchberg, 2010*, edited by Richard Heinrich, Elisabeth Nemeth, Wolfram Pichler and David Wagner, 59–84. Frankfurt/Main et al.: De Gruyter. doi:10.1515/9783110330496.59.

Nettel, Ana Laura and Georges Roque. 2012. "Persuasive Argumentation Versus Manipulation." *Argumentation* 26 (1): 55–69. doi:10.1007/s10503-011-9241-8.

Neurath, Otto. 1991a. "Statistik und Proletariat [1927]." In *Gesammelte bildpädagogische Schriften*, by Otto Neurath, edited by Rudolf Haller, 76–84. Vienna: Hölder-Pichler-Tempsky.

Neurath, Otto. 1991b. "Bildstatistik und Arbeiterbildung [1929]." In *Gesammelte bildpädagogische Schriften*, by Otto Neurath, edited by Rudolf Haller, 139–43. Vienna: Hölder-Pichler-Tempsky.

Neurath, Otto. 1991c. "Bildstatistik nach Wiener Methode [1931]." In *Gesammelte bildpädagogische Schriften*, by Otto Neurath, edited by Rudolf Haller, 180–91. Vienna: Hölder-Pichler-Tempsky.

Neurath, Otto. 1991. *Gesammelte bildpädagogische Schriften*. Edited by Rudolf Haller. Vienna: Hölder-Pichler-Tempsky.

Neurath, Otto. 2010. *From Hieroglyphics to Isotype: A Visual Autobiography*. Edited by Matthew Eve and Christopher Burke. London: Hyphen Press.

Newall, Michael. 2011. *What Is a Picture? Depiction, Realism, Abstraction*. Basingstoke et al.: Palgrave Macmillan.

Nida-Rümelin, Julian, Jakob Steinbrenner and Dominic McIver Lopes, eds. 2012. *Fotografie zwischen Dokumentation und Inszenierung*. Ostfildern: Hatje Cantz.

Nida-Rümelin, Martine. 2015. "Qualia: The Knowledge Argument." In *The Stanford Encyclopedia of Philosophy*, edited by Edward N. Zalta, Summer 2015 Edition. The Metaphysics Research Lab, Center for the Study of Language; Information, Stanford University. http://plato.stanford.edu/archives/sum2015/entries/qualia-knowledge.

Niiniluoto, Ilkka. 2015. "Scientific Progress." In *The Stanford Encyclopedia of Philosophy*, edited by Edward N. Zalta, Summer 2015 Edition. The Metaphysics Research Lab, Center for the Study of Language; Information, Stanford University. http://plato.stanford.edu/archives/sum2015/entries/scientific-progress/.

Nortmann, Ulrich and Christoph Wagner, eds. 2010. *In Bildern denken?: Kognitive Potentiale von Visualisierung in Kunst und Wissenschaft*. Paderborn et al.: Fink.

Norton, John D. 1996. "Are Thought Experiments Just What You Thought?" *Canadian Journal of Philosophy* 26 (3): 333–66. doi: 10.1080/00455091.1996.10717457.

Norton, John D. 2004. "On Thought Experiments: Is There More to the Argument?" *Philosophy of Science* 71 (5): 1139–51. doi:10.1086/425238.

Origgi, Gloria. 2004. "Is Trust an Epistemological Notion?" *Episteme* 1 (1): 61–72. doi:10.3366/epi.2004.1.1.61.

Origgi, Gloria. 2012. "A Social Epistemology of Reputation." *Social Epistemology* 26 (3–4): 399–418. doi: 10.1080/02691728.2012.727193.

Ortony, Andrew, ed. 1998. *Metaphor and Thought*. 2nd ed., reprinted. Cambridge et al.: Cambridge University Press.

Ottino, Julio M. 2003. "Is a Picture Worth 1,000 Words? Exciting New Illustration Technologies Should Be Used with Care." *Nature* 421: 474–6. doi: 10.1038/421474a.

Pang, Alex Soojung-Kim. 1997. " 'Stars Should Henceforth Register Themselves': Astrophotography at the Early Lick Observatory." *The British Journal for the History of Science* 30 (2): 177–202. doi: 10.1017/S0007087497003002.

Peacocke, Christopher. 1987. "Depiction." *The Philosophical Review* 96 (3): 383–410. doi: 10.2307/2185226.

Peacocke, Christopher. 1994. "Reply. Non-Conceptual Content: Kinds, Rationales and Relations." *Mind & Language* 9 (4): 419–30. doi: 10.1111/j.1468-0017.1994. tb00316.x.

Peacocke, Christopher. 2001a. "Does Perception Have a Nonconceptual Content?" *The Journal of Philosophy* 98 (5): 239–64. doi:10.2307/2678383.

Peacocke, Christopher. 2001b. "Phenomenology and Nonconceptual Content." *Philosophy and Phenomenological Research* 62 (3): 609–15. doi:10.1111/j.1933-1592.2001.tb00077.x.

Peirce, Charles S. 1983. *Phänomen und Logik der Zeichen.* Edited by Helmut Pape. Suhrkamp-Taschenbuch Wissenschaft; 425. Frankfurt/Main: Suhrkamp.

Perini, Laura. 2005a. "Explanation in Two Dimensions: Diagrams and Biological Explanation." *Biology and Philosophy* 20 (2–3): 257–69. doi:10.1007/s10539-005-2562-y.

Perini, Laura. 2005b. "The Truth in Pictures." *Philosophy of Science* 72 (1): 262–85. doi:10.1086/426852.

Perini, Laura. 2005c. "Visual Representations and Confirmation." *Philosophy of Science* 72 (5): 913–26. doi:10.1086/508949.

Perini, Laura. 2010. "Scientific Representation and the Semiotics of Pictures." In *New Waves in Philosophy of Science*, edited by P. D. Magnus and Jacob Busch, 131–54. Basingstoke et al.: Palgrave Macmillan.

Perini, Laura. 2012a. "Depiction, Detection, and the Epistemic Value of Photography." *The Journal of Aesthetics and Art Criticism* 70 (1): 151–60. doi:10.1111/j.1540-6245.2011.01506.x.

Perini, Laura. 2012b. "Image Interpretation: Bridging the Gap from Mechanically Produced Image to Representation." *International Studies in the Philosophy of Science* 26 (2): 153–70.

Perini, Laura. 2012c. "Truth-Bearers or Truth-Makers?" *Spontaneous Generations: A Journal for the History and Philosophy of Science* 6 (1): 142–7. doi: 10.4245/sponge. v6i1.17357.

Platon. 1994a. *Sämtliche Werke Band 1: Apologie, Kriton, Ion, Hippias II, Theages, Alkibiades I, Laches, Charmides, Euthyphron, Protagoras, Gorgias, Menon, Hippias I, Euthydemos, Menexenos.* Edited by Friedrich Schleiermacher, Walter F. Otto and Ursula Wolf. Reinbek/Hamburg: Rowohlt-Taschenbuch-Verlag.

Platon. 1994b. *Sämtliche Werke Band 3: Kratylos, Parmenides, Theitetos, Sophistes, Politikos, Philebos, Briefe.* Edited by Friedrich Schleiermacher, Walter F. Otto and Ursula Wolf. Reinbek/Hamburg: Rowohlt-Taschenbuch-Verlag.

Platon. 2006. *Sämtliche Werke Band 2: Lysis, Symposion, Phaidon, Kleitophon, Politeia, Phaidros.* Edited by Friedrich Schleiermacher, Walter F. Otto and Ursula Wolf. 31st ed. Reinbek/Hamburg: Rowohlt-Taschenbuch-Verlag.

Popper, Karl R. 2002. *Conjectures and Refutations: The Growth of Scientific Knowledge.* Routledge Classics. London et al.: Routledge.

Pörksen, Uwe. 1997. *Weltmarkt der Bilder: eine Philosophie der Visiotype.* Stuttgart: Klett-Cotta.

Poser, Hans. 2001. *Wissenschaftstheorie: eine philosophische Einführung.* Stuttgart: Reclam.

Pritchard, Duncan and John Turri. 2014. "The Value of Knowledge." In *The Stanford Encyclopedia of Philosophy*, edited by Edward N. Zalta, Spring 2014 edition.

The Metaphysics Research Lab, Center for the Study of Language; Information, Stanford University. http://plato.stanford.edu/archives/spr2014/entries/knowledge-value/.

Purves, William Kirkwood, David Sadava, Andreas Held and Jürgen Markl. 2011. *Purves Biologie*. 9th ed. Heidelberg: Spektrum, Akademie Verlag.

Ratzka, Thorsten. 2012. "Die Fenster zum Himmel." In *Visualisierung und Erkenntnis: Bildverstehen und Bildverwenden in Natur- und Geisteswissenschaften*, edited by Dimitri Liebsch and Nicola Mößner, 237–64. Cologne: Herbert von Halem.

Reichenbach, Hans. 1961. *Experience and Prediction: An Analysis of the Foundations and the Structure of Knowledge*. Chicago, Ill. et al.: University of Chicago Press.

Reicher, Maria E. 2013. "Wie aus Gedanken Dinge werden. Eine Philosophie der Artefakte." *Deutsche Zeitschrift für Philosophie* 61 (2): 219–32. doi:10.1524/dzph.2013.0018.

Reichert, Uwe. 2016. "Eine neue Ära der Astrophysik. Das Zeitalter der Gravitationswellen-Astrononomie hat begonnen." *Sterne und Weltraum* 4: 24–35.

Renkl, Alexander and Katharina Scheiter. 2015, online first. "Studying Visual Displays: How to Instructionally Support Learning." *Educational Psychology Review*. doi:10.1007/s10648-015-9340-4.

Reydon, Thomas. 2013. *Wissenschaftsethik: eine Einführung*. UTB 4032. Stuttgart: Ulmer.

Richter, Marianne Ina. 2014. *Scientific Visualisation: Epistemic Weight and Surpluses*. Frankfurt/Main et al.: Peter-Lang-Edition.

Ritchin, Fred. 2010. *After Photography*. 1st paperback ed. New York, NY: Norton.

Robin, Harry. 1992. *The Scientific Image: From Cave to Computer*. New York: H.N. Abrams.

Rollins, Mark. 2008. "Pictorial Representation." In *The Routledge Companion to Aesthetics*, edited by Berys Nigel Gaut and Dominic McIver Lopes, 2nd ed., reprint, 383–97. London et al.: Routledge.

Roque, Georges. 2015. "Should Visual Arguments Be Propositional in Order to Be Arguments?" *Argumentation* 29 (2): 177–95. doi:10.1007/s10503-014-9341-3.

Roskies, Adina L. 2007. "Are Neuroimages Like Photographs of the Brain?" *Philosophy of Science* 74 (5): 860–72. doi:10.1086/525627.

Roskies, Adina L. 2008. "Neuroimaging and Inferential Distance." *Neuroethics* 1: 19–30. doi: 10.1007/s12152-007-9003-3.

Russell, Bertrand. 1910–1911. "Knowledge by Acquaintance and Knowledge by Description." *Proceedings of the Aristotelian Society* 11 (New Series): 108–28.

Russell, Bertrand. 1912. *Problems of Philosophy – Online Edition*. Edited by Andrew Chrucky. Oxford University Press. www.ditext.com/russell/russell.html.

Sachs-Hombach, Klaus, ed. 2003. *Was ist Bildkompetenz?: Symposium "Was ist Bildkompetenz?"* Wiesbaden: Deutscher Universitäts-Verlag.

Sachs-Hombach, Klaus, ed. 2005a. *Bildwissenschaft: Disziplinen, Themen, Methoden*. Suhrkamp-Taschenbuch Wissenschaft 1751. Frankfurt/Main: Suhrkamp.

Sachs-Hombach, Klaus, ed. 2005b. *Bildwissenschaft zwischen Reflexion und Anwendung*. Cologne: Herbert von Halem.

Sachs-Hombach, Klaus. 2005c. "Konzeptionelle Rahmenüberlegungen zur interdisziplinären Bildwissenschaft." In *Bildwissenschaft: Disziplinen, Themen, Methoden*, edited by Klaus Sachs-Hombach, 11–20. Suhrkamp-Taschenbuch Wissenschaft 1751. Frankfurt/Main: Suhrkamp.

Sachs-Hombach, Klaus. 2006. *Das Bild als kommunikatives Medium: Elemente einer allgemeinen Bildwissenschaft.* 2nd ed. Cologne: Herbert von Halem.

Sachs-Hombach, Klaus, ed. 2009. *Bildtheorien. Anthropologische und kulturelle Grundlagen des Visualistic Turn.* Suhrkamp-Taschenbuch Wissenschaft 1888. Frankfurt/Main: Suhrkamp.

Sachs-Hombach, Klaus. 2012. "Bilder in der Wissenschaft." In *Visualisierung und Erkenntnis: Bildverstehen und Bildverwenden in Natur- und Geisteswissenschaften,* edited by Dimitri Liebsch and Nicola Mößner, 31–42. Cologne: Herbert von Halem.

Sachs-Hombach, Klaus and Eva Schürmann. 2005. "Philosophie." In *Bildwissenschaft: Disziplinen, Themen, Methoden,* edited by Klaus Sachs-Hombach, 109–23. Suhrkamp-Taschenbuch Wissenschaft 1751. Frankfurt/Main: Suhrkamp.

Sacks, Oliver W. 2010. *The Mind's Eye.* New York: Alfred A. Knopf.

Sady, Wojciech. 2012. "Ludwik Fleck." In *The Stanford Encyclopedia of Philosophy,* edited by Edward N. Zalta, Summer 2012 Edition. The Metaphysics Research Lab, Center for the Study of Language; Information, Stanford University. http://plato. stanford.edu/archives/sum2012/entries/fleck/.

Salmon, Wesley C. 1983. *Logik.* Edited by Joachim Buhl. Stuttgart: Reclam.

Salmon, Wesley C. 1993. "The Value of Scientific Understanding." *Philosophia* 51 (1): 9–19.

Savedoff, Barbara. 2008. "Documentary Authority and the Art of Photography." In *Photography and Philosophy: Essays on the Pencil of Nature,* edited by Scott Walden, 111–37. Malden, MA: Blackwell.

Savigny, Eike von. 1983. *Zum Begriff der Sprache: Konvention, Bedeutung, Zeichen.* Stuttgart: Reclam.

Sazdanovic, Radmila. forthcoming. "Visualization and Visual Thinking in Mathematics." In *Visualization: A Critical Survey of the Concept,* edited by Erna Fiorentini. Kultur: Forschung Und Wissenschaft. Berlin, Münster, Vienna, Zurich, London: LIT-Verlag.

Schäfer, Lothar and Thomas Schnelle. 1980. "Ludwik Flecks Begründung der soziologischen Betrachtungsweise in der Wissenschaftstheorie." In *Entstehung und Entwicklung einer wissenschaftlichen Tatsache: Einführung in die Lehre vom Denkstil und Denkkollektiv,* by Ludwik Fleck, edited by Lothar Schäfer and Thomas Schnelle, VII–XLIX. Suhrkamp-Taschenbuch Wissenschaft 312. Frankfurt/Main: Suhrkamp.

Schantz, Richard, ed. 2009. *Wahrnehmung und Wirklichkeit.* Frankfurt/Main: Ontos Verlag.

Schawelka, Karl. 2007. *Farbe: warum wir sie sehen, wie wir sie sehen.* Weimar: Verlag der Bauhaus-Universität.

Schickore, Jutta. 1999. "Sehen, Sichtbarkeit und empirische Forschung." *Journal for General Philosophy of Science* 30 (2): 273–87. doi:10.1023/A:1008374032737.

Schirra, Jörg R. J. 2005. "Computational Visualistics: Dealing with Pictures in Computer Science." In *Bildwissenschaft zwischen Reflexion und Anwendung,* edited by Klaus Sachs-Hombach, 494–509. Cologne: von Halem.

Schirra, Jörg R. J. 2012. "Sind Bilder ein Gegenstand der Informatik? Überlegungen zur Computervisualistik." In *Visualisierung und Erkenntnis: Bildverstehen und Bildverwenden in Natur- und Geisteswissenschaften,* edited by Dimitri Liebsch and Nicola Mößner, 329–59. Cologne: Herbert von Halem.

Schirra, Jörg R. J. and Klaus Sachs-Hombach. 2013. "The Anthropological Function of Pictures." In *Origins of Pictures: Anthropological Discourses in Image Science*, edited by Klaus Sachs-Hombach, 132–59. Cologne: Herbert von Halem.

Schlaudt, Oliver. 2009. *Messung als konkrete Handlung: eine kritische Untersuchung über die Grundlagen der Bildung quantitativer Begriffe in den Naturwissenschaften.* Würzburg: Königshausen & Neumann.

Schnotz, Wolfgang. 2002. "Commentary: Towards an Integrated View of Learning from Text and Visual Displays." *Educational Psychology Review* 14 (1): 101–20. doi:10.1023/A:1013136727916.

Scholz, Martin. 2000. *Technologische Bilder: Aspekte visueller Argumentation.* Weimar: VDG.

Scholz, Oliver R. 1993. "When Is a Picture?" *Synthese* 95 (1): 95–106. doi: 10.1007/BF01064669.

Scholz, Oliver R. 2001. "Das Zeugnis anderer: Prolegomena zu einer sozialen Erkenntnistheorie." In *Erkenntnistheorie: Positionen zwischen Tradition und Gegenwart*, edited by Thomas Grundmann, 354–75. Paderborn: Mentis.

Scholz, Oliver R. 2004. "Quellen der Erkenntnis: Metapher, Begriff und Sache." In *Beiheft zur Zeitschrift für Deutsche Philologie: „Quelle" – Zwischen Ursprung und Konstrukt. Ein Leitbegriff in der Diskussion* 12, edited by Thomas Rathmann and Nikolaus Wegmann, 40–65. Berlin: Erich Schmidt Verlag.

Scholz, Oliver R. 2009a. "Abbilder und Entwürfe: Bilder und die Strukturen der menschlichen Intentionalität." In *Bildtheorien. Anthropologische und kulturelle Grundlagen des Visualistic Turn*, edited by Klaus Sachs-Hombach, 146–62. Suhrkamp-Taschenbuch Wissenschaft 1888. Frankfurt/Main: Suhrkamp.

Scholz, Oliver R. 2009b. *Bild, Darstellung, Zeichen: philosophische Theorien bildlicher Darstellung.* 3rd ed. Klostermann Rote Reihe. Frankfurt/Main: Klostermann.

Scholz, Oliver R. 2009c. "Das Zeugnis der Sinne und das Zeugnis anderer." In *Wahrnehmung und Wirklichkeit*, edited by Richard Schantz, 183–209. Frankfurt/Main: Ontos Verlag.

Scholz, Oliver R. 2010. "Aus Bildern lernen." In *In Bildern denken? kognitive Potentiale von Visualisierung in Kunst und Wissenschaft*, edited by Ulrich Nortmann and Christoph Wagner, 43–52. Paderborn et al.: Fink.

Scholz, Oliver R. 2011. "Bildspiele." In *Image and Imaging in Philosophy, Science, and the Arts: Proceedings of the 33rd International Wittgenstein-Symposium in Kirchberg, 2010*, edited by Richard Heinrich, Elisabeth Nemeth, Wolfram Pichler and David Wagner, 365–82. Frankfurt/Main et al.: De Gruyter. doi:10.1515/9783110330519.365.

Scholz, Oliver R. 2013. "Wissenschaftstheorie, Erkenntnistheorie und Metaphysik – Klärungen zu einem ungeklärten Verhältnis." *Philosophia Naturalis* 50 (1): 5–24. doi:10.3196/003180213809359792.

Schürmann, Eva. 2006. "Was will die Bildwissenschaft?" *Philosophische Rundschau* 53: 154–68. doi: 10.1628/003181506783353278.

Scruton, Roger. 2008. "Photography and Representation." In *Photography and Philosophy: Essays on the Pencil of Nature*, edited by Scott Walden, 138–66. Malden, MA: Blackwell.

Searle, John Rogers. 1983. *Sprechakte: ein sprachphilosophischer Essay.* Edited by R. Wiggershaus. Suhrkamp-Taschenbuch, Wissenschaft 458. Frankfurt/Main: Suhrkamp.

Seidel, Markus. 2011. "Relativism or Relationism? A Mannheimian Interpretation of Fleck's Claims About Relativism." *Journal for General Philosophy of Science* 42 (2): 219–40. doi:10.1007/s10838-011-9163-z.

Shah, Priti and James Hoeffner. 2002. "Review of Graph Comprehension Research: Implications for Instruction." *Educational Psychology Review* 14 (1): 47–69. doi:10.1023/A:1013180410169.

Shin, Sun-Joo. 2015. "The Mystery of Deduction and Diagrammatic Aspects of Representation." *Review of Philosophy and Psychology* 6 (1): 49–67. doi:10.1007/s13164-014-0212-5.

Shin, Sun-Joo, Oliver Lemon and John Mumma. 2014. "Diagrams." In *The Stanford Encyclopedia of Philosophy*, edited by Edward N. Zalta, Winter 2014 Edition. The Metaphysics Research Lab, Center for the Study of Language; Information, Stanford University. http://plato.stanford.edu/archives/win2014/entries/diagrams/.

Siegel, Susanna. 2015. "The Contents of Perception." In *The Stanford Encyclopedia of Philosophy*, edited by Edward N. Zalta, Spring 2015 Edition. The Metaphysics Research Lab, Center for the Study of Language; Information, Stanford University. http://plato.stanford.edu/archives/spr2015/entries/perception-contents/.

Sober, Elliott. 1976. "Mental Representations." *Synthese* 33 (2–4): 101–48. doi:10.1007/BF00484711.

Steinbrenner, Jakob. 2009. "Bildtheorien der analytischen Tradition." In *Bildtheorien. Anthropologische und kulturelle Grundlagen des Visualistic Turn*, edited by Klaus Sachs-Hombach, 284–315. Suhrkamp-Taschenbuch Wissenschaft 1888. Frankfurt/Main: Suhrkamp.

Stenning, Keith. 2002. *Seeing Reason: Image and Language in Learning to Think*. Oxford: Oxford University Press.

Suhm, Christian. 2004. "Theoretische Entitäten und ihre realistische Deutung: Realismus einer Strategie zur Verteidigung des wissenschaftlichen Realismus." In *Was ist wirklich? Neuere Beiträge zu Realismusdebatten in der Philosophie*, edited by Christian Suhm and Christoph Halbig, 139–81. Berlin: De Gruyter.

Suhm, Christian. 2005. *Wissenschaftlicher Realismus: eine Studie zur Realismus-Antirealismus-Debatte in der neueren Wissenschaftstheorie*. Frankfurt/Main: Ontos-Verlag.

Tufte, Edward R. 2010. *Visual Explanations: Images and Quantities, Evidence and Narrative*. 8th ed., with revisions. Cheshire, Conn.: Graphics Press.

Tufte, Edward R. 2011. *Envisioning Information*. 13th ed. Cheshire, Conn.: Graphics Press.

Tversky, Amos. 1977. "Features of Similarity." *Psychological Review* 84 (4): 327–52. doi:10.1037/0033-295x.84.4.327.

Tversky, Barbara. 2011. "Visualizing Thought." *Topics in Cognitive Science* 3 (3): 499–535. doi:10.1111/j.1756-8765.2010.01113.x.

Tversky, Barbara. 2015. "The Cognitive Design of Tools of Thought." *Review of Philosophy and Psychology* 6 (1): 99–116. doi:10.1007/s13164-014-0214-3.

van Fraassen, Bas C. 1980. *The Scientific Image*. Clarendon Paperbacks. Oxford: Clarendon Press.

van Fraassen, Bas C. 2010. *Scientific Representation: Paradoxes of Perspective*. Oxford: Clarendon Press.

Vekiri, Ioanna. 2002. "What Is the Value of Graphical Displays in Learning?" *Educational Psychology Review* 14 (3): 261–312. doi:10.1023/A:1016064429161.

Verdi, Michael P. and Raymond W. Kulhavy. 2002. "Learning With Maps and Texts: An Overview". *Educational Psychology Review* 14 (1): 27–46. doi: 10.1023/A:1013128426099.

Vertesi, Janet. 2015. *Seeing Like a Rover: How Robots, Teams, and Images Craft Knowledge of Mars*. Chicago: The University of Chicago Press.

Vögtli, Alexander and Beat Ernst. 2007. *Wissenschaftliche Bilder: eine kritische Betrachtung*. Basel: Schwabe.

Walden, Scott, ed. 2008. *Photography and Philosophy: Essays on the Pencil of Nature*. Malden, MA: Blackwell.

Walton, Kendall L. 1990. *Mimesis as Make-Believe: On the Foundations of the Representational Arts*. Cambridge, Mass. et al.: Harvard University Press.

Wartenberg, Thomas E. 2011a. "Film as Philosophy." In *The Routledge Companion to Philosophy and Film*, edited by Paisley Livingston and Carl Plantinga, 549–59. Routledge Philosophy Companions. London et al.: Routledge.

Wartenberg, Thomas E. 2011b. "On the Possibility of Cinematic Philosophy." In *New Takes in Film-Philosophy*, edited by Havi Carel and Greg Tuck, 9–24. Basingstoke et al.: Palgrave Macmillan.

Watzlawick, Paul, Janet H. Beavin and Don D. Jackson. 1974. *Menschliche Kommunikation. Formen, Störungen, Paradoxien*. 4th ed. Bern, et al.: Huber.

Weiss, Dieter G. 2012. "Das neue Bild der Zelle: Wechsel der Sichtweisen in der Zellbiologie durch neue Mikroskopieverfahren." In *Visualisierung und Erkenntnis: Bildverstehen und Bildverwenden in Natur- und Geisteswissenschaften*, edited by Dimitri Liebsch and Nicola Mößner, 295–328. Cologne: Herbert von Halem.

Wiegand, Wilfried, ed. 1981. *Die Wahrheit der Photographie: klassische Bekenntnisse zu einer neuen Kunst*. Frankfurt/Main: Fischer.

Wiesing, Lambert. 2006. *Artifizielle Präsenz: Studien zur Philosophie des Bildes*. Suhrkamp-Taschenbuch Wissenschaft 1737. Frankfurt/Main: Suhrkamp.

Wilder, Kelley E. 2009. *Photography and Science*. London: Reaktion Books.

Wilder, Kelley E. 2011. "Visualizing Radiation: The Photographs of Henri Becquerel." In *Histories of Scientific Observation*, edited by Lorraine Daston and Elizabeth Lunbeck, 349–68. Chicago et al.: University of Chicago Press.

Wittgenstein, Ludwig. 1995. *Tractatus logico-philosophicus, Tagebücher 1914–1916, Philosophische Untersuchungen*. Edited by Joachim Schulte. 10th ed. Suhrkamp-Taschenbuch Wissenschaft 501. Frankfurt/Main: Suhrkamp.

Wollheim, Richard. 1998. "On Pictorial Representation." *The Journal of Aesthetics and Art Criticism* 56 (3): 217–26. doi: 10.2307/432361.

Woodward, James. 2005. *Making Things Happen: A Theory of Causal Explanation*. Oxford et al.: Oxford University Press.

Zimmermann, Anja. 2009. *Ästhetik der Objektivität: Genese und Funktion eines wissenschaftlichen und künstlerischen Stils im 19. Jahrhundert*. Bielefeld: transcript.

Zittel, Claus. 2012. "Ludwik Fleck and the Concept of Style in the Natural Sciences." *Studies in East European Thought* 64 (1–2): 53–79. doi:10.1007/s11212-012-9160-8.

Zittel, Claus. 2014. "Ludwik Flecks Gestaltbegriff und sein Blick auf die Gestaltpsychologie seiner Zeit." *N. T. M. Zeitschrift für Geschichte der Wissenschaften, Technik und Medizin* 22 (1–2): 9–29. doi:10.1007/s00048-013-0106-0.

Index

Page numbers in *italics* denote figures, those in **bold** denote tables. End of chapter notes are indicated by a letter n between page number and note number.